Mineral Biofortification in Crop Plants for Ensuring Food Security

Mirza Hasanuzzaman •
Muhammad Suleman Tahir • Mohsin Tanveer •
Adnan Noor Shah
Editors

Mineral Biofortification in Crop Plants for Ensuring Food Security

Springer

Editors
Mirza Hasanuzzaman
Department of Agronomy
Sher-e-Bangla Agricultural University
Dhaka, Bangladesh

Muhammad Suleman Tahir
Department of Chemical Engineering
Khwaja Fareed University of Engineering and Information Technology
Rahim Yar Khan, Pakistan

Mohsin Tanveer
University of Tasmania
Tasmanian Institute of Agriculture
Hobart, TAS, Australia

Xinjiang Institute of Ecology and Geography
Chinese Academy of Sciences
Urumqi, China

Adnan Noor Shah
Department of Agricultural Engineering
Khwaja Fareed University of Engineering and Information Technology
Rahim Yar Khan, Punjab, Pakistan

ISBN 978-981-99-4089-9 ISBN 978-981-99-4090-5 (eBook)
https://doi.org/10.1007/978-981-99-4090-5

This Springer imprint is published by the registered company Springer Nature Singapore Pte Ltd.
The registered company address is: 152 Beach Road, #21-01/04 Gateway East, Singapore 189721, Singapore

Preface

The world population is expected to reach 8.6 billion in 2030, 9.8 billion in 2050, and 11.2 billion by 2100, but agricultural lands will not be able to follow at the same speed. According to the United Nations, every year, 80 million people are added to the world's population. Already one in nine people around the world suffer from hunger, and the only way to feed them is by doubling food production in a sustainable way. To meet this challenge, we must increase food production by 70% to feed the increasing population by the year 2050. Current climate change is putting a great deal of pressure on increasing crop production, which is also a great concern for not only increasing crop yield but to produce grains and vegetables enriched with minerals and essential nutrients. The cause of nutritional deficiencies, apart from an insufficient quantitative and qualitative supply of food, may also include food allergies and intolerances. Considering the global challenges to prevent nutritional deficiencies in different environmental settings, various methods of food fortification may provide, at least partially, solutions to the issue of latent hunger caused by a deficiency of minerals or vitamins in the diet.

Biofortification of essential micronutrients into crop plants can be achieved through three main approaches, namely transgenic, conventional, and agronomic, involving the use of biotechnology, crop breeding, and fertilization strategies, respectively. Higher numbers of crops have been targeted by transgenic means, while the practical utilization of biofortification is higher by breeding methods. Cereals being staple crops have been targeted by all three approaches. The same is the case for legumes and vegetables. Interestingly, oil seed biofortification has been achieved through transgenic means because limited availability of genetic diversity for the targeted component, low heritability, and linkage drags in the targeted crop. To alleviate nutritional issues and nutrient deficiencies, the biofortification of edible plants is considered the most appropriate approach. By contrast, biofortification focuses on improving the nutritional content of the region's current agricultural biodiversity, preserving its habits and customs.

This book, *Mineral Biofortification in Crop Plants for Ensuring Food Security*, contains 14 chapters on the most recent information with up-to-date literature, which provides a comprehensive overview of the recent advances on the aspects of the need of biofortification in plants under climate changes. This book aims to bring together a galaxy of eminent experienced scientists to present the latest developments in this field. While many books cover the phytonutrients of crops, this reference book reports the latest methodologies, techniques, and environmental changes used for biofortification. Moreover, the comprehensive mechanisms of the plant responses to genetically induced biofortification, production, and response of fortified plants under climate change, and its effect on food security are also highlighted in this book.

We would like to give special thanks to the authors for their outstanding and timely work in producing such fine chapters. Our profound thanks also to Ayesha Siddika and Md. Rakib Hossain Raihan, Department of Agronomy, Sher-e-Bangla Agricultural University, for their valuable support in formatting and incorporating all editorial changes in the manuscripts. I am highly thankful to Ms. Momoko Asawa, Editorial Assistant, Medicine and Life Sciences, Books Editorial Services, Springer, Japan, and Priyanga Kabali, Project Editor, Springer Nature, India, for their prompt responses during the production. I am also thankful to all other editorial staff for their precious help in formatting and incorporating editorial changes in the manuscripts.

Dhaka, Bangladesh — Mirza Hasanuzzaman
Rahim Yar Khan, Punjab, Pakistan — Muhammad Suleman Tahir
Hobart, TAS, Australia — Mohsin Tanveer
Rahim Yar Khan, Punjab, Pakistan — Adnan Noor Shah

Contents

About the Editors

Mirza Hasanuzzaman is a Professor of Agronomy at Sher-e-Bangla Agricultural University, Dhaka, Bangladesh. He received his Ph.D. from the United Graduate School of Agricultural Sciences, Ehime University, Japan, with a Japanese Government (MEXT) Scholarship. Later, he completed his postdoctoral research in the Center of Molecular Biosciences (COMB), University of the Ryukyus, Okinawa, Japan, with a 'Japan Society for the Promotion of Science (JSPS)' postdoctoral fellowship. Subsequently, he became an Adjunct Senior Researcher at the University of Tasmania with an Australian Government's Endeavour Research Fellowship. Dr. Mirza Hasanuzzaman is one of the Highly Cited Researchers recognized by Clarivate Analytics. He published over 250 articles and edited over 20 books on important aspects of plant physiology, plant stress tolerance, and crop production. According to Scopus, Prof. Hasanuzzaman's publications have already received over 19000 citations with an *h*-index of 73. Prof. Hasanuzzaman supervised the dissertations of 41 master's and 3 Ph.D. students. Dr. Hasanuzzaman is an editor and reviewer of more than 100 international journals and was a recipient of the 'Publons Global Peer Review Award 2017, 2018 and 2019' which is managed by Web of Science. He is a Fellow of Bangladesh Academy of Sciences, The Linnean Society of London, Royal Society of Biology, and International Society of Environmental Botanists. He received the World Academy of Science (TWAS)

Young Scientist Award, 2014; University Grants Commission (UGC) Gold Medal, 2018; Global Network of Bangladeshi Biotechnologists (GNOBB) Award, 2021; Bangladesh Academy of Sciences (BAS) Gold Medal Award-2022 (Senior Group) and Society for Plant Research Young Scientist Award (Agriculture), 2023.

Muhammad Suleman Tahir is working as the Vice Chancellor at Khwaja Fareed University of Engineering and Information Technology in Rahim Yar Khan, Punjab, Pakistan. He joined KFUEIT after his previous role as the Dean of the Engineering Faculty at the University of Gujrat and brings with him vast experience of 23 years in the public and private sectors. Besides holding a Post Doc and a Ph.D. in Chemical Engineering from Graz University of Technology, Austria, he is a renowned national and international speaker. He has published several books and book chapters approved by Higher Education Commission (HEC) of Pakistan and several national and international research articles. His rich experience in developmental projects includes the completion of 15 construction projects at the University of Gujrat worth PKR 2 billion, 03 community training projects worth PKR 35 million under the umbrella of NAVTTC, and 05 research projects worth PKR 28 million funded by HEC.

Mohsin Tanveer has received his Ph.D. in Plant Physiology from the University of Tasmania, Australia. Since 2021, he has been working as Associate Researcher in Plant Physiology Lab, University of Tasmania, Australia. He has recently joined the Xinjiang Institute of Ecology and Geography, Chinese Academy of Sciences as an Associate Professor and currently leading two projects on salinity tolerance in halophytes as co-PI. His research interests include Plant Stress Physiology, Nutrient Metabolism, Ion Transport and Agricultural Sustainable Development. In 2021, he has been listed among top 40 Young Researchers of

Australia and first one in Botany field. According to Google scholars, Dr. Mohsin's publications have already received over 7265 citations with a h-index of 43. He is also among the Top 3% Scientists in the specified subject among the scientists in the AD Scientific Index in the World.

Adnan Noor Shah obtained a Ph.D. (Agronomy) from Huazhong Agriculture University, China, in 2017. His specialization includes plant nutrition, plant–soil interaction, crop physiology, crop cultivation, farming system, crop ecology, and plant abiotic stress responses, in addition to this, he is working on plant genomics and epigenetics. He served as Assistant Professor at Gomal University Dera Ismail Khan, Pakistan, from 2018 to 2019. In addition, he completed his 2 years postdoctorate fellowship from Anhui Agricultural University, China, during 2019–2021. Now, he has been working as an Assistant Professor in the Department of Agricultural Engineering, Khwaja Fareed University of Engineering and Information Technology, Rahim Yar Khan 64200, Pakistan, since March 2022. He is part of the editorial board in different international and national journals, BMC Plant Biology, International Journal of Plant Production, Functional plant biology, Frontiers in plant science, Frontiers in Agronomy, Journal of Pure and Applied Agriculture, Agriculture-MDPI.

Chapter 1
Agronomic Biofortification: An Effective Tool for Alleviating Nutrient Deficiency in Plants and Human Diet

K. S. Karthika, I. Rashmi, S. Neenu, and Prabha Susan Philip

Abstract Nutrient deficiencies in soils result in malnutrition in human which is a widespread problem. Malnutrition affects humans health rendering them susceptible to various diseases, and significantly imbalance the socio-economic structure of any country by decreasing the Gross Domestic Product. Attaining nutritional security is the way to deal with this serious issue and so far it has not been realized in developing countries. To achieve nutritional security and to establish a hunger-free world, it is inevitable to alleviate malnutrition. This could be by following a balanced diet which can be accomplished by overcoming dietary deficiencies through dietary diversification, supplementation, food fortification, agronomic fortification and crop improvement. We discuss on agronomic biofortification as an efficient tool in alleviating nutrient deficiency in plants and human diet through explaining the essential nutrients, their role and the effects of their deficiencies in humans. In this chapter, the importance of nutrients in food web and human health, are explained along with the different aspects of biofortification, factors affecting biofortification of essential minerals in soil-plant system, challenges and future prospects of agronomic biofortification.

Keywords Agronomic biofortification · Essential nutrients · Micronutrient malnutrition · Nutritional security

K. S. Karthika (✉)
ICAR-National Bureau of Soil Survey and Land Use Planning, Regional Centre, Bangalore, Karnataka, India

I. Rashmi
ICAR-Indian Institute of Soil and Water Conservation, Regional Station, Kota, Rajasthan, India

S. Neenu
ICAR-Central Plantation Crops Research Institute, Kasaragod, Kerala, India

P. S. Philip
Radio Tracer Laboratory, Kerala Agricultural University, Thrissur, India

M. Hasanuzzaman et al. (eds.), *Mineral Biofortification in Crop Plants for Ensuring Food Security*, https://doi.org/10.1007/978-981-99-4090-5_1

1.1 Introduction

Malnutrition resulting from nutrient deficiency is a serious cause of concern of human health in many parts of the world, especially in developing and underdeveloped countries. The major reason for essential mineral deficiency is due to lack of human diet diversity, consuming one or two staple foods on a regular basis. The human diet is mainly based on cereals and animal-based thus leading to deficiency of one or more essential nutrients. Almost one third world population (about 2 billion) suffers from deficiency of one or more essential nutrients (Wakeel et al. 2018) and evidences of ill effects of under nourishments have been reported in about 850 million population round the globe as per recent estimates by United Nations Millennium Goal Report (2006). Deficiencies are mainly due to micronutrients and micronutrient malnutrition is associated with trace elements like zinc (Zn), selenium (Se) and iron (Fe). Insufficient intake of micronutrients and vitamins is extensively found especially in those following vegetarian diet (Connolly 2008).

The dream of nutritional security has not yet realised at global level, particularly in developing countries (Stein et al. 2008). Nutritional security can be achieved only when nutrition is provided through a balanced diet i.e., in right quantity and right time provision of carbohydrates, proteins, fat, minerals and vitamins by means of diverse and wide variety of food sources. There are several approaches for addressing dietary deficiencies, and those being adopted are dietary diversification, supplementation, food fortification, agronomic fortification and crop improvement (genetic fortification). Earlier adopted conventional methods like mineral supplements, diversified diets including varieties of fruits, vegetables and other intervention have not been sufficient to address the issue of undernourishment as many of these programmes did not reach all target groups (Welch and Graham 2004; White and Broadley 2005) and the cost involved in this is too high to afford for low income generating countries. Though nutrient deficiency in humans could be reduced by increasing nutrient supplements, in reality, it is not within the reach of many people.

It is at this juncture, biofortification provides a solution to this question. The approaches like biofortification are now highly preferred, despite the fact that significant technological advancements were able to attain the food security across the world. In this chapter, the details on essential nutrients, their role in plants and human, malnutrition due to their deficiency, factors affecting biofortification of essential minerals in soil-plant system and its effect on human health, plant traits improvement for nutrient accumulation in grain, are discussed along with explaining challenges/constraints and future prospects of agronomic biofortification.

1.2 Essential Nutrients for Plants and Human, their Role and Malnutrition Caused by their Deficiency

1.2.1 Essential Nutrients for Plants

Essential nutrients are those required for the normal life of an organism and their functions cannot be substituted by any other element or compound. Plants can absorb more than 90 elements from the soil and other sources and about 64 elements are identified in plant parts by tissue analysis. But all these elements are not essential for completing the life cycle of the plants. The essentiality of the element is decided based on the Criteria of Essentiality propounded by Arnon and Stout (1939). According to these criteria:

(a) The element must be absolutely essential for supporting normal growth and reproduction. In the absence of the particular element the plants cannot complete their vegetative or reproductive stage of the life cycle.
(b) The requirement of the element must be particular and not replaceable by another element. To put it another way, deficiency of any one element cannot be corrected by supplying another element.
(c) The element must have a direct role in the metabolism of the plant.

As per the criteria of essentiality to complete the life cycle of the plants, they need 18 essential nutrients including carbon (C), hydrogen (H), oxygen (O), nitrogen (N), phosphorus (P), potassium (K), calcium (Ca), magnesium (Mg), sulfur (S), iron (Fe), manganese (Mn), copper (Cu), zinc (Zn), boron (B), molybdenum (Mo), nickel (Ni) and chlorine (Cl). These are presented in Table 1.1 along with their major forms of uptake. Among these nutrients, some are called macro nutrients which are required in higher quantity (in excess of 10 mmoles kg^{-1} of dry matter) *viz.* C, H, O, N, P, K, Ca, Mg and S, and some are micronutrients which are required in small quantity (less than 10 mmoles kg^{-1} of dry matter) viz., Fe, Mn, Cu, Zn, B, Mo, Ni and Cl. The macronutrients are again classified into primary nutrients (C, H, O, N, P and K) and secondary nutrients (Ca, Mg and S). Still there are some other nutrients which are known as beneficial elements that are not essential to complete the life cycle of the plants but required to enhance the growth of the plants (Marschner 2012). Among the primary nutrients, plants receive C, H, and O from water and air and all other nutrients are obtained from soil and other fertilizer sources. Deficiency of any one of these essential elements may lead to loss of yield and reduction in quality of the product.

Table 1.1 Essential Nutrients required for plant growth and the major forms of uptake

Nutrient	Chemical Symbol	Principal forms for uptake	Mobility in plants	Mobility in soil
Carbon	C	CO_2	Mobile	Mobile
Hydrogen	H	H_2O	Mobile	Mobile
Oxygen	O	H_2O, O_2	Mobile	Mobile
Nitrogen	N	NH_4^+, NO_3^-	Mobile	Mobile as NO_3^-, immobile as NH_4^+
Phosphorus	P	$H_2PO_4^-$, HPO_4^{2-}	Somewhat mobile	Immobile
Potassium	K	K^+	Very mobile	Somewhat mobile
Calcium	Ca	Ca^{2+}	Immobile	Somewhat mobile
Magnesium	Mg	Mg^{2+}	Somewhat mobile	Immobile
Sulfur	S	SO_4^{2-}, SO_2	Mobile	Mobile
Iron	Fe	Fe^{2+}, Fe^{3+}	Immobile	Very mobile
Manganese	Mn	Mn^{2+}	Immobile	Immobile
Boron	B	H_3BO_3	Immobile	Immobile
Zinc	Zn	Zn^{2+}	Immobile	Mobile
Copper	Cu	Cu^{2+}	Immobile	Immobile
Molybdenum	Mo	MoO_4^{2-}	Immobile	Somewhat mobile
Chlorine	Cl	Cl^-	Mobile	Mobile
Nickel	Ni	Ni^{2+}	Immobile	Somewhat mobile
Cobalt	Co	Co^{2+}	Mobile	Somewhat mobile

Table 1.2 Critical leaf concentrations for sufficiency of major mineral elements in crop plants (%)

Macronutrients	Sufficiency (%)
Nitrogen (N)	1.5–5.0
Phosphorus (P)	0.2–0.5
Potassium (K)	0.5–6.0
Calcium (ca)	0.05–2.0
Magnesium (mg)	0.15–1.0
Sulfur (S)	0.1–0.5

1.2.2 Role of Nutrients in Plants

Nutrients essential to plants are classified into macro-nutrients and micro-nutrients based on their contribution to plant growth and requirement. The critical leaf concentrations for sufficiency of major and micronutrients in crop plants are presented in Tables 1.2 and 1.3 respectively. Macronutrients comprise primary and secondary nutrients which are required in large quantities by the plants. Primary nutrients are C, H, O, N, P, K. Secondary nutrients are those generally required in smaller amounts compared to the primary nutrients. The secondary nutrients include Ca, Mg and S. The functions of these nutrients in plants are different. Those nutrients

Table 1.3 Critical leaf concentrations for sufficiency and toxicity of micronutrient elements in crop plants (%)

Micronutrients	Sufficiency (mg kg^{-1})	Toxicity (mg kg^{-1})
Iron (Fe)	5–150	>500
Manganese (Mn)	10–20	200–5300
Copper (cu)	1–5	15–30
Zinc (Zn)	15–30	100–300
Boron (B)	5–100	100–1000
Chlorine (cl)	10–60	40–70
Molybdenum (Mo)	0.1–1.0	1000
Nickel (Ni)	0.1	20–30
Cobalt (co)	0.1	0.5

that are of huge importance in human health are Ca, S, Fe, Zn and Se. Hence, the roles of these nutrients in plants and their deficiencies are described here.

1.2.2.1 Role of Calcium in Plants

Calcium is the main constituent of cell wall of plants. It plays a major role in the formation of cell wall membrane and its plasticity affecting the cell division. It promotes early root development. Calcium helps in activating many enzyme systems involved in protein synthesis and transfer of carbohydrates. It is involved in the development of root hairs and hence helps in improving the nutrient absorbing capacity of plants. Calcium supports seed production and enhances the stiffness of straw. Calcium acts as a base for neutralizing the organic acids which are toxic to plants. It helps in better absorption of nitrogen and micronutrients including Cu, Zn, B and Mn by correcting acidity. Calcium also influences the water economy of the plants.

1.2.2.2 Deficiency Symptoms of Calcium in Plants

Calcium deficiency first appears in younger leaves and leaf tips as it is immobile in plant system (Karthika et al. 2018). The characteristic foliar symptoms of calcium deficiency are necrotic lesions on leaf margins and tips, brownish leaf veins and leaf deformities. Due to calcium deficiency, young leaves appear smaller in shape than the normal leaves. The growing tips of roots and leaves get distorted; turn brown and show die back symptoms at the tips and margins. Necrotic spots appear on leaves and finally the terminal bud dies. Root system development will be poor and often lead to rotting. Small and lesser number of nodules will form in legumes due to calcium deficiency. In citrus plants, the leaves turn green to light green towards the margins when calcium is deficient in the plants. Due to the lack of adequate calcium, newly emerging leaves may stick together at the margins which lead to tearing as the leaves uncurl and expand. This also leads to weakening of stem structure. Young leaves may be cupped and crinkled along with terminal bud damage in some plants. Premature falling of flower and flower buds occur due to calcium deficiency in some plants.

1.2.2.3 Role of Sulfur in Plants

Sulfur is a constituent of certain essential amino acids. It is involved in the metabolism of biotin, thiamin and co-enzyme A. Even though sulfur is not a constituent of chlorophyll, it helps in chlorophyll formation process. Sulfur improves the vegetative and root growth. In legumes, sulfur promotes the formation of root nodules for nitrogen fixation. Sulfur also stimulates the formation of seeds in onion and garlic. Sulfur is an important constituent of some volatile compounds in mustard oil and many proteins and enzymes. It is actively involved in the oxidation-reduction systems in respiration process of plants.

1.2.2.4 Deficiency Symptoms of Sulfur in Plants

Plants grown under sulfur deficient conditions display stunted growth, chlorosis, anthocyanosis, premature and reduced flowering and reduced seed setting. Plants with sulfur deficiency show chlorotic symptoms and stay chlorotic for an extensive period until necrosis. Since sulfur is immobile in plants the symptoms appear first in younger leaves. Younger leaves turn chlorotic with evenly lighter colored veins. The deficiency symptom does not commonly appear in many plants. The overall growth rate is retarded due to a decrease in protein formation thus delaying the maturity. The plants appear as thin stemmed, stiff and woody. Symptoms may be similar to nitrogen deficiency symptoms. Sulfur limitation affects the iron uptake mechanisms and hence shows iron deficiency symptoms (Courbet et al. 2019). Sulfur deficiency in plants restricts their ability to utilize nitrogen and consequently leads to nitrogen deficiency symptoms also (Kaur et al. 2011). Sulfur deficiency in oilseeds causes disorders in protein metabolism thereby inducing ‘white blooming’ in which white flower color is produced from a surplus of petal cells with carbohydrates. This occurs as a result of the formation of leuco-anthocyanins during the periods of intense photosynthetic activity (Schnug and Haneklaus 2005). Excess concentration of sulfur may lead to interveinal yellowing of leaves and also the leaf tissues bum.

1.2.2.5 Role of Iron in Plants

Iron is a part of the heme enzyme system which takes part in oxidation-reduction reactions in plant metabolic processes like photosynthesis and respiration. It is a component of protein ferridoxin which is required for nitrate and sulfate reductions. It is essential for the formation of chlorophyll and required for the protein synthesis in chloroplast. Iron is involved in protein metabolism and it acts as oxygen carrier in respiration process. It regulates respiration and photosynthesis in plants.

1.2.2.6 Deficiency Symptoms of Iron in Plants

Since iron is immobile in plant system the deficiency symptoms first appear in younger leaves. The typical iron deficiency symptom is interveinal chlorosis with a clear distinction between veins and chlorotic areas especially in young leaves, as iron is involved in chlorophyll synthesis. As the deficiency progresses the youngest leaves appear as white. Plants also show retarded growth and the development of fruits and seeds are affected. Iron toxicity leads to hindered growth in plants. In rice, brown spots appear on upper corners of lower leaves and it spreads towards the base of the leaf.

1.2.2.7 Role of Zinc in Plants

Zinc is a constituent of many enzymes which regulate plant metabolic processes. It acts as a catalyst in the chlorophyll formation. Zinc is involved in the synthesis of RNA and protein. It influences the synthesis of growth hormones like indole acetic acid (IAA) in plants. It has a role in the reproduction of some plants and also it is associated with plant water uptake.

1.2.2.8 Deficiency Symptoms of Zinc in Plants

Since zinc is immobile in plants, the deficiency symptoms first appear on younger leaves as interveinal chlorosis analogous to iron deficiency. The difference from iron deficiency is that zinc deficiency symptoms appear as banding at the basal part of the leaf but interveinal chlorosis due to iron deficiency is exhibited along the entire length of the leaf. The plant shows stunted growth and severely dwarfed due to short internodes. There will be peculiar yellow stripping between the veins of the leaves. The young leaves are abnormally small, thick, narrow and mottled. The older leaves turn yellow and drop prematurely. Due to the lack of sufficient auxin production, leaves show bushy appearance. Zinc deficiency leads to less production of seeds and deformation of fruits.

1.2.2.9 Role of Selenium in Plants

Selenium is a beneficial element for higher plants, enhancing the photosynthesis, antioxidant metabolism, carbohydrates and secondary metabolites in plant leaves (Andrade et al. 2018). Selenium application at low concentration augments photosynthesis in plants. Availability of sulfur has a major effect on selenium build up due to antagonism effects of the two oxyanions. It can mitigate stress in plants due to its capacity to induce the synthesis of sulfur and nitrogen compounds. It can also stimulate antioxidants, enzymes and metabolites. Selenium in low doses can improve phytoremediation potential by augmenting photosynthesis and enhancing the capacity of plants to tolerate stress. Selenium increases plant growth by carbohydrate accumulation and delaying senescence in plants (Nawas et al. 2013; Kaur et al. 2018; Hajiboland et al. 2019). Selenium is found to mitigate abiotic stresses due to UV-B radiation (Golob et al. 2017), cadmium toxicity (Shahid et al. 2019), drought (Nawas et al. 2015) and fungal infections (Kornaś et al. 2019).

1.2.2.10 Toxicity Symptoms of Selenium in Plants

High levels of selenium lead to chlorosis in leaves. Later leaves dry and wither. Some plants safely accumulate very high amounts of selenium without any toxicity while others cannot.

1.2.3 Essential Nutrients in Human and their Role in Human Nutrition

1.2.3.1 Role of Calcium in Human

Calcium is one of the essential and abundant nutrients required for many vital functions in human body. It is mainly stored in teeth and bones of the body. Calcium is used in small amounts throughout the body. It is involved in muscle functions, nerve transmission, vascular contraction, vasodialation, intracellular signaling, hormone secretion etc. Changes in serum calcium concentration affect any of these functions. Low calcium is linked to higher risk of seizures as it is related to nerve transmission and intracellular signaling process. In pregnant ladies, more calcium is required for the growth of the fetus. Calcium acts as a co-factor of many enzymes.

The condition of low calcium levels in body is known as hypocalcemia. Low calcium may lead to memory loss or confusion, numbness and tingling in the hands, feet and face, muscle spasm, depression, hallucination, muscle cramps, weak and brittle nails, easy fracturing of the bones, slow growth of hair, fragile and thin skin, osteoporosis etc. Calcium has a very important role in neurotransmitter release and muscle contractions. Hence, calcium deficiency leads to seizures in otherwise healthy people.

High calcium in body leads to calcification of soft tissues and blood vessels. It may also lead to kidney problem and stone formation in kidney. High calcium may result in constipation in many.

1.2.3.2 Role of Sulfur in Human

Sulfur is an essential element for human body and is needed for most of the body tissues. Human body contains around 3–6% sulfur containing amino acids like cysteine, methionine, taurine etc. Sulfur is used for dermatological issues like wound healing and acute exposure to radioactive material. Sulfur is involved in metal transport, regulation of gene expression, free radical scavenging, protein stabilization and synthesis, enzyme functionality, tissue integrity and protection, remodeling of extracellular matrix components, DNA methylation and repair, lipid metabolism, detoxification of xenobiotics and signaling molecules etc. (Palego 2015). Sulfur plays a vital role in proper cross-linking of the connective tissues such as skin, tendon and ligaments and extracellular proteins like GAGs and hyaluronic acids. Sulfur, as a component of glutathione, plays a key role in the liver as a phase 2 detoxification in the liver. Sulfur is involved in stress response and exercise recovery.

For the formation of collagen and keratin, sulfur is required and hence the deficiency of sulfur can lead to problems in hair, nail and skin. The shortage of methionine leads to the build-up of hydrogen peroxide in hair follicles and a gradual loss of hair colour (Wood et al. 2009). Deficiency of thiamine, a sulfur containing vitamin, results in malfunctioning of central nervous system and circulatory system which cause beriberi (Shils et al. 2006, 2010).

1.2.3.3 Role of Iron in Human

Iron acts as a co-factor for a large number of enzymes engaged in oxidative phosphorylation which is the metabolic pathway that converts nutrients to energy. Iron is a major component of hemoglobin which transfers oxygen from the lungs to the tissues. It also plays a role in the metabolism as part of some proteins and enzymes. The iron from diet is stored in a protein complex called ferritin. Another iron protein made in the liver transport iron within the blood to other locations for storage. Iron is a part of another protein myoglobin which stores oxygen in the muscle tissues. Iron is an important nutrient required for brain development and growth of children. It is essential for the normal production and function of several cells and hormones.

Deficiency of iron leads to anaemia. Under mild deficiency conditions anaemia can be associated with some functional impairment affecting the cognitive development, work capacity and immunity mechanisms. During pregnancy period iron deficiency may lead to a variety of undesirable outcomes for both mother and child like increased risk of sepsis, perinatal mortality, maternal mortality and low birth weight. Anaemia due to iron deficiency reduces learning ability and is linked with increased rates of morbidity.

Toxic iron (Fe) levels in the body may be a result of genetic or metabolic disorders, recurrent blood transfusions or too much intake. Excess Fe levels over an elongated period could lead to liver and heart damage, skin changes and diabetes (Fraga and Oteiza 2002).

1.2.3.4 Role of Zinc in Human

More than one-third of the world's population is affected by zinc deficiency. Zinc is mandatory to carry out a number of physiological functions in human body. Zinc is very much essential for the proper working of the immune system of the body. It plays a key role in cell division, cell growth, healing of wounds, normal vision, immunity, fertility and carbohydrate breakdown process. Zinc is required for sensing smell and taste. Zinc is essential for growth and development for the child during the pregnancy period, infancy and childhood. For the proper functioning of insulin, zinc is very much needed. Zinc is necessary for the proper functioning of several enzymes that help in nerve function, metabolism, digestion and many other processes. For DNA synthesis zinc is required.

Zinc deficiency leads to poor appetite, delayed wound healing, loss of taste or smell, depressed mood, diarrhoea, hair loss etc. Zinc toxicity leads to nausea, vomiting, abdominal pain, headache etc.

1.2.3.5 Role of Selenium in Human

The trace nutrient element selenium has an important role in human health. Selenium acts as an antioxidant and catalyst for thyroid hormone production and DNA synthesis. It is a constituent of selenoproteins which protects the body from oxidative stress which can lead to cardiovascular diseases. About 30 selenoproteins were identified in mammals and in humans it is around 25. For the proper functioning of immune system selenium is needed and it is found to counteract with development of virulence due to HIV. Selenium is very much required for proper sperm motility and helps to reduce the possibility of miscarriage. Deficiency of selenium may lead to increased risk of cardiovascular diseases in human. It also leads to down regulate the cholesterol levels in blood plasma. In China, keshan disease, a widespread cardiomyopathy developed due to selenium deficiency. In human body, low selenium levels cause a reduced immune response to polio vaccination and suboptimal levels are directly correlated with a wide variety of diseases like cystic fibrosis, heart diseases, alzheimer's disease, cancer, cognitive decline, oxidative stress related disorders, impairment in immune functions, hypothyroidism, reduced fertility etc. (Rayman 2012). Another disease reported due to the severe deficiency of selenium in human is Kaschin-Beck disease characterized by the deterioration of cartilage and joints. The optimum intake level of selenium for human body is 400 micrograms daily for adults.

Excess selenium results in several symptoms including gastrointestinal upset, fatigue, hair loss and mild nerve damage (Guerrero-Romero and Rodríguez-Morán 2005). Long term consumption of high doses of selenium may lead to a specific disease called selenosis. Excessive consumption also damages the cardiovascular, neurological, gastrointestinal and hematopoietic systems (Yang et al. 1983; Raisbeck 2000). Typical symptoms of selenosis are hair loss, brittle hair, stratified and thickened nails, gastrointestinal disturbances and garlic odor from breath and skin.

1.3 Importance of Minerals in Plants and Food System/Web

Quality of food results from subjective and objective quality traits of food crops (Schupan 1961). Here we describe the importance of minerals in plants and food web and their role in maintaining the quality of food. We try to look at how the deficiency of a particular nutrient affects the quality of food and how it causes imbalance in human diet and health.

1.3.1 Calcium

In case of calcium, characteristic calcium deficiency symptoms carry specific name in specific crops like bitter pit in apple, blossom end rot in tomato, pepper and zucchini etc. Calcium is reported to influence the crop quality especially in fruits like apple. Increased calcium application reduced the bitter pit in apple in which the fruit skin develops pits and brown spots appear on the skin resulting in bitter taste. In carrots, calcium deficiency leads to cavity spot with typical symptoms of oval necrotic spots developing into craters. Calcium pectate in the cell wall is very important for resistance against fungal and bacterial pathogens and hence calcium is important in storage and post-harvest quality of fruits (Winkler and Knoche 2019).

1.3.2 Sulfur

Deficiency of sulphur in plants significantly affect the quality of plant-based food and feed. The S containing essential amino acid methionine is often a limiting nutritional factor in seed rich diets. In the cereal protein composition, S has a prominent effect and it may affect the protein content and quality. Sulfur deficiency in wheat leads to the accumulation of low sulfur storage proteins and high molecular weight subunits of glutenin. This influences baking quality of wheat since the disulphide bridging in dough preparation needs polymerisation of glutelin fraction to form optimum loaf volume (Liu et al. 2011). In Brassicales, sulfur fertilization results in increased content of S-rich metabolite called glucosinolates which are important in the defence mechanisms of the plants against pests and diseases. In cruciferous vegetables like onion and garlic, S is responsible for the smell and taste

(Karthika et al. 2018). Increased sulfur application improves the quality characteristics like pungent smell in onions. As a flavour component, crop bio-fumigants and cancer prevention agents, S compounds are important for human also.

1.3.3 Iron

Deficiency of Fe is one of the most important abiotic stress factors which cause a reduction the yield and quality of fruits and vegetables (Abadía et al. 2011). Lack of sufficient quantity of Fe lead to significant yield reduction in leafy vegetables and also affect quality in terms of colour, hardness and acidity (Alvarez-Fernandez et al. 2003). In Pakchoi, Fe deficiency reduced the nutrient content, soluble proteins and vitamin C leading to undesirable quality (Ding et al. 2007). In cabbage, lack of Fe nutrition stunts the growth and limits the yield (Karthika et al. 2020). In rape seed, Fe intake increased the yield and quantity, and also increased the height of the plant, amount of nitrate reductase activity and photosynthesis. In strawberry, fruit quality increased with foliar Fe fertilization (Karp et al. 2002). Application of Fe in tomato and citrus species improved the fruit quality (Sanchez-Sanchez et al. 2002, 2005). Deficiency of Fe decreased fruit quality and yield, and ultimately led to plant death (Alvarez-Fernandez et al. 2003, 2006). Iron treatment has significant positive effect on grain yield; yield components, quality as well as carotenoids contents of bread wheat (*Triticum aestivum* L.) (Ghafari and Raazmjoo 2013). Foliar application of ferrous sulphur significantly improved the growth and yield parameters of tomato plants in green house (Zarghamnejad et al. 2015); Foliar application of Fe improved the Fe content in fenugreek seed (Chhibba et al. 2007); protein and Fe contents of wheat grain (Zeidan et al. 2010).

1.3.4 Zinc

Deficiency of Zn may result in low productivity of the crop. A low concentration of Zn in seeds will lead to low dietary intake of Zn while consumption (Pathak et al. 2012). In rice, foliar application of Zn resulted in enhanced yield and grain Zn content (Phattarakul et al. 2012). Zinc fertilizers improved the dry matter content, grain yield and grain Zn concentration in rice. Application of micronutrients enhanced the Zn content in rice grain and increased the straw yield (Bana et al. 2022; Haefele et al. 2022). In rice, Zn deficiency was identified as *Khaira* disease in India. Application of Zn in wheat improved the kernel weight, grain yield and protein content (Malakouti 2000). Foliar Zn application in wheat increased the grain Zn concentration and decreased the concentration of anti-nutrients like phytate (Cakmak and Kutman 2018a, b). In Zn deficient soils, application of Zn fertilizers resulted in increased yield and crop quality in cereals. Also, Zn application improved the fruit quality in horticultural crops like apple, grape, apricot, citrus etc. (Malakouti 2001).

1.3.5 Selenium

Selenium (Se) in plants is found to improve the secondary metabolism by increasing tocopherol, flavonoids, phenolic compounds, ascorbic acid and vitamin A (Hartikainen et al. 2000; Xu et al. 2003; Ríos et al. 2008; Businelli et al. 2015). These secondary metabolites are health promoting phytochemicals which prevent a series of human diseases and also used as active ingredients in medicines (El-Nakhel et al. 2019). But it was reported that high concentration of Se in plants are phytotoxic by inhibiting growth and modifying the nutritional characters of plants (Hartikainen et al. 2000). Selenium has an effect on the quality of fruits and vegetables. In *Camelia oleifera* selenium application resulted in higher cellular content of linoleic acid and sterol and a lower content of oleic acid (Song et al. 2015). In broccoli, Se treatment was found to have a positive effect on retaining the sensory and postharvest quality by means of lowering the respiratory intensity and ethylene production (Lv et al. 2017). In chicory and lettuce, the quality improvement was achieved by lowering the content of phenylalanine ammonia lyase activity and ethylene production (Malorgio et al. 2009). Hu et al. (2003) reported that selenium increased the plant yield and total amino acid and vitamin C content in green tea. In tomato, the post-harvest gray mould disease caused by *Botritis cinerea* was effectively controlled by the application of selenium (Wu et al. 2016).

1.4 Factors Affecting Biofortification of Essential Minerals in Soil-Plant System

Biofortification is an efficient method to improve the nutrient content of crops especially in cereals, herbages, corn etc. But in conventional fortification artificial additives are required. Compared to conventional fortification, biofortification is more effective and economic as it improves the micronutrient content of the crops. There are many factors affecting the efficiency of biofortification of mineral elements. They include plant species: genotypes and phenotypes, soil characteristics, nature of application, rate/form of application and climatic conditions (Ebrahimi et al. 2019; El-ramady et al. 2021; Izydorczyk et al. 2021).

Factors controlling bioavailability of micronutrients in crops and human can be classified into two, soil factors and plant factors. The soil factors include soil pH, soil moisture, soil temperature, soil texture, soil organic matter content, nutrient interactions, rhizospheric environment, clay content, oxidation states etc. Plant factors like genetic variability, antinutritional compounds etc. Other factors include the food matrix, processing and storage of food also affect the bioavailability of nutrients.

1.4.1 Plant Factors

There is genetic variability among plants which greatly influence the bioavailability of various nutrients to human. Hence developing plant genotypes containing high content of micronutrients and an improved bioavailability of these micronutrients can improve the human nutrition and health. The edible portion of a plant contains anti-nutrients and the concentration of which depends on the genetic and environmental factors, can lower the bioavailability of dietary iron, zinc and other nutrients. Like an anti-nutrient is phytic acid which fixes iron and zinc and hence reduces the bioavailability of iron and zinc in human.

1.4.1.1 Genetic Factors

Genetic diversity is a major factor which influences the bioavailability of various nutrients. Exploiting genetic disparity in crop plants is one of the most prevailing tools to improve the level of nutrients in a given diet on a large scale. Modern agriculture not only depends on diversity of crops but also diversity of nutrition. Concentrating on high yielding cereal crops like wheat, rice, maize etc. could drastically reduce the production of nutritionally rich grains. Reliance on a few crops is the main reason for widespread deficiencies of iron and zinc. Breeding programmes must make use of the genetic diversity from crop wild relatives, landraces and old cultivars to improve the nutrient content of modern varieties.

1.4.1.2 Anti-Nutrients

Dietary phytate is one of the major anti-nutrients which prevent the absorption of nutrients to the plant and human system. Phytates are insoluble complexes which cannot be digested or absorbed because of the absence of intestinal phytate enzymes in humans. A diet with high phytic acid and phenols like wheat is the major source of human calories.

1.4.2 *Soil Factors*

1.4.2.1 pH

Soil pH is one of the main factors influencing the availability of nutrients to plants. In general, soil pH controls solubility, mobility and concentration of nutrient ions in soil solution and also the acquisition of nutrients by plants (Fageria et al. 1997). As the soil pH increases, micronutrient availability in soil decreases except for Mo. Between pH 5 and 7, cations are strongly held in soil solution. At lower pH, B compounds are soluble and stay available to plants (Deb et al. 2009). With increasing pH the solubility of free Fe decreases and which in turn reduces the availability of iron in calcareous soils. As the soil pH increased from 5 to 8, the soil solution Zn^{2+} concentration reduced 1000 times and became almost 10^{-10} M. As a result, a raise in soil pH is linked with acute decline in concentrations of Zn in plant tissues (Marschner 1995). Molybdenum availability is a severe problem in acid soils as it gets fixed with Fe and Al compounds and also with silicates, hence unavailable for plant use (Choudhary and Suri 2014). In loamy soils with relatively low organic matter, an increase in soil pH from 5.9 to 6.4 resulted in an increase in Se availability.

1.4.2.2 Soil Moisture and Temperature

Soil moisture plays an important role in providing a suitable medium for diffusion of nutrients to roots. Wet soil conditions at low temperature decrease the availability of micronutrient cations (Choudhary and Suri 2014; Kumar et al. 2016). Zinc deficiency often appears when the soil temperature goes down during the winter season especially due to freezing. This leads to decreased Zn solubility of Zn present in the soil. Mineralization of micronutrients from soil organic matter is greatly influenced by the soil moisture content. Water stagnation for long time increases the soil pH and reduces the redox potential which in turn decreases the availability of S, B, Cu and Zn in such soils. Low temperature affects the micronutrient availability resulting in decreased root activity, nutrient diffusion, dissolution rate etc. As the soil temperature increases, Cu content increases but manganese concentration decreases with soil temperature rise. In Poaceae family low soil temperature reduces the production of phytosiderophores and mobilization and uptake of soil Fe but elevated soil temperature reduces Fe uptake by intensifying the microbial decomposition of phytosiderophores and enhancing CO_2 production (Marschner et al. 1986).

1.4.2.3 Soil Texture

Nutrient availability is strongly influenced by soil texture. Sandy soils (coarse textured soils) are deficient in many nutrients but clayey soils are rich in plant nutrients as the organic matter in clayey soil can hold nutrients in a better way than sandy soils. Leaching in sandy soil often results in loss of plant nutrients along with the leachate and hence unavailable to plants. Mineral soils with low organic matter also have low plant available nutrients as many of the parent materials are originally deficient of nutrient elements.

1.4.2.4 Nutrient Interactions

Interaction of nutrients may occur in plants and soil. The factors influencing the nutrient acquisition depends not only on the capacity of the soil to supply nutrients to plants but also with the balance among other nutrients in the soil (Deb et al. 2009).

In soils with high P concentration, Zn absorption by plants may reduce and hence the yield. Excess application of K decreases the concentration of Mn and Fe content in rice plants. Addition of dolomite to improve the soil pH often reduces the availability of Zn at higher pH levels. It also decreases uptake of heavy metals by the root system due to the physiological antagonism of Ca with other metal ions. As the concentration of Fe increases, Mn and Zn become deficient whilst high levels of Mn can induce Fe and Zn deficiencies. Similarly, higher rates of Zn application induce Cu deficiency in cereals especially in wheat and barley because of antagonistic relationship between Cu and Zn. In citrus, high Cu content in soil results in Fe chlorosis. Copper toxicity can be eliminated by the application of Fe fertilizer in soil. Copper is having an antagonistic interaction with Mo also. With Fe, Mo can precipitate iron molybdate in roots and can lead to Fe deficiency.

1.4.2.5 Organic Matter

Organic matter is a store house of many nutrients, especially micronutrients. Organic matter reduced the availability of nutrients by forming insoluble compounds through complexation with humic acid, lignin and other organic compounds of high molecular weight. Organic matter can improve the availability of nutrients by solubilization and mobilization of nutrients by the action of organic ligands like amino acids, short chain organic acids and other organic compounds (Mortvedt 2000).

1.4.2.6 Clay Content

Clay surfaces are having negatively charged particles and hence surrounded by a cloud of positively charged cations. The cation exchange capacity of soil generally increased with increase in pH and with increased cation exchange capacity metal cations are attracted towards negative charged sites on soil colloids and hence the soil solution is exhausted resulting in decreased metal availability to plants.

1.4.2.7 Oxidation State

In soil, the oxidation–reduction reactions occur at various conditions. Reducing conditions occur as result of water logging and the soil becomes anaerobic, while oxidized conditions are found in well drained aerobic soils. At neutral pH, the reduced states Fe, Mn and Cu are more soluble than the oxidized states. At anerobic conditions with low pH, micronutrient cations are more available at limited drainage and hence the flooded soils usually show higher availability than aerated soils. In well drained calcareous soils at high pH, deficiency of Fe, Mn, Zn etc. are reported in plants even the concentration of these elements are adequate in the soil.

1.4.2.8 Rhizosphere

Rhizospheric microenvironment plays an important role in the availability of nutrients. Availability of nutrients in rhizosphere depends on the combined effects of soil interaction between plant roots and microbes in the nearby soils. Rhizospheric chemical environment is entirely different from the bulk of the soil. Microbes in the rhizosphere constantly produce chelating agents by decomposing the plant and animal residues. These chelates help to transform solid phase micronutrient cations into soluble metal complexes and thus increase the availability of insoluble micronutrients to plants. The arbuscular mycorrhizal fungi (AMF) present in the rhizosphere changes the rhizospheric environment by way of exudation or secretion of organic acids or chelating agents which can solubilize and mobilize plant nutrients from organic and inorganic complexes. Similar to AMF, in rhizosphere there are numerous microbes that play very important role in nutrient regulations.

1.4.3 Other Factors

Other factors include the cultural practices, food processing and storage practices. Cultural practices like dehulling, milling, fermenting and cooking are usually practiced to improve texture and add quality to the final food product. But these processes are known to take out the important minerals, vitamins and other micronutrients thereby resulting in a noticeable reduction of their concentration in the final processed food. The aleurone layer and embryo of wheat seeds contain around ten-fold Zn concentration compared to its endosperm concentration and removing the zinc rich aleurone layer during the time of processing will lead to considerable reduction in zinc content of the final product.

1.5 Agronomic Biofortification of Essential Mineral Nutrients for Improving Crop Quality and Ensuring Food Security

Agronomic biofortification is an effective mechanism of enriching the edible portion or seed with essential nutrient for better human nutrition. The enrichment of staple crops with essential nutrients through fertilization techniques has received worldwide attention. Addition of chemical fertilizer and organic amendments to increase crop production is well known. Fertilizers, natural or industrial one either supply one or more essential nutrient for crop growth and development. Comparing various essential nutrients, N, P and K are the macronutrients which are largely applied by farmers. However, secondary nutrients, Ca, Mg, S, though essential are often less applied by crop producers. Similarly, micronutrients, Fe, Mn, Cu, Zn and B etc. are essential micronutrients that exhibit widespread deficiency in soils and in crops thus directly affecting human health (Ritchie and Roser 2019). However, present agricultural practices only aim to increase crop yields rather than nutritional quality, which is a major challenge for enhancing nutrient content in grains. Among nutrients, most common deficiency is associated with micronutrients like Fe, Zn, Se etc. Recently FAO highlighted that proper nutrition is vital for good health and could be achieved by sustainable agriculture practices. Of many reasons, poor micronutrient absorption is mainly due to presence of phytic acid and polyphenols that form complex compounds with dietary minerals. Phytates present in seeds is a major concern for adsorption of micronutrients like Fe and Zn, influences bioavailability of nutrients and their absorption in human body. However, ascorbic acid and other organic compounds act as enhancers in mineral intake, promoting nutrient absorption in gastrointestinal track.

Biofortification is an effective management practice to enrich seeds with nutrient content, enhance nutrient uptake and improve human health. Agronomic, conventional breeding and genetic engineering are the three common methods of biofortification. Amongst these, agronomic biofortification is the cheapest, fastest and easiest approach of nutrient enrichment of seeds. Agronomic biofortification methods improve the solubility and availability of essential nutrients in soil solution, thus enhancing its crop uptake and accumulation in grain/edible portion. Moreover, in the present world economy, with high inflation rates post covid could further push down the poor from affording good quality food and other nutrient supplements. Therefore, agronomic biofortification is the best way to reach the poorest population to improve desired nutrient content in the diet. Some of the common agronomic techniques for biofortification are listed below.

1.5.1 Addition of Fertilizers

Fertilizer application is the easiest agronomic approach to supply essential nutrients to crop nutrition. Nutrient concentration in grains is strongly linked with the native/inherent soil fertility, which at optimum moisture contents and nutrient cycling increases its uptake by crops. In general, fertilizers when applied either to roots or leaves, is effective in enhancing its bioavailability to crops. Once the fertilizers are applied to soil, it is directly absorbed by crops and subsequently translocated to various tissues within plants. However, soil application of fertilizers is subjected to leaching/runoff losses, immobilization or fixation of nutrients, microbial consumption etc. Therefore, foliar application is considered as easy technique for biofortification of plant parts though modifications in soil properties by integrated nutrient management approaches, including manures, compost, biofertilizers etc. are important strategies to enhance nutrient bioavailability in soil and crop uptake. However, foliar pathway is considered advantageous over non-foliar, soil application.

1.5.2 Foliar Spray of Fertilizers

Unlike root application, foliar spray of nutrients is an effective approach where inorganic fertilizers are not easily translocated to edible parts. Foliar fertilization is generally known as the effective way of enriching grains with essential nutrients. Foliar spraying transports nutrients and thus is a handy approach to enhance N, P, K, S, Fe, Zn etc. in edible plant parts. Soluble fertilizers of macro and micronutrients are mixed with water to make a formulation which is directly sprayed upon crops especially at fruiting or seed setting stages. Sometimes, lime is also applied with foliar fertilizer spray, where lime acts a neutralizing agent that augment nutrient accumulation and translocation within plant parts. During foliar spray, micronutrients especially Fe and Zn can be easily absorbed by leaves, stems, or tissues and translocated through phloem or xylem. Cakmak and Kutman (2018a, b) and Kiran et al. (2021) highlighted that foliar application of micronutrients at specific growth stages of crops, can improved Fe or Zn content in grain, enhance crop yield in micronutrient deficient soils. It is interesting to note that effectiveness of foliar application depend upon, crop varieties/species, crop stages, site specific conditions governs efficacy of nutrients or minerals. Prom-U-Thai et al. (2020) reported that foliar application of Zn on rice resulted in difference in Zn use efficiency that varied from site to site depending upon crop types. Similar foliar spray on wheat crop enhanced grain Zn concentration. This can be explained by increased Zn concentration (95%) and its bioavailability (74%) was observed in wheat crop subjected to 0.5% zinc sulphate foliar spray (Hussain et al. 2012). Increased translocation of nutrients from leaves to seeds by foliar spraying of Fe and Zn has been reported by other workers (Peck et al. 2008; Velu et al. 2014).

Timing of foliar spray is an important factor determining the efficacy of applied essential nutrients. Cakmak et al. (2010) reported that in wheat, milk to grain filling stage is the best for foliar application to enrich grains with Zn content especially in aleurone and embryo parts (Yang et al. 2011). Velu et al. (2014) opined that foliar spraying during grain physiological maturity stage nearing harvest, is considered efficient method to accumulate nutrients in seeds. Another study by Palmer et al. (2014) reported that optimum time of foliar application of Zn in wheat crops can be 9–12 days post anthesis, when phloem enriches wheat caryopsis with essential nutrients.

1.5.3 Seed Coating and Priming

Seed coating with essential nutrients or micronutrients, or similar materials is effective in seed embryo germination, which later help in growth and development of plant. This method of seed priming provides initial boost for seeds to fight against deficiency or hidden hunger, similar to animal vaccination. Seed priming is a technique in which the seeds of crops/plant parts are added to a nutrient solution or water for efficient absorption of essential nutrients and prevent its deficiency. Coating of seeds with fertilizers or nutrients, other materials, nanoparticles, micromolecules injected with essential nutrients, are a few novel approaches of biofortifying seed materials with nutrient or minerals.

1.5.4 Addition of Biofertilizers

Microorganism or microbial consortia improves nutrient mineralization, increases nutrient uptake, utilization efficiency, directly influencing crop yield. Plant growth promoting rhizobacteria (PGPR) are beneficial microbes reducing fertilizer and other chemical input in cropping system. Microbes secrete organic acids, exudates that enhance mineralization of nutrients. Phytosiderophores secreted by these organisms in rhizosphere improve uptake of Zn, Fe and other micronutrients. The PGPR has many benefits of enhancing nutrient availability, improving root functions, suppress plant diseases, augment crop growth and development. Among PGPR, Bacillus is the most commonly used for micronutrient biofortification. Some of the studies carried out with *Bacillus subtilis* and *Bacillus aryabhattai* increases Zn enrichment in maize grains (Mumtaz et al. 2018), another species of *Bacillus pichinotyi*-YAM2, *Bacillus cereus*-YAP6 etc. strains are used to wheat fortification with Fe and Se (Yasin et al. 2015). *Rhizobium*, blue green algae, *Azolla*, *Azotobacter* etc. are a few important biofertilizers commonly utilized to improve nitrogen fixation and assimilation by crops. Mycorrhiza is another group of organisms that are effective biofortification techniques to reduce human malnutrition (Wang and Qiu 2006). The AM (arbuscular mycorrhizae) fungi is a versatile species, which more effectively and efficiently utilizes the mineral nutrients such as P, Fe, Zn, Cu and improve water availability to plant roots. Almost all the pulse crops are associated with AM that enhances nutrient uptake and subsequently its accumulation in grains. Combining various sources of biofertilizers, PGPR, AM etc. are ecofriendly techniques for not only enhancing crop yield but improving nutritional quality of food grains.

1.5.5 *Nano Fertilizers*

Nano fertilizers techniques decreasing the particle size into nanoscale in the form of nano particles, containing macro and micronutrients applied in control mode. The reduction in particle size increases surface area of particles leading to more contact of fertilizers/nutrients with plant parts that will enhance nutrient uptake and efficiency of applied fertilizers (Szerement et al. 2022). Thus, nano fertilizers are an opportunity of providing "right amount of fertilizers" at "right time" and "right place" to crop-soil-environment. Uptake, mobilization and accumulation of essential nutrients in crops vary according to crop species and type, chemical composition, and concentration of nano particles. For instance, Di et al. (2019) reported decline in wheat germination when seeds were treated with ZnO based nanoparticles. However, another study by Zhang et al. (2015) reported no effect of similar nanoparticles on growth of cucumber and corn plants. Application of nanofertilizers will not only decrease the losses, but amount could be reduced to half than conventional fertilizers. Application of Fe nanofertilizers significantly decreased the dosage of conventional Fe fertilizers without affecting growth and metabolism of crops (Elanchezhian et al. 2017). Similarly, Tarafder et al. (2020) reported that application of nano fertilizers with a combination of urea –hydroxyapaptite and Fe, Zn, Cu nanoparticles improved total uptake of Cu, Fe and Zn by 16, 3 and 146 times respectively.

1.5.6 *Effect of Nitrogen Fertilizers*

Nitrogen fertilizers play a dominant role in improving crop growth and development. Grain quality and nutrient content is directly influenced by N rate, time of application and method. Nitrogen is a critical component of amino acids, energy compounds viz., ATP etc. and involved in various plant metabolic activities. Nutrient uptake and utilization efficiency of all the essential nutrients is directly influenced by nitrogen assimilation in crops. Cakmak et al. (2010) reported an improved accumulation of Zn in seeds due to higher Zn uptake and translocation with increased N levels in durum wheat. Nitrogen supply enhances release of phytosiderophores from crop roots that improve Fe uptake and translocation in wheat crops grown in Fe deficient soils (Aciksoz et al. 2011).

1.5.7 Crop Diversification/Crop Rotation

Monocropping system such as cereal based cropping system usually causes soil degradation, nutrient deficiencies, decline in crop productivity and increasing pest-disease menace in many regions. However, crop diversification or crop rotation with different crop types improves soil health, nutrient status, increases crop yield and provide livelihood security to farmers. Legumes are the best option for crop rotation which not only enhances soil fertility by increasing N content, but also improve subsequent crop yield. Pulse crops in crop rotation improve nutrient cycling and mineralization enhances the bioavailability of macro and micronutrients in soil. Addition of N by legumes thus could enhance Zn and Fe content in crop grains.

1.6 Human Health *Vis-a-Vis* Essential Nutrients/Minerals

Deterioration in soil health either due to nutrient imbalances or other factors lead to reduction in soil quality (Kumar and Karthika 2020) which in turn affect the human health. Soil health is thus essential in maintaining human health and essential for nutritional security. Cereal crops such as rice, wheat, maize are the major staple foods consumed in South Asian and African countries. Human nutrient deficiency or malnutrition is a worldwide concern due to consumption of plant-based diet which contains insufficient nutrients, vitamins, and minerals. In a recent study, it is estimated that globally 2 billion people suffer from one or other micronutrient deficiency (Prom-U-Thai et al. 2020). Deficiencies of essential nutrients or minerals are commonly observed in pregnant women, lactating mothers, sportsmen, and labourers. Most often, hidden hunger is the common phenomenon borne out of nutrient deficiency in humans. Agronomic biofortification technique increases nutrient content in seeds or edible parts for better utilization by human body. Some of the deficiency symptoms in human health due to poor intake of essential elements are given in Table 1.4.

Table 1.4 Effect of mineral deficiency on human health

Essential nutrient	Deficiency symptoms in human
Iron	Fatigue, weakness, dizziness, reduced work capacity, drastic decline in haemoglobin, maternal mortality
Zinc	Low weight gain, diarrhoea, pneumonia in infants; dwarfing among toddlers and children, taste sensitivity, chronic non-healing ulcers, recurring infections and eyesight problems
Selenium	Foggy memory, fatigue and frequent cold and flu
Calcium	Affect bones, teeth, osteoporosis, fracture
Sulfur	Tendonitis, arthritis and muscle and joint stiffness

1.6.1 Iron Deficiency

About 1.62 billion people suffer from anaemia due to Fe deficiency (Prasad 2014). It has been found that polished rice has very low Fe concentration (0.5–0.6 mg kg^{-1}), whereas, wheat and maize have 34–38 mg kg^{-1} Fe content. Studies at IRRI thus found that iron deficiency causes more than 85% of maternal mortality and severely affect mental health of children. Blanket dose of Fe as foliar spray on cereal crops is recommended by Prasad and Shivay (2018a, b) as biofortification of cereal grains could reduce anaemia, improve haemoglobin content, and reduce mortality during pregnancy. This approach could benefit more people living below poverty line. In addition, including green leafy vegetables which are good source of Fe in cereal diet on regular basis can alleviate Fe deficiency. Application of nitrogen fertilizers along with foliar spray of Fe enhances Fe content in spinach leaves (Yasin et al. 2015). Similarly, Ghaly et al. (2017) reported that recommended N doses improved Fe content in leafy vegetables as compared to control treatment. Biofortified foods with essential nutrients and minerals have been tested for years. Positive results of biofortification were observed on humans (Rosado et al. 2009).

1.6.2 Zinc Deficiency

Another nutrient is Zn, deficiency of which is commonly reported in many countries during the last decade. Prasad (2013) reported dwarfism and hypogonadism in adolescents' group in countries of Iran and Egypt due to Zn deficiency. Compared to Fe content in soil, Zn is 50 mg kg^{-1} which is almost one thousand parts of Fe or even less. This results in Zn deficiency in soils under cropping system worldwide (Alloway 2008). Foliar application and Zn coated urea are the most common mechanisms utilized to improve Zn content in grains of rice, wheat, chickpea crops (Shivay and Prasad 2014). Presence of phytate or phytic acid acts as an inhibitor for Zn absorption in human body. Hotz and Brown (2004) claimed that phytic acid to Zn molar ratio is an important criterion for determining bioavailability of Zn. They reported, phytic acid to Zn molar ratio in most diets were < 5, 5–15 and > 15 and represented high (50%), medium (30%) and low (15%) nutrient absorption levels respectively. However, Zn absorption by human gut is reported as 15–30% only, resulting in potential risk of growing Zn deficiency in humans (Clemens 2014).

1.6.3 Selenium Deficiency

Selenium mineral is important for animal health. However, famous keshan disease in humans was caused due to Se deficiency and mutation of Coxsack virus reported in

China (Ge and Yang 1993). In human body, Se-methionine and Se-cysteine are the common bioavailable forms of Se. Thus, nutrient is required to maintain glutathionase enzyme activity in human beings. On the other hand, Se toxicity causes loss of hair and nails in humans (Yang et al. 1983). Selemium bioavailability depends upon dietary intake factors such as lipids, metals which can form complex with Se reducing its absorption by human body. Wang et al. (2013) reported that biofortification of Se in crops enhance crop yield and grain Se content.

1.6.4 Calcium

Calcium is the most abundant nutrient in human body associated with healthy teeth and bones that plays an important role in blood clotting, nerve and heart functions. Calcium deficiency leads to hip fracture and osteoporosis. Calcium deficiency is mostly observed in acidic soil, which is ameliorated with lime, dolomites amendment. However, neutral to alkaline soils have abundant calcium content, optimum for plant growth and development. Groundnut is an oilseed, legume crop with high demand for Ca and sulfur nutrients. Calcium plays an important role in pegging stage, pod haulm and grain yield, determines oil quality and yield. The main source of Ca to humans is milk, which can improve Ca intake.

1.6.5 Sulfur

Sulfur is yet another essential nutrient which plays an important role in synthesis of protein in human body. Sulfur is a major constituent of methionine and cystine amino acid, which makes 3–6% of proteins in human body (Prasad and Shivay 2018a, b) and is an essential dietary intake. Sulfur is an essential component of thiamine, deficiency of which causes beri-beri disease. The recommended dietary intake of S is 1.4 mg kg^{-1} (Prasad 2014). Gypsum, elemental sulfur, pyrite, organic manures etc. are the most common sulfur fertilizers used in agriculture. Oilseed crops generally have a higher demand of S which is directly associated with seed yield and oil quality. According to Nimin et al. (2007), sulfur deficiency shows symptoms such as arthritis, muscle and joint stiffness, whereas reduced cystine level can cause renal failure and psychiatric disorders (Tapel et al. 2000; Bark et al. 2013). Garlic, mustard greens, onion, broccoli, eggs, cabbage and cauliflowers are common sulfur rich foods.

Biofortification approach is beneficial for farmers, as nutrient fortified food grains are in high demand and would improve the overall livelihood security. This approach will provide nutrient rich seeds for crop production. Improved crop yield and better nutritional quality of seed will improve land productivity. Foliar application of nutrients will reduce the amount of nutrients and prevent their losses. More field experiments on agronomic biofortification of essential nutrients should be performed under variable soil-crop-environment conditions to ensure success of fortified food grains. Impact of soil properties, nutrient cycling, release of nutrients/micronutrients on long term application and its accumulation in edible parts/grain, synergistic-antagonistic effect of nutrients etc. can improve the understanding of agronomic biofortification. Impact of nanoparticles-based fertilizers, biofertilizers on the release pattern and availability of nutrients needs to be explored under varied soil and climatic conditions. Future research is required to identify more rhizospheric microbes, explore rhizosphere properties that could provide crop-nutrient/mineral-microbial interaction studies for improving agronomic biofortification. There is still scarcity of the efficient methods of nutrient application to enhance nutrient uptake and its accumulation in seeds. Besides, both plant breeding and microbial based approaches with agronomic methods of fortification might provide long term solutions, which needs detailed investigation.

1.7 Plant Traits Improvement for Nutrient Accumulation in Grain

Nutrient accumulation in grain can be improved by biofortification either by conventional breeding approaches or by the application of biotechnology. Breeding can be carried out either directly to increase mineral concentration or to increase bioavailability. These breeding approaches are described in brief below.

1.7.1 Germplasm Evaluation and Screening

Nutrient efficient cultivars are essential in resulting in improved yield and nutrient concentration in crops. Germplasm evaluation is essential in understanding genetic variability as these genetic variations between cultivars play an important role in their response to nutrient efficacy. More nutrient efficient plant genotypes need to be screened to decrease nutrient deficiency. Efficiency could be by improved root uptake, nutrient translocation and utilization. When the soils are deficient in nutrients, the cultivars with efficient genotypes need to be screened for better efficiency and increase nutrient accumulation in them. This could be helpful in overcoming nutrient deficiency. For micronutrient enriched traits international genomes of different crops such as rice, wheat, beans, cassava and sweet potato were studied by Graham et al. 2001. Screening of cultivars for efficient and responsive genotypes can be one of the breeding techniques in achieving nutritional security.

1.7.2 Conventional Breeding

This technique is another breeding approach which has proved to be successful in achieving nutritional security. Conventional breeding has succeeded in developing more than 290 varieties of biofortified crops. These crops include rice, wheat, maize, cassava, beans, sweet potato etc. which were fortified for iron, zinc and vitamin A. This has helped in overcoming the nutritional deficiency of iron, zinc and vitamin A.

1.7.3 Biotechnological Interventions

There are many limitations to conventional breeding approach, which demands the interventions through biotechnological aspects/transgenics to carry out biofortification (Dhaliwal et al. 2022). Yield and nutrient concentration exist in an indirect relationship which is one of the main limitations of conventional breeding indicating the need for new gene editing techniques. Transgenic approaches have been found successful in the release of biofortified varieties of many crops. The methods involved in biofortification by transgenics are synthesis of transgenes, variations in transporters expression and suppressing the anti-nutrient concentration to improve the uptake and nutrient efficiency of crops. For staple crops like rice, genetic modifications were carried out to improve the Fe concentration in grain (Trijatmiko et al. 2016).

1.8 Challenges/Constraints in Agronomic Biofortification

Nutritional security of human can be achieved when food sources like plants and animals are nourished well. Intensified agricultural practices using high yielding varieties and high analysis fertilizers has led to increased food grain production over the years. This has no doubt ensured food security but at the same time, nutrient imbalances started to be a rising concern for soils. The same has reflected in the nutritional content on the food crops produced from such soils.

1.8.1 Challenges/Constraints

Agronomic biofortification is a frontier area of research and is a crop based strategy. Agronomic fortification, otherwise, called as crop biofortification, can be achieved through applying fertilizers those meet the inadequate supply of the lacking nutrient.

1.8.1.1 Restricted Research

Research on biofortification has not explored much and is restricted to a few major staple food crops. There are not enough studies to tell about the bioavailability of the micronutrients in the food crop and distribution of these within the plant system. The interaction studies of the trace elements like Zn, Fe and Se are meagre. So far research studies in this direction are carried out either as laboratory/pot trial or as hydroponics experiment where limited factors in a controlled way act (Nawaz et al. 2017). The conclusions from such experiments are not sufficient to predict the behaviour of numerous factors acting together in an uncontrolled manner and hence hardly any comparison is possible with the laboratory experiments.

1.8.1.2 Fixing Fertilizer Dosage of Application Is Tough

Global scenario of widespread micronutrient deficiencies often referred as 'hidden hunger' (Prom-u-thai et al. 2020) in food crops are resulted where soils are reported to be deficient in certain micronutrients, due to low inherent concentration of micronutrients and/or inadequate supply of micronutrients, improper management of acidity or other problem conditions like calcareous nature of soils etc. (Fageria and Baligar 2008). Since the hidden hunger is associated with elements required in trace amounts that are in a narrow range between toxicity and deficiency, utmost care is required to be taken for fixing the doses of application. Standardised information on crop specific, site specific and nutrient specific doses to be applied for the purpose of biofortification is lacking in this area, which would be helpful to growers as well as future researchers who can bring about required modifications from time to time.

1.8.1.3 Soil Conditions

The problematic soil health conditions as aforementioned in this chapter would further hinder the plant intake of the most essential plant micronutrients even when adequate inputs are provided by transforming the nutrients into unavailable forms for plant roots (Fageria et al. 2002).

1.8.1.4 Cost

The extra cost involved in micronutrient fertilizer application is a true concern for farmers. Proper awareness needs to be provided for making them understand the public health benefits and possible economic returns achievable through increased yield levels are much more in comparison with the additional cost involved. Agronomic biofortification is reported to be a cost-effective strategy by several researchers as it is on par with premium price of fortified grain produces (Harris et al. 2007; Shivay et al. 2008; Manzeke et al. 2014; Joy et al. 2016). Strategic planning is highly demanded from the end of policy makers for the wide acceptability of agronomically fortified foods among the farmers and consumers and in turn this would help in future interventions.

1.9 Future Prospects of Biofortification

Agronomic biofortification is found to be offering a low cost as well as rapid solution as compared to that of breeding approaches for alleviating hidden hunger. Staple food crops like rice, wheat and other cereals or millets are thought to be better choice for agronomic fortification thus making it available even for weaker economic section (Waters and Sankaran 2011). Micronutrient deficiency in crops is due to the inadequacy of soils with respect to the nutrients and this can be corrected by timely application of micronutrient containing fertilizers. It is observed that high response to micronutrient fertilizers would be expected under such nutrient stressed conditions and better uptake can also be expected. In comparison with soil application, response of plants to foliar application of micronutrient formulation has proven to yield better results. Recent advancements on development of nano- sized fertilizer products are promising as they are efficient even at low application doses. Biofertilizers that can solubilise or mobilise insoluble forms of micronutrients for plant uptake also play a pivotal role in agronomic biofortification.

Studies by Joy et al. (2015) revealed the benefits of agronomic fortification over fortification of finished products at the time of value addition processes. Zinc application to wheat crop through foliar application has much cost and health benefits over flour fortification. Zinc has got strong competition with cadmium which is considered to be non-essential and toxic element from plants' perspective. Applications of zinc through fertilizers formulations provide an opportunity to block the entry and accumulation of toxic elements in the plant system (Welch et al. 1999; Revees and Chaney 2008; Cakmak 2009). Dietary cadmium levels were reported in human beings consuming durum wheat and rice cultivated in places having elevated cadmium concentration.

Further, several reports suggest that biofortification with Zn, Se and Fe enable plants with a better defence against stress conditions (Morales Espinoza et al. 2019; Rizwan et al. 2019; Noreen et al. 2020) as in case of barley, salinity stress related effect was found mitigated by application of Zn along with ascorbic acid; Se application as nano-sized formulation increased the antioxidant content in tomato fruits under salinity conditions; Zn as well as ascorbic acid application has contributed decreased Cd levels in wheat grain.

1.10 Conclusion

Nutrient imbalances in soil have direct influence on soil and plant health. Thus, it affects the nutritional security. To overcome the ill effects of malnutrition in human, agronomic biofortification could be resorted to. It could be an efficient tool in alleviating nutrient deficiency and thus provide a strong base for enhancing human health thereby ensuring nutritional security. A wide knowledge gap exists in this crop biofortification research as enough data are lacking to predict the field responses under variable conditions; forms in which micronutrients are distributed and found as compounds in edible parts; different antagonistic and competitive interactions possible in the soil as well as within plant system in relation to the nutrients and also the behaviour and impact on micronutrient elements in soils with heavy metal contamination. Future research can be very well oriented in a manner to address the issue with a planned exploration on all relevant areas.

References

Abadía J, Vazquez S, Rellan-Alvarez R, El-Jendoubi H, Abadia A, Alvarez-Fernandez A, Lopez-Millan AF (2011) Towards a knowledge-based correction of iron chlorosis. Plant Physiol Biochem 49:471–482. https://doi.org/10.1016/j.plaphy.2011.01.026

Aciksoz SB, Ozturk L, Gokmen OO, Romheld V, Cakmak I (2011) Effect of nitrogen on root release of phytosiderophores and root uptake of Fe (III)- phytosiderophore in Fe-deficient wheat plants. Physiol Plant 142(3):287–296. https://doi.org/10.1111/j.1399-3054.2011.01460.x

Alloway BJ (ed) (2008) Micronutrient deficiencies in global crop production. Springer, Dordrecht

Álvarez-Fernández A, Paniagua P, Abadía J, Abadía A (2003) Effects of Fe deficiency chlorosis on yield and fruit quality in peach (*Prunus persica* L. *Batsch*). J Agric Food Chem 51:5738–5744. https://doi.org/10.1021/jf034402c

Alvarez-Fernandez A, Abadia J, Abadia A (2006) Iron deficiency, fruit yield and quality. In: Barton LL, Abadia J (eds) Iron nutrition in plants and rhizospheric microorganisms. Springer, Dordrecht, The Netherlands, pp 85–101

Andrade G, Gorgulho B, Lotufo P, Bensenor I, Marchioni D (2018) Dietary selenium intake and subclinical hypothyroidism: a cross-sectional analysis of the ELSA-Brasil study. Nutrients 10(6):693

Arnon DI, Stout PR (1939) The essentiality of certain elements in minute quantity for plants with special reference to copper. Plant Physiol 14:371–375

Bana RC, Gupta AK, Bana RS, Shivay YS, Bamboriya SD, Thakur NP et al (2022) Zinc-coated urea for enhanced zinc biofortification, nitrogen use efficiency and yield of basmati rice under Typic fluvents. Sustainability 14:104. https://doi.org/10.3390/su14010104

Bark M, Malhi GS, Gay LJ, Dean OM (2013) The promise of N-acetylcysteine in neuropsychiatry. Trends Pharmacol Sci 34:166–173

Businelli D, D'Amato R, Onofri A, Tedeschini E, Tei F (2015) Se-enrichment of cucumber (*Cucumis sativus* L.), lettuce (*Lactuca sativa* L.) and tomato (*Solanum lycopersicum* L. karst) through fortification in pre-transplanting. Sci Hortic 197:697–704. https://doi.org/10.1016/j.scienta.2015.10.039

Cakmak I, Kutman UB (2018a) Agronomic biofortification of cereals with zinc: a review. Soil Sci 69:172–180. https://doi.org/10.1111/ejss.12437

Cakmak I, Pfeiffer WH, McClafferty B (2010) Review: biofortification of durum wheat with zinc and iron. Cereal Chem 87:10–20. https://doi.org/10.1094/CCHEM-87-1-0010

Cakmak I (2009) Enrichment of fertilizers with zinc: an excellent investment for humanity and crop production in India. J Trace Elements Medicine Biol 23:281–289

Cakmak I, Kutman UB (2018b) Agronomic biofortification of cereals with zinc: a review. European J Soil Sci 69(1):172–180. https://doi.org/10.1111/ejss.12437

Chhibba IM, Nayyar VK, Kanwar JS (2007) Influence of mode and source of applied iron on fenugreek (*Trigonella corniculata* L.) in a Typic Ustochrept in Punjab, India. Int J Agric Biol 9(2):254–256

Choudhary AK, Suri VK (2014) Integrated nutrient management technology for direct-seeded upland rice (*Oryza sativa*) in North–Western Himalayas. Commun Soil Sci Plant Anal 45(6): 777–784

Clemens S (2014) Zn and Fe biofortification: the right chemical environment for human bioavailability. Plant Sci 225:52–57

Connolly EL (2008) Raising the bar for biofortification: enhanced levels of bioavailable calcium in carrots. Trends Biotechnol 26:401–403

Courbet G, Gallardo K, Vigani G, Brunel-Muguet S, Trouverie J, Salon C, Ourry A (2019) Disentangling the complexity and diversity of crosstalk between Sulphur and other mineral nutrients in cultivated plants. J Exp Bot 70:4183–4196

Deb DL, Sakal R, Datta SP (2009) Fundamentals of soil science. Indian society of soil science, Cambridge Printing Works, New Delhi, p 728

Dhaliwal SS, Sharma V, Shukla AK, Verma V, Kaur M, Shivay YS, Nisar S, Gaber A, Brestic M, Barek V, Skalicky M (2022) Biofortification—a frontier novel approach to enrich micronutrients in field crops to encounter the nutritional security. Molecules 27:1340. https://doi.org/10.3390/molecules27041340

Di Gioia F, Petropoulos SA, Ozores-Hampton M, Morgan K, Rosskop EN (2019) Zinc and iron agronomic biofortification of *Brassicaceae* microgreens. Agronomy 9:677. https://doi.org/10.3390/agronomy9110677

Ding F, Wang XF, Shi QH, Wang EL, Feng JF (2007) Effects of iron deficiency on photosynthetic characteristics and quality of Chinese cabbage (*brassica Chinensis* L.). Shandong Agric Sci 6: 51–53. https://doi.org/10.3969/j.issn.1001-4942.2007.06.016

Ebrahimi N, Stoddard FL, Hartikainen H, Seppanen MM (2019) Plant species and growing season weather influence the efficiency of selenium biofortification. Nutr Cycl Agroecosyst 114:111–124

Elanchezhian R, Kumar D, Ramesh K, Biswas AK, Guhey A, Patra AK (2017) Morpho-physiological and biochemical response of maize (*Zea mays* L.) plants fertilized with nano-iron (Fe_3O_4) micronutrient. J Plant Nutr 40:1969–1977. https://doi.org/10.1080/01904167.2016.1270320

El-Nakhel C, Pannico A, Kyriacou MC, Giordano M, De Pascale S, Rouphael Y (2019) Macronutrient deprivation eustress elicits differential secondary metabolites in red and green-pigmented butterhead lettuce grown in closed soilless system. J Sci Food Agric 99:6962–6972. https://doi.org/10.1002/jsfa.9985

El-ramady H, Abdalla N, Elbasiouny H, Elbehiry F, Elsakhawy T (2021) Nano-biofortification of different crops to immune against COVID-19: a review. Ecotoxicol Environ Saf 222:12500

Fageria NK, Baligar VC (2008) Ameliorating soil acidity of tropical Oxisols by liming for sustainable crop production. Adv Agron 99:345–399

Fageria NK, Baligar VC, Clark RB (2002) Micronutrients in crop production. Adv Agron 77:85–268

Fageria NK, Baligar VC, Wright RJ (1997) Soil environment and root growth dynamics of field crops. Recent Res Dev Agron 1:15–58

Fraga CG, Oteiza PI (2002) Iron toxicity and anti-oxidant nutrients. Toxicology 180:23–32. https://doi.org/10.1016/S0300-483X(02)00379-7

Ge K, Yang G (1993) The epidemiology of selenium deficiency in the etiological study of endemic diseases in China. Am J Clin Nutr 57(2):259S–263S. https://doi.org/10.1093/ajcn/57.2.259S

Ghafari H, Razmjoo J (2013) Effect of foliar application of nano-iron oxidase, iron chelate and iron sulphate rates on yield and quality of wheat. Int J Agron Plant Prod 4:2997–3003

Ghaly FM, Baddourand GA, El-Azazy HM (2017) Nitrate accumulation and oxalate formation in spinach (*Spincea oleracea* L) as affected by nitrogen fertilization levels and iron foliar application. J Soil Sci Agric Eng 8(11):571–576. https://doi.org/10.21608/jssae.2017.38092

Golob A, Kavčič J, Stibilj V, Gaberščik A, Vogel-Mikuš K, Germ M (2017) The effect of selenium and UV radiation on leaf traits and biomass production in *Triticum aestivum* L. Ecotoxicol Environ Saf 136:142–149

Graham RD, Welch RM, Bouis HE (2001) Addressing micronutrient malnutrition through enhancing the nutritional quality of staple foods: principles, perspectives and knowledge gaps. Adv Agron 70:77–142

Guerrero-Romero F, Rodriguez-Moran M (2005) Complementary therapies for diabetes: the case for chromium, magnesium, and antioxidants. Arch Med Res 36:250–257

Haefele SM, Thomas CL, Saito K (2022) Long-term fertility experiments for irrigated rice in the west African Sahel: effect on macro-and micronutrient concentrations in plant and soil. Field Crops Res 275:108357. https://doi.org/10.1016/j.fcr.2021.108357

Hajiboland R, Rahmat S, Zeinalzadeh N, Farsad-Akhtar N, Hosseinpour-Feizi MA (2019) Senescence is delayed by selenium in oilseed rape plants. J Trace Elem Med Bio 55:96–106

Harris D, Rashid A, Miraj G, Arif M, Shah H (2007) On-farm seed priming with zinc sulphate solution–a cost-effective way to increase the maize yields of resource poor farmers. Field Crops Res 102:119–127

Hartikainen H, Xue T, Piironen V (2000) Selenium as an anti-oxidant and pro-oxidant in ryegrass. Plant Soil 225:193–200. https://doi.org/10.1023/A:102651292

Hotz C, Brown KH (2004) International zinc nutrition consultative group (IZiNCG) technical document #1. Assessment of the risk of zinc deficiency in populations and options for its control. Food Nutr Bull 25:99–199

Hu Q, Xu J, Pang G (2003) Effect of selenium on the yield and quality of green tea leaves harvested in early spring. J Agric Food Chem 51:3379–3381

Hussain S, Maqsood MA, Rengel Z, Aziz TJ (2012) Biofortification and estimated human bioavailability of zinc in wheat grains as influenced by methods of zinc application. Plant Soil 361: 279–290. https://doi.org/10.1007/s11104-012-1217-4

Izydorczyk G, Ligas B, Mikula K, Witek-Krowiak A, Moustakas K, Chojnacka K (2021) Biofortification of edible plants with selenium and iodine–a systematic literature review. Sci Total Environ 754:141983

Joy EJM, Ahmad W, Zia MH, Kumssa DB, Young SD, Ander EL, Watts MJ, Stein AJ, Broadley MR (2016) Valuing increased zinc (Zn) fertilizer-use in Pakistan. Plant Soil 411:139–150

Joy EJM, Stein AJ, Young SD, Ander EL, Watts MJ, Broadley MR (2015) Zinc-enriched fertilizers as a potential public health intervention in Africa. Plant Soil 389:1–24

Karp K, Starast M, Kaldmae H (2002) Influence of the age of plants and foliar fertilization on the yield of strawberry cultivar Jonsok under plastic mulch. Acta Hortic 567:459–562

Karthika KS, Philip PS, Neenu S (2020) Brassicaceae plants response and tolerance to nutrient deficiencies. In: Hasanuzzaman M (ed) The plant family brassicaceae. Springer Nature Singapore Pte Ltd, Singapore, pp 337–361

Karthika KS, Rashmi I, Parvathi MS (2018) Biological functions, uptake and transport of essential nutrients in relation to plant growth. In: Hasanuzzaman M, Fujita M, Oku H, Nahar K, Hawrylak Nowak B (eds) Plant nutrients and abiotic stress tolerance. Springer Nature Singapore Pte Ltd, Singapore, pp 1–49

Kaur G, Chandna R, Pandey R, Abrol YP, Iqbal M, Ahmad A (2011) Sulfur starvation and restoration affect nitrate uptake and assimilation in rapeseed. Protoplasma 248:299–311

Kaur M, Sharma S, Singh D (2018) Influence of selenium on carbohydrate accumulation in developing wheat grains. Comm Soil Sci Plant Anal 49:1650–1659

Kiran A, Wakeel A, Sultana R, Khalid A, Qurrat-ul-Ain MR, Shahzad AN, Khan S, Noor M (2021) Concentration and localization of Fe and Zn in wheat grain as affected by its application to soil and foliage. Bull Environ Contam Toxicol 106(5):852–858. https://doi.org/10.1007/s00128-021-03183-x

Kornaś A, Filek M, Sieprawska A, Bednarska-Kozakiewicz E, Gawrońska K, Miszalski Z (2019) Foliar application of selenium for protection against the first stages of mycotoxin infection of crop plant leaves. J Sci Food Agric 99:482–485

Kumar A, Choudhary AK, Pooniya V, Suri VK, Singh U (2016) Soil factors associated with micronutrient Acquisition in Crops-Biofortification Perspective. In: Singh U, Praharaj C, Singh S, Singh N (eds) Biofortification of food crops. Springer, New Delhi, pp 159–176. https://doi.org/10.1007/978-81-322-2716-8_13

Kumar KSA, Karthika KS (2020) Abiotic and biotic factors influencing soil health and/or soil degradation. In: Giri B, Varma A (eds) Soil health, soil biology 59. Springer Nature, Switzerland AG, pp 145–161

Liu Y, Ohm J-B, Hareland G, Wiersma J, Kaiser D (2011) Sulfur, protein size distribution, and free amino acids in flour mill streams and their relationship to dough rheology and bread making traits. Cereal Chem 88:109–116

Lv J, Wu J, Zuo J, Fan L, Shi J, Gao L, Li M, Wang Q (2017) Effect of se treatment on the volatile compounds in broccoli. Food Chem 216:225–233

Malakouti MJ (2001) The effect of balanced fertilization and zinc application on improving apple yield, quality and reducing browning incidence. Acta Hortic 564:153–158

Malakouti MJ (2000) Balanced nutrition of wheat: an approach towards self-sufficiency and enhancement of national health- a compilation of papers. Ministry of Agriculture, Karaj, Iran, p 544p

Malorgio F, Diaz KE, Ferrante A, Mensuali-Sodi A, Pezzarossa B (2009) Effects of selenium addition on minimally processed leafy vegetables grown in a floating system. J Sci Food Agric 89:2243–2251

Manzeke GM, Mtambanengwe F, Nezomba H, Mapfumo P (2014) Zinc fertilization influence on maize productivity and grain nutritional quality under integrated soil fertility management in Zimbabwe. Field Crops Res 166:128–136

Marschner H, Römheld V, Horst W, Martin P (1986) Root induced changes in the rhizosphere: importance for the mineral nutrition of plants. Zeitschriftfür Pflanzenernährung und Bodenkunde 149:441–456

Marschner H (2012) Marschner's mineral nutrition of higher plants. Elsevier, London

Marschner H (1995) Mineral nutrition of higher plants. Academic, New York

Morales-Espinoza MC, Cadenas-pliego G, Marissa P, Delia A, Cabrera M, Fuente D (2019) Se nanoparticles induce changes in the growth, antioxidant responses, and fruit quality of tomato developed under NaCl stress. Molecules 24:3030

Mortvedt JJ (2000) Bioavailability of micronutrients. In: Sumner ME (ed) Handbook of soil science. CRC Press, Boca Raton, FL, p 2148

Mumtaz MZ, Ahmad M, Jamil M, Asad SA, Hafeez F (2018) Bacillus strains as potential alternate for zinc biofortification of maize grains. Int J Agric Biol 20:1779–1786

Nawaz F, Ahmad R, Ashraf MY, Waraich EA, Khan SZ (2015) Effect of selenium foliar spray on physiological and biochemical processes and chemical constituents of wheat under drought stress. Ecotoxicol Environ Saf 113:191–200

Nawaz F, Ashraf MY, Ahmad R, Waraich EA, Shabbir RN, Hussain RA (2017) Selenium supply methods and time of application influence spring wheat (*Triticum aestivum* L.) yield under water deficit conditions. J Agric Sci 155:643–656

Nawaz F, Ashraf MY, Ahmad R, Waraich EA (2013) Selenium (se) seed priming induced growth and biochemical changes in wheat under water deficit conditions. Biol Trace Elem Res 151: 284–293

Nimin ME, Han B, Corodoba F (2007) Are we getting enough sulfur in our diet? Nutr Metab 4:24–32. https://doi.org/10.1186/1743-7075-4-24

Noreen S, Sultan M, Akhter MS, Shah KH, Ummara U, Manzoor H, Ulfat M, Alyemeni MN, Ahmad P (2020) Foliar fertigation of ascorbic acid and zinc improves growth, antioxidant enzyme activity and harvest index in barley (*Hordeum vulgare* L.) grown under salt stress. Plant Physiol Biochem 158:244–254

Palego L, Betti L, Giannaccini G (2015) Sulfur metabolism and sulfur-containing amino acids: I-molecular effectors. Biochem Pharmacol 4(7):1000158

Palmer LJ, Palmer LT, Rutzke MA, Graham RD, Stangoulis JCR (2014) Nutrient variability in phloem: examining changes in K, mg, Zn and Fe concentration during grain loading in common wheat (*Triticum aestivum*). Physiol Plant 152:729–737. https://doi.org/10.1111/ppl.12211

Pathak GC, Gupta B, Pandey N (2012) Improving reproductive efficiency of chickpea by foliar application of zinc. Braz J Plant Physiol 24(3):173–180

Peck AW, McDonald GK, Graham RD (2008) Zinc nutrition influences the protein composition of flour in bread wheat (*Triticum aestivum* L.). J Cereal Sci 47:266–274. https://doi.org/10.1016/j.jcs.2007.04.006

Phattarakul N, Rerkasem B, Li L, Wu L, Zou C, Ram H et al (2012) Biofortification of rice grain with zinc through zinc fertilization in different countries. Plant Soil 361:131–141. https://doi.org/10.1007/s11104-012-1211-x

Prasad R, Shivay YS (2018a) Sulfur in soil, plant and human nutrition. Proc Nat Acad Sci, India Section B: Biol Sci 88(2):429–434. https://doi.org/10.1007/s40011-016-0769-0

Prasad R (2013) Fertilizer nitrogen, food security, health and environment. Proc Indian Nat Sci Acad 79(4):997–110

Prasad R (2014) Major sulfur compounds in plants and their role in human nutrition and health–an overview. Proc Indian Nat Sci Acad 80(5):1045–1054

Prasad R, Shivay YS (2018b) Iron fertilization of cereal crops for human health-a new concept. Indian J Fertilisers 14(5):68–77

Prom-U-Thai C, Rashid A, Ram H, Zou C, Guilherme L, Corguinhan A, Guo S, Kaur C, Naeem A, Yamuangmorn S, Ashraf MY, Sohu VS, Zhang Y, Martins F, Jumrus S, Tutus Y, Yazici MA, Cakmak I (2020) Simultaneous biofortification of rice with zinc, iodine, iron and selenium through foliar treatment of a micronutrient cocktail in five countries. Front Plant Sci 11:589835. https://doi.org/10.3389/fpls.2020.589835

Raisbeck MF (2000) Selenosis. Vet Clin North Am Food Anim Pract 16:465–480. https://doi.org/10.1016/S0749-0720(15)30081-5

Rayman MP (2012) Selenium and human health. Lancet 379:1256–1268. https://doi.org/10.1016/S0140-6736(11)61452-9

Reeves PG, Chaney RL (2008) Bioavailability as an issue in risk assessment and management of food cadmium: a review. Sci Total Environ 398:13–19

Ríos JJ, Blasco B, Cervilla LM, Rubio-Wilhelmi MM, Ruiz JM, Romero L (2008) Regulation of sulfur assimilation in lettuce plants in the presence of selenium. Plant Growth Regul 56:43–51. https://doi.org/10.1007/s10725-008-9282-7

Ritchie H, Roser M (2019) Micronutrient deficiency. Published on-line. https://ourworldindata.org/Micronutrient

Rizwan M, Ali S, Ali B, Adrees M, Arshad M, Hussain A, Rehman MZU, Waris AA (2019) Zinc and iron oxide nanoparticles improved the plant growth and reduced the oxidative stress and cadmium concentration in wheat. Chemosphere 214:269–277

Rosado JL, Hambidge KM, Miller LV, Rosado JL, Hambidge KM, Miller LV, Garcia OP, Westcott J, Gonzalez K, Conde J, Hotz C, Pfeiffer W, Ortiz-Monasterio I, Krebs NF (2009) The quantity of zinc absorbed from wheat in adult women is enhanced by biofortification. J Nutr 139(10):1920–1925. https://doi.org/10.3945/jn.109.107755

Sanchez-Sanchez A, Juarez M, Sanchez-Andreu J, Jorda J, Bermudez D (2002) Humic substances and amino acids improve the effectiveness of chelate FeEDDHA in lemon trees. J Plant Nutr 25(11):2433–2442

Sanchez-Sanchez A, Juarez M, Sanchez-Andreu J, Jorda J, Bermudez D (2005) Use of humic substances and amino acids to enhance iron availability for tomato plants from applications of the chelate FeEDDHA. J Plant Nutr 28:1877–1886. https://doi.org/10.1080/01904160500306359

Schnug E, Haneklaus S (2005) Sulfur deficiency symptoms in oilseed rape (*brassica napus*L.)–the aesthetics of starvation. Phyton Annales Rei Botanicae 45:79–95

Schuphan W (1961) Nutritional values in crops and plants. Problems for producers and consumers. Faber and Faber Limited, London

Shahid MA, Balal RM, Khan N, Zotarelli L, Liu GD, Sarkhosh A, Fernández-Zapata JC, Nicolás JJM, Garcia-Sanchez F (2019) Selenium impedes cadmium and arsenic toxicity in potato by modulating carbohydrate and nitrogen metabolism. Ecotoxicol Environ Saf 180:588–599

Shils ME, Shike M, Ross AC, Caballero B, Cousins RJ (eds) (2006) Modern nutrition in health and disease, 10th edn. Lippincott Williams, Baltimore

Shivay YS, Prasad R (2014) Effect of source and methods of zinc application on corn productivity, nitrogen and zinc concentrations and uptake by high quality protein corn. Egypt J Biol 16:72–78. https://doi.org/10.4314/ejb.v16i1.10

Shivay YS, Kumar D, Prasad R (2008) Effect of zinc-enriched urea on productivity, zinc uptake and efficiency of an aromatic rice–wheat cropping system. Nutr Cyc Agroecosyst 81:229–243

Song YR, Jiang XG, Peng SF, Li FC, Liu J, Chen D (2015) Effect of selenium content on the quality and functional components of selenium-riched *Camellia oleifera* oil. J Chin Inst Food Sci Technol 15:142–149

Spinazzi M, Angelini C, Patrini C (2010) Sub-acute sensory ataxia and optic neuropathy with thiamine deficiency. Nat Rev Neuro 16(5):288–293

Stein AJ, Meenakshi JV, Qaim M, Nestel P, Sachdev HPS, Bhutta ZA (2008) Potential impacts of iron biofortification in India. Social Sci Med 66:1797–1808

Szerement J, Kloc AS, Mokrzycki J, Hersztek MM (2022) Agronomic biofortification with se, Zn, and Fe: an effective strategy to enhance crop nutritional quality and stress defense—a review. J Soil Sci Plant Nutr 22:1129–1159

Tapel M, van der Giet M, Schwartzfeld C, Laufer U, Lierman D, Zidek W (2000) Prevention and radiographic contrast-agent induced reductions in renal function by acetyl cysteine. New England J Medic 343:180–184. https://doi.org/10.1056/NEJM200007203430304

Tarafder C, Daizy M, Alam MM, Ali MR, Islam MJ, Islam R, Ahommed MS, AlySaadAly M, Khan MZH (2020) Formulation of a hybrid nanofertilizer for slow and sustainable release of micronutrients. ACS Omega 5:23960–23966. https://doi.org/10.1021/acsomega.0c03233

Trijatmiko KR, Dueñas C, Tsakirpaloglou N, Torrizo L, Arines F, Adeva C, Balindong J, Oliva N (2016) Biofortified indica rice attains iron and zinc nutrition dietary targets in the field. Sci Rep 6:19792

United Nations (2006) The millennium development goals report. Department of Economic and Social Affairs

Velu G, Ortiz-Monasterio I, Cakmak I, Hao Y, Singh RJ (2014) Biofortification strategies to increase grain zinc and iron concentrations in wheat. J Cereal Sci 59:365–372

Wakeel A, Farooq M, Khurram B, Levent O (2018) Micronutrient malnutrition and biofortification: recent advances and future perspectives. Plant Micronutrient Use Efficiency 225-243. https://doi.org/10.1016/B978-0-12-812104-7.00017-4

Wang B, Qiu YL (2006) Phylogenetic distribution and evolution of mycorrhizas in land plants. Mycorrhiza 16:299–363

Wang Y, Wang X, Wong Y (2013) Generation of selenium-enriched rice with enhanced grain yield, selenium content and bioavailability through fertilisation with selenite. Food Chem 141(3): 2385–2393

Waters BM, Sankaran RP (2011) Moving micronutrients from the soil to the seeds: genes and physiological processes from a biofortification perspective. Plant Sci 180:562–574

Welch RM, Graham R (2004) Breeding for micronutrients in staple food crops from a human nutrition perspective. J Exp Bot 55:353–364

Welch RM, Harti JJ, Norvell WA, Sullivan LA, Kochian LV (1999) Effects of nutrient solution zinc activity on net uptake, translocation, and root export of cadmium and zinc by separated sections of intact durum wheat (*Triticum turgidum* L. *vardurum*) seedling roots. Plant Soil 208:243–250

White PJ, Broadley MR (2005) Biofortifying crops with essential mineral elements. Trends Plant Sci 10:586–593

Winkler A, Knoche M (2019) Calcium and the physiology of sweet cherries: a review. Sci Hortic 245:107–115

Wood JM, Decker H, Chavan B, Rokos H, Spencer JD, Hess S, Thornton MJ, Paus R, Schallreuter KU (2009) H_2O_2 mediated stress affects human hair colour by blunting methionine sulfoxide repair. FASEB J 23(7):2065–2070

Wu Z, Yin X, Bañuelos GS, Lin ZQ, Zhu Z, Liu Y, Yuan L, Li M (2016) Effect of selenium on control of postharvest gray mold of tomato fruit and the possible mechanisms involved. Front Microbiol 6:1–11

Xu J, Yang F, Chen L, Hu Y, Hu Q (2003) Effect of selenium on increasing the antioxidant activity of tea leaves harvested during the early spring tea producing season. J Agr Food Chem 51:1081–1084. https://doi.org/10.1021/jf020940y

Yang GQ, Wang SZ, Zhou RH, Sun SZ (1983) Endemic selenium intoxication of humans in China. Am J Clin Nutr 37(5):872–881. https://doi.org/10.1016/S0140-6736(83)90575-5

Yang XW, Tian XH, Gale WJ, Cao YX, Lu XC, Zhao AQ (2011) Effect of soil and foliar zinc application on zinc concentration and bioavailability in wheat grain grown on potentially zinc-deficient soil. Cereal Res Commun 39:535–543. https://www.jstor.org/stable/23792319

Yasin M, El-Mehdawi AF, Anwar A, Pilon-Smits EAH, Faisal M (2015) Microbial-enhanced selenium and iron biofortification of wheat (*Triticum aestivum* L.)–applications in phytoremediation and biofortification. Int J Phytoremediation 17:341–347. https://doi.org/10.1080/15226514.2014.922920

Zarghamnejad Z, Kordlaghari KP, Keshavarzi K (2015) Efficacy of foliar application of ferrous and manganese sulfate on wet and dry biomass of tomato (*Lycopersicon esculentum*) in greenhouse. Int J Biosciences 6:437–444

Zeidan MS, Manal F, Hamouda HA (2010) Effect of foliar fertilization of Fe, Mn and Zn on wheat yield and quality in low sandy soils fertility. World J Agric Sci 6:696–699

Zhang R, Zhang H, Tu C, Hu X, Li L, Luo Y, Christie P (2015) Phytotoxicity of ZnO nanoparticles and the released Zn (II) ion to corn (*Zea mays* L.) and cucumber (*Cucumis sativus* L.) during germination. Environ Sci Pollut Res 22:11109–11117. https://doi.org/10.1007/s11356-015-4325-x

Chapter 2
The Role of Biofortification in Enhancing Plant Growth, Development, Yield, and Quality

Nusrat Jabeen

Abstract Biofortification is to get better crops with better nutrition. It is a feasible and cost-effective means of crops breeding with higher levels of minerals, proteins and healthier fats to deliver essential nutrients to population that may have limited access to different healthy diets. This can be achieved through transgenic, conventional, and agronomic approaches, involving the use of biotechnology, crop breeding, and nutrients fertilization strategies, respectively. Biofortification differs from food fortification because it has increased the amount of nutrients set in the crop being grown to make plant foods naturally more nutritive instead of adding nutrient to the foods during processing. This chapter reviews how bio fortification through these different strategies work and could help to improve growth and developmental status of crops to achieve high quality and level of protein, oil, micronutrients, minerals, vitamins and total yield of staple crop to alleviate micronutrient malnutrition.

Keywords Biofortification · Cost-effective · Strategies · Malnutrition · Plant yield

2.1 Introduction

The essential mineral nutrients are vital for healthy survival of all human being in this planet. Plants are the only source of all these minerals. No plant alone contain all minerals, someplants are rich for some mineral and deficient to others, limiting the nutrients in plants and ultimately in humans. Most of the agriculture systems do not focus human health but to increase crop yield and its productivity to meet the requirements. Besides, micronutrient deficient soil has been observed across many parts of the world, thus limiting the uptake of essential nutrients in plants and ultimately insufficient intake negatively affects human biology. Around the world more than 2 billion people are reported to be affected by the deficiency of key micronutrients such as iron (Fe), zinc (Zn), selenium (Se), iodine (I), and vitamins in

N. Jabeen (✉)
Biosaline Laboratory, Department of Botany, University of Karachi, Karachi, Pakistan

M. Hasanuzzaman et al. (eds.), *Mineral Biofortification in Crop Plants for Ensuring Food Security*, https://doi.org/10.1007/978-981-99-4090-5_2

the daily diet (Hodge 2016; Huang et al. 2020) despite increased food crop production (Gould 2017). Deficiency of essential mineral may lead to stunted growth, reduced immunity, fatigue, perinatal complications, irritability, weakness, degeneration of tissues and muscles, sterility, and high risk of morbidity and mortality (Bailey et al. 2015). Food nutritional security is a growing matter of concern to human health for people who live especially in developing world where daily diet comprises mainly of micronutrient-poor staple food crops (Maertens and Velde 2017). This global threat addresses the need to shift agriculture from producing more quantity of food crops to producing nutrient rich food crops in an adequate amount to combat "hidden hunger" or "micronutrient malnutrition".

There are different approaches used for the improvement of food nutritional profile to tackle micronutrient deficiency such as-.

Dietary diversification; a food based approach aims to utilize foods with a high content of micronutrients throughout the year. It needs to change food production practices, food selection patterns, and traditional household methods for preparing and processing local foods. It is constrained by resource availability for poor households and seasonal availability of fruits and vegetables. It is difficult to sustain on large scale due to limited availability of land and reduced resources for poor household.

Food supplementation; food supplements are concentrated sources of nutrients may be natural or synthetic like vitamins, minerals, and essential fatty acids, consumed in dose form (e.g., pills, tablets, capsules, liquids in measured doses). Intakes of supplements in an adequate amount correct nutritional deficiencies and support some physiological functions but these are not medicinal products to treat diseases or modify any physiological functions in humans and may not solve the root cause of micronutrient deficiencies. The vitamins consumption is more common practice in developed countries than non-developed countries because of their economic status and low-income consumers (Wiltgren et al. 2015).

Food fortification; Fortification is the addition of vitamins, minerals, and micronutrients to foods during processing to boost their nutritional value. Food makers add synthetic vitamins and minerals to food with no taste, texture or smell unlike natural enriched foods. Several foods fortified during processing with nutrients such as iron, ferrous sulfate, ferrous fumarate, ferric pyrophosphate, electrolytic iron powder compounds, Iodine and vitamins etc. to overcome nutritional deficiencies and provide health benefits with nominal risk (Jha and Warkentin 2020). Fortified foods must be recommended one part of an overall healthy lifestyle because these foods are heavily processed, high in sugar, fats, sodium, vitamins and minerals. Intake in excess amount may be harmful and cause some health problem like obesity, so the efficiencies of these strategies are low. This is a better option to get nutrients from unprocessed foods like fruits and vegetables to reduce health risk factor. Taking it into consideration a cost-effective, long-term, and sustainable approach, biofortification was introduced, it is an act of breeding nutrients into food crops.

Biofortification is a practical way to address countryside people who may not have an access to commercially available fortified foods and supplements. Their diet based on cereal crops with less protein and vitamins because their soils are low in essential micronutrients like Zn, Fe and I. Most of the soil is affected by salinity and alkalinity (Fuge and Johnson 2015). Micro-nutrient deficiencies affect crop growth, its development like seed formation, flowering, yield, and its quality. Some micronutrients like boron (B), magnesium (Mg), and copper (Cu) give stability and strength to cell wall and increase plant pathogen resistance (Chaudhary et al. 2022).

Biofortification aimed to rectify nutrients imbalances by not only breeding for improved agronomic performance, but also redirecting efforts to improve the essential nutrient contents of new crop cultivars being developed (Stangoulis and Kanez 2022). Furthermore, World Health Organization (WHO) and the Consultative Group on International Agricultural Research (CGIAR) aimed for the development of nutritionally enhanced high-yielding biofortified crops to achieve high food security in future (Bouis 2000). A successful biofortified crop must be high yielding, profitable for the farmers, effective in reducing micronutrient malnutrition in humans and acceptable to both farmers and consumers in target areas (Hotz et al. 2007). The present review discusses gains through crop plants biofortification in past and effective strategies for future.

2.2 Key Micronutrients Uptake and Function in Plants

A sufficient amount of micronutrients present in soil, are available to plants in different forms with different uptake mechanisms. Their availability in soil and uptake by plant could be affected by different factors including soil pH, competing cations, anions, similarities in micronutrients uptake mechanisms, soil geomorphology, organic matter, and its microbiology. Some micronutrients in soil react with some compound like phosphates and carbonates to form precipitates; some may interact with clay colloids and other mineral complexes, and become unavailable to the crop (Marschner 2012; Dimkpa and Bindarban 2016). Though micronutrients are required in less quantity but without it many of the processes that drive plant metabolism of N, P, K, Mg, Ca, and S, as well as crop responses to ecological alarming situations, would not function at optimum level.

Micronutrients such as Fe, Cu, Mn, and Cl are involved as cofactor in plant photosynthesis. Beside this Fe, Mn, Zn, Cu, Ni, Mo, and Cl all involved different enzymes function like DNA/RNA polymerases, N-metabolizing enzymes, superoxide dismutases, catalases, dehydrogenases, oxidases, ATPases, and numerous other redox enzymes. A specific role of Zn is in the enzymatic processes involved in the biosynthesis of auxin which results enhanced root growth and its ability to absorb nutrients and water. Boron is required for maintenance of cell membrane and involved in cell wall formation, pollination and seed set. Copper is required for carbohydrate metabolism and synthesis of lignin necessary for cell wall strengthening. Nickel is involved with urease enzyme which converts urea to ammonia for nitrogen metabolism in plants. Molybdenum is a component of the nitrogenase enzyme system which fix nitrogen in soil by symbiotic and free-living N-fixing.

Deficiency of any required micronutrients could result in poor seed, root and shoot development, reduced tolerance to any kind of stress which badly affect plant yield, or lead plant death. Improvement in micronutrients supplementation to crops in a balanced fertilizer regime could mitigate the adverse effects of all deficiencies. The positive response of crops to micronutrient supplementations must not aim only in terms yield stimulation but also in terms of enhanced nutrient uptake into edible parts which is important for human consumption.

2.3 Biofortification Approaches

Biofortification has been advocated as a long-term alternative of traditional therapies for increasing mineral nutrition (Zhu et al. 2007). This method results in the improvement of both mineral content and bioavailability in the edible parts of staple crops through agronomic intervention, plant breeding, or genetic engineering but mineral bioavailability can only be influenced through plant breeding and genetic engineering (Chaudhary et al. 2022).

2.3.1 Agronomic Approach

The agronomic method of biofortification is simple and inexpensive process of physical application of nutrients by means of adopting proper agronomic practices to improve the nutritional and health status of crops which positively affect the human nutritional status (Cakmak and Kutman 2017).

The organic minerals are more available and can be solubilized and mobilized from the soil in the edible parts of plants more easily than inorganic forms of minerals. They are less excreted and toxic to the crop. Application of macro nutrients fertilizers like nitrogen, phosphorus, and potassium (NPK) are important and necessary to increase crop yield and save the human population from hunger (Graham et al. 2007). Application of micronutrient as ferlitilzer like Fe, Zn, Cu, Mn, I, Se, Mo, Co, and Ni are found to contribute not only to increase crop nutritional value and yield but also to decrease human malnutrition (Cakmak 2008). When mineral elements become unavailable in soil or cannot readily translocate to the targeted areas of crops growing in soils, the application of soluble inorganic fertilizers to the roots or to the leaves are practiced by different approaches. Agronomic approach of biofortification can be considered as an effective strategy for application/supplementation of micronutrients to the growing crops with major advantages (Chaudhary 2022) i.e.,

- It is applied on already being cultivated crop cultivars by the farmers with suitable produce.

- With low quantity of micronutrient high quality of grains and yield can be achieved in the same year.
- It always creates a win–win approach for developing countries.

Different species of microorganisms like *Bacillus*, *Pseudomonas*, *Rhizobium*, *Azotobacter* etc. are plant growth-promoting soil microorganisms. Besides fertilizers these microorganisms can also be utilized to enhance the nutrient mobility from soil to edible parts of plants and improve their nutritional status (Rengel et al. 1999; Smith and Read 2007). Maintenance of soil physical, chemical, and biological properties is also required for successful biofortification. Loose, crumby, and granular soils are porous with high nutrient, and moisture-holding capacity. Applications of lime, gypsum and elemental sulfur have also been practiced to improve the soil chemical properties. Organic amendments, less use of pesticides, addition of legume crops in the cropping system increase biological activity in the soil and enhance the micronutrient bio-availability for plant uptake. Most important is a judicious use of macronutrients like N, P, K, Mg, and Ca in soil which help in the proper uptake of other nutrients (Marcelle 1995). Studies have shown that micronutrient fertilization is most effective when combined with nitrogen–phosphorus–potassium (NPK) and organic fertilizers (Sakellariou and Mylona 2020). Different crops are targeted through agronomical biofortification for the improvements in their growth yield and nutritional status is tabulated (Table 2.1).

2.3.2 *Breeding Approach*

Conventional breeding is thought to be the most reliable approach of biofortification which offer sustainable, cost-effective alternative to transgenic- and agronomic approaches. It involves older techniques and natural processes for the development of new varieties of crops. In conventional plant breeding, high nutrients parent lines are crossed with desired trait of recipient line over several generations to produce plants with variable desired nutrient concentration and agronomic traits found in different germplasm thus improve the levels of minerals and vitamins in crops. All crops don't have the genetic potential to meet desired micronutrient levels with conventional breeding, and therefore, to achieve sufficient improvements breeding strategies have to rely on the limited genetic variation present in the gene pool (Mulualem 2015; Garg et al. 2018). Genetic engineering is an alternative approach of breeding when variable desired traits is not available naturally in the available germplasm, a specific micronutrient does not naturally exist in crops, and/or modifications cannot be done by conventional breeding. This approach also target the exclusion of antinutrient and inclusion of promoters that can increase micronutrients bioavailability (Jha and Warkentin 2020). Genetically modified crops such as corn, rice, wheat, and soybeans have the potential for increased agriculture productivity

Table 2.1 Summary of studies on different crops quality and productivity through agronomical biofortification

Crops	Nutrients supply	Types of agronomic biofrotification	Adequate supply results	References
Cereals				
Barley	Zinc sulphate	Soil application	$ZnSO_4$ fertilization applied at 15 kg ha^{-1} significantly improved quality traits, dry-matter yield, crude protein, acid detergent fibre (ADF) and shoot Zn concentration in crop harvested at 50% heading stage.	Sher et al. (2022)
	NPK + Sulphur (i.e. ammonium sulphate, potassium sulphate, and Wigor S (80% elemental Sulphur and 20% bentonite)	Soil application	Addition of Sulphur in fertilization positively affects the content of manganese, iron, zinc, and copper in the grain. Ammonium sulphate has the greatest impact on the content and uptake of all micronutrients, except for zinc to manage the quality of crop.	Barczak et al. (2019)
	Sodium selenate and sodium selenite	Foliar application	Sodium selenate has been reported to be effectively taken by plants than sodium selenite and significantly increased grains yield and weight.	Rodrigo et al. (2013)
	Biofertlizers + NPK fertilizers + vermicompost	Soil application	Barley grain quality and quantity has been improved by increased Fe and Zn concentrations through vermicompost application and Mn concentration by chemical fertilizer.	Maleki et al. (2011)
Maize	Zinc sulphate and iron sulphate	Soil and foliar application both	Soil application improves the yield attributes of crop	Anwar et al. (2022)

(continued)

Table 2.1 (continued)

Crops	Nutrients supply	Types of agronomic biofrotification	Adequate supply results	References
			whereas foliar spray is better for nutrients availability, optimum crop development and growth with improved grain quality.	
	Vermicompost, zinc and iron	Soil application	Superior crop yield (grain and stalk) and quality (protein, starch, amylose) was achieved with the combined effect of Zn and iron along with vermicompost	Pandey et al. (2022)
	Phosphate solubilizing bacteria (PSB)	Soil application	Significantly maximize plant growth and nutritional status of the maize crop and has been proved to be used as promising bio-fertilizer	Ahmad et al. (2019)
	Farmyard manure+ $ZnSO_4$ + $FeSO_4$	Soil and foliar application both	Soil application of FYM enriched $ZnSO_4$ and $FeSO_4$ each@25 and 15 kg ha^{-1} and foliar spray of $ZnSO_4$ and $FeSO_4$ each@0.5% recorded the higher values in all growth parameters, grain yield, Stover yield and all yield parameters.	Nikhil and Salakinkop (2018)
	Potasium iodide (KI) Potassium iodate (KIO_3)	Soil and foliar application both	Soil application did not affect grain yield and its iodine content whereas foliar application enhanced grain iodine concentrations up to 5- to ten-fold without affecting grain yield.	Cakmak et al. (2017)
	Mn	Soil application	Nutrient uptake (Mn) was increased by 69% and total yield of crop was increased by 16%.	Dimkpa and Bindarban (2016)

(continued)

Table 2.1 (continued)

Crops	Nutrients supply	Types of agronomic biofrotification	Adequate supply results	References
Rice	Chelated se	Foliar application	The results showed higher seed-setting rate, grain weight, grain yield, grain protein, total se and Fe content and net photosynthetic rate. Reduced both the chalky rice rate and chalkiness. Increased grain 2-acetyl-1-pyrroline (2-AP) content and activities of all enzymes related to biosynthesis of 2-AP	Lou et al. (2020)
	Zinc solubilizing rhizobacteria	Soil application	The results provide a strong base for developing biofertilizer from various bacillus strains which were gram positive. They have ability to solubilize indigenous ZnO through diverse plant growth promoting traits. The inoculation with these strains alone and in combination improved the growth, productivity and nutrient quality of rice seedlings with increased bacterial population in the rice rhizosphere.	Naseer et al. (2020)
	Potasium iodide (KI) Potassium iodate (KIO_3)	Soil and foliar application both	Soil application did not affect grain yield and its iodine content whereas foliar application enhanced not only iodine concentrations of grains up to 5- to ten-fold but also in endosperm part of polished rice without affecting grain yield.	Cakmak et al. (2017)
	Se		Soil selenate application was more	Boldrin et al. (2013)

(continued)

Table 2.1 (continued)

Crops	Nutrients supply	Types of agronomic biofrotification	Adequate supply results	References
		Soil and foliar application both	effective for shoot dry matter production (by 6%) and grain se accumulation than selenite. Foliar application of both selenate and selenite increased grain yield. In both forms of application selenate resulted in higher contents of se in rice grains, soil application yield nearly 450% higher se contents than foliar application. Soil application of selenite resulted in a higher percentage of se in roots than in shoots and grains.	
Sorghum	Zn, iron and se	Soil application	Application of all micronutrients (Zn, Fe and se) enhanced the growth, quality and uptake of nutrients in sorghum accessions. Se results highest plant height, stem diameter, 1000-grain weight and Zn produced the maximum protein, oil and starch contents.	Qureshi et al. (2021)
	Ca, Zn and iron	Foliar application	Combined foliar application of nutrients (ca at 3% + Zn at 2% + Fe at 1%) improved yield and quality of sorghum by increasing plant height, number of leaves, stem diameter, leaf area per plant, fresh biomass, dry matter yield, dry matter contents, crude proteins, ash contents	Asif et al. (2020)

(continued)

Table 2.1 (continued)

Crops	Nutrients supply	Types of agronomic biofrotification	Adequate supply results	References
			and highest plant contents of Zn.	
	NPK+ biocompost+ biofertilizer	Soil application	Recorded maximum green and dry fodder yield, plant height and stem girth.	Aditi et al. (2019)
Wheat	Phosphate glass fertilizers (GF)	Soil application	Soil application of the GF formulas improved wheat growth, yield, grain quality and photosynthetic parameters as compare to the NPK treatment, especially with N supplementation. Its implementation also effects positively on grain mineral, sugar and protein contents.	Ait-El-Mokhtar et al. (2022)
	Zinc sulphate ($ZnSO_4$) And iron sulphate ($FeSO_4$)	Soil and foliar application both	Application of $FeSO_4$ and $ZnSO_4$ as alone or in combination both soil and foliar application improved growth and quality attributes of wheat. It increased plant height, number of tillers, spike length, number of spikelets per spike, number of grains per spike, 1000 grain weight, economical and biological yield, harvesting index, calcium, magnesium, iron, zinc, copper and protein contents. Combine foliar spray of 0.5% $ZnSO_4$ and 1% $FeSO_4$ is recommended to enhance wheat crop productivity with good quality grains.	Ramzan et al. (2020)

(continued)

Table 2.1 (continued)

Crops	Nutrients supply	Types of agronomic biofrotification	Adequate supply results	References
	Zn	Soil application, foliar spray and seed priming	Results showed improved grain yield, grain Zn and grain quality in both zero and plough tillage systems. Zn seed priming method was found to be most effective in improving the wheat grain yield. Soil Zn application in zero tillage and foliar applications in plough tillage were the most effective for grain Zn biofortification.	Nadeem et al. (2020)
	Zn and iron	Foliar application	Foliar application of various levels of Zn and Fe significantly improved all growth and yield parameters. The plants fertilized with 0.3% Zn and 1% Fe showed higher grain yield. Zn content of grain was observed higher in plants fertilized with 0.4% Zn and 0.5% Fe. Grain Fe content was noted higher in the plants fertilized with 0.1% Zn and 2% Fe.	Jalal et al. (2020)
	Se (sodium selenate)	Soil and foliar application both	Foliar and soil application of se increased 2.6- to 4.6-fold se concentration in grains. Foliar applied se has been found to increase the yield of two wheat cultivar Divana and Simonida in the second year of experiment.	Manojlović et al. (2019)
	Potasium iodide (KI) Potassium iodate (KIO_3)	Soil and foliar application both	Soil application did not affect grain yield and its iodine content	Cakmak et al. (2017)

(continued)

Table 2.1 (continued)

Crops	Nutrients supply	Types of agronomic biofrotification	Adequate supply results	References
			whereas foliar application enhanced grain iodine concentrations up to 5- to ten-fold without affecting grain yield. This increase was also observed in the iodine concentration of the endosperm part of wheat grains.	
Legumes / pulses				
Common bean	Nitrogen and silicon	Soil application (nitrogen); foliar application (Si)	Common bean result maximum grain yield, 20% higher than the control, when fertilize with the split dose of N (60 + 60 kg ha^{-1}) accompanied with the 3.0 g Si L^{-1}, foliar spray irrespective of seed inoculation.	Bueno et al. (2022)
	Zinc + magnesium	Foliar application	Foliar application of zinc plus magnesium increased plant height and biomass yield in common bean cultivars.	Yeboah et al. (2021)
	Zinc oxide nanoparticles +zinc nitrate + chitosan	Foliar application	Foliar application of ZnO nanoparticles at 25 ppm and zinc nitrate at 50 ppm complexed with chitosan improved the growth and biomass production of crop. ZnO nanoparticles as fertilizer are more effective than conventional fertilizer, it accelerates maturation of the crop and reduces the amount of fertilizer to be used without affecting crop yield and maximize the crop productivity.	Palacio-Márquez et al. (2021)

(continued)

Table 2.1 (continued)

Crops	Nutrients supply	Types of agronomic biofrotification	Adequate supply results	References
Chickpea	Mineral, chelated and nanoforms of Zn, Fe i.e., $ZnSO_4$, EDTA-Zn, nano-ZnO $FeSO_4$, EDTA-Fe nano-Fe_2O_3	Foliar application	The foliar application of 0.5% of each of nano-fertilizers of Zn and Fe proved higher translocation over the mineral and chelated forms of nutrient fertilizers and significantly improved grain yield and its nutrient content to a greater extent.	Dhaliwal et al. (2021a, 2021b)
	Zinc sulphate + Ferrus sulphate + urea	Soil and foliar application both	At the time of flowering and pod formation $ZnSO_4$ (0.5%), $FeSO_4$ (0.5%) and urea (2%) application result highest grain Zn and Fe content. Crop grain yield and amount of protein were significantly increased over control with these treatments.	Vajinder et al. (2021)
	Arbuscular mycorrhizal fungi (AMF) + se + I	Soil application	The soil inoculation with AMF improved the concentration of I and se both in single and joint supply of these elements. It is found to be very effective improving the chickpea total dry weight seed yield and quality with the increased levels of antioxidants, protein, and macro- and microelements.	Golubkina et al. (2020)
Soybean	$FeSO_4$	Foliar application	Among various treatments maximum grain Fe concentration, maximum Fe uptake in grain and straw, maximum grain and straw yield was obtained with 0.5% $FeSO_4$ foliar	Dhaliwal (2022)

(continued)

Table 2.1 (continued)

Crops	Nutrients supply	Types of agronomic biofrotification	Adequate supply results	References
			application at 30, 60, and 90 days after sowing over the control.	
	Li (LiOH–li hydroxide and Li_2SO_4- li sulfate)	Foliar application	Supplementation of li significantly effects morphology, yield components and nutritional status of soybean plants. The highest grain yield was recorded with the use of approx. 45.7 mg kg^{-1} Li_2SO_4. The highest concentrations of li in grains were obtained with the application of 120 mg kg^{-1}.	Martins dos Santos et al. (2019)
	Fe compounds Fe-citrate, Fe-phosphate, humic acid, humic acid with $FeCl_3$, nano-Fe	Foliar application	Results showed enhanced growth and yield of soybean with Fe fortification at the time of flowering and pod filling. Application of Fe in any form (organic or inorganic) increased Fe concentration in vegetative parts and in seeds. The highest yield and seed Fe concentration in soybean was found with combined application of humic acid (25 mg) + Fe (2 mM).	Sharma et al. (2019)
Pea	Glass fertilizers (GF) Biofertilzers	Soil application	Addition of biofertilizers with glass fertilizer helped to release components in soil for high quality pea production. It has been found that 60 kg fed^{-1} GF + biofertilizers increased pea pod yield by 55.6% and	Sayed & Ouis (2022)

(continued)

Table 2.1 (continued)

Crops	Nutrients supply	Types of agronomic biofrotification	Adequate supply results	References
			63.23% as compare to control treatment in the first and second seasons respectively.	
	Phosphorous and zinc	Soil apllication	Application of 65 kg P/ha + 20 kg Zn/ha recorded highest plant height, growth yield, reproductive yield, seed index, seed yield, stover yield and plant dry weight.	Subbarao & Dawson (2022)
	Compost, chemical fertilizers $[(NH_4)_2SO_4]$ Nutrients mix. (aqua cool)	Soil and foliar application	Chemical fertilizer, ammonium sulphate @243 K/fed showed better result than compost in increasing total growth, yield and pod quality of pea plant with the spray of aqua cool @ 2 cm/L	Eata et al. (2020)
	Zinc sulphate ($ZnSO_4$), copper sulphate (Cu_2SO_4), boric acid (H_3BO_3)	Foliar application	Foliar sprays of $ZnSO_4$ + Cu_2SO_4 and $ZnSO_4$ + H_3BO_3 result higher growth and reproductive yield of pea.	Safina et al. (2021)
Oilseeds				
Canola	Gibberellic -Acid3 (GA3)	Foliar application	Foliar application of GA3 @ 5 g ha^{-1} result maximum growth and yield of canola variety SURHRAN-2012.	Buriro et al. (2022)
	Zinc	Soil and foliar application both	Foliar spray of Zn @ 5 g/L result highest oil and seed yield. It also increased quality of seed i.e., oil, protein and Zn content in seed and plant.	Afsahi et al. (2020)
	Supplementary irrigation+ selenium	Soil and foliar application both	The results showed that supplementary irrigation at flowering and grain filling stages and foliar application of	Mohtashami et al. (2020)

(continued)

Table 2.1 (continued)

Crops	Nutrients supply	Types of agronomic biofrotification	Adequate supply results	References
			selenium @1.5 mg L^{-1} increased canola yield, yield components, oil percentage, and water productivity in subtropical dryland conditions.	
	Nitrogen, Sulfur, boron	Soil and foliar application both	Sidedress N efficiently consumed by crop and lead to highest yield of crop. Same result was obtained with soil-applied S. soil-applied B did not bring any change in yield, but the foliar application of boron at early flowering was observed to increase yields up to 10%.	Ma et al. (2015)
Mustard	Zn, Fe and urea	Foliar application	The combined N, Fe and Zn foliar effect (recommended dose of fertilizer +0.5% $FeSO_4$ + 0.5% $ZnSO_4$ + 1% urea @ 45 and 65 days after sowing) was found to be the most effective for enhancing total yield, quality and nutritional powers of Indian mustard	Dhaliwal et al. (2021a, 2021b)
	Zinc and boron	Soil application	Effect of boron and zinc levels (i.e., 1 and 2 kg/ha; 5 and 10 kg/ha respectively) along with NPKS @ 80: 40:40:40 kg/ha on growth and yield of yellow mustard revealed maximum no. of silique/plant, no. of seeds/siliqua, test weight, seed yield, Stover yield and harvest index.	Yanthan and Singh (2021)

(continued)

Table 2.1 (continued)

Crops	Nutrients supply	Types of agronomic biofrotification	Adequate supply results	References
	Zinc and boron	Soil application	Soil application of $ZnSO_4$ @ 20 kg/ha along with borax @ 2 kg/ha recorded maximum seed yield, oil content, oil yield, uptake of zinc and boron.	Nadaf and Chandranath, (2019)
	Phosphorus, Sulphur, Phosphate solubilizing bacteria (PSB)	Soil application	The maximum yield was obtained with the application of phosphorus and Sulphur @ 40 kg/ha each along with seed inoculation with PSB biofertilizer.	Solanki et al. (2018)
Sunflower	Four slow release nitrogenous fertilizers (SRNF) (bacterial, neem and sulfur-coated urea and N loaded biochar), Three N levels (100%, 80% & 60%)	Soil application	Among all four SRNF neem-coated urea @148 kg N ha^{-1} with N @100% and 80% significantly increased growth rate, leaf area index, total dry matter, achene yield, biological yield and harvest index of the sunflower crop.	Waqar et al. (2022)
	Potassium	Soil application	Soil application of potassium @150 kg ha^{-1} with normal irrigation resulted in the highest values for growth, yield, and quality attributes.	Dar et al. (2021)
	Zinc	Foliar application	2.0% concentration of Zn applied through foliage increased yield and oil content of sunflower.	Keerio et al. (2020)
	Sulfur	Soil application	Soil application of Sulphur @20 kg ha^{-1} was found to increase overall sunflower plant growth, achene yield and oil content.	Saleem et al. (2019)

(continued)

Table 2.1 (continued)

Crops	Nutrients supply	Types of agronomic biofrotification	Adequate supply results	References
Fruit				
Tomato	NPK fertilizers	Soil application	The application of NPK fertilizer with nitrogen content 31.7% increased the vegetative growth, yield as well as chemical content of tomato plants.	Bekbayeva et al. (2021)
	Sodium selenite (Na_2SeO_3)	Soil application	The soil application of sodium selenite in low concentration improved crop yield and the nutritional quality of the tomato. 53% se accumulation was observed in the fruits under the 5 mgL^{-1} treatment.	Foroughbakhch et al. (2020)
	Conventional fertilizer with urea, chicken manure compost, vermicompost	Soil application	Vermicompost led to greater improvement in growth, fruit quality, yield and soil quality as well.	Wang et al. (2017)
Vegetables				
Broccoli	Organic fertilizer (Tecamine max) chemical fertilizer with NPK	Soil application	100 kg chemical fertilization/ ha + 7 ml/L of organic extract (Tecamine max) proved to be the best fertilizer combination to increase head weight and the heads yield.	Al-Bayati et al. (2021)
	Phosphorus sources (rock phosphate calcium superphosphate, phosphoric acid monoammonium phosphate combined with biofertilizers (i.e., inoculation with mycorrhiza)	Soil application	It has been observed that phosphorus affects positively growth, yield and chemical parameters in broccoli plants but it depends on the source of phosphate fertilizer. Inoculation with mycorrhiza (biofertilizers) may also have a beneficial impact on the growth and yield parameters.	Mohamed et al. (2021)

(continued)

Table 2.1 (continued)

Crops	Nutrients supply	Types of agronomic biofrotification	Adequate supply results	References
Cabbage	Nitrogen nutrition levels and foliar treatments (se, Si, and SA—Salicylic acid)	Soilless and foliar application	Among two cultivation system the plants of hydroponic system were characterized by higher yields than those grown in pot cultivation (mix of peat and sand) due to stimulating effect of N nutrition. Foliar application of Si and se resulted high content of Chl a, Chl b, carotenoids, and high antioxidant activity whereas foliar spray with the SA significantly influenced the plants quality.	Liu et al. (2022)
	Licorice extract microelements, calcium,	Foliar application	Licorice extract and some nutrients spray led to a significant increase in plant height, leaf area, total leaf chlorophyll, anthocyanin content, curl percentage, head weight, total yield, nitrogen, phosphorous and potassium percent in leaves.	Sarhan and Mahmood (2021)
	Yeast suspension, Nitrogen fertilizer	Foliar application	Application of yeast suspension at 6 g L^{-1} and nitrogen fertilizer at 3 g L^{-1} recorded the highest values of plant height, leaf length and width, head weight and its diameter and total yield of cabbage.	Alsaady et al. (2020)
Carrot	Poultry mortality compost (PMC)	Soil application	It has been concluded from the studies that soil application of PMC not only improves the yield but also biofortifies vegetables with	Mubarak et al. (2022)

(continued)

Table 2.1 (continued)

Crops	Nutrients supply	Types of agronomic biofrotification	Adequate supply results	References
			micronutrients such as Zn, Fe, and Mn.	
	Compost, chicken manure, NPK fertilizer	Soil application	Chicken manure or compost used alone or with NPK fertilizer increased agronomic and nutritional quality of carrots.	Maxime Merlin et al. (2020)
Potato	Poultry mortality compost (PMC)	Soil application	Soil application of PMC not only improves the yield but also biofortifies potato vegetable with micronutrients such as Zn, Fe, and Mn.	Mubarak et al. (2022)
	Potassium solubilizing bacteria (PSB) (*Bacillus cereus*)	Soil application	The bio-fertilization of potato with *Bacillus cereus* significantly increased the total yield of potato with the increase in concentration and uptake of leaf N, P, and K. the application of K-feldspar @ 240 kg K_2O ha^{-1} to potato inoculated with PSB yield 40 ton ha^{-1} potato tuber	Ali et al. (2021)
	Boron, potassium	Foliar application	Foliar spray with 1000 ppm potassium sources improve growth, number of tubers, yield per plant and total yield whereas combined effect of potassium sources with 100 ppm boron has been found to be highly effective in improving potato size, tuber weight, total yield, dry matter, and other quality parameters.	Ewais et al. (2020)

(continued)

Table 2.1 (continued)

Crops	Nutrients supply	Types of agronomic biofrotification	Adequate supply results	References
Sweet potato	NPK, Seaweed extract (*Ulva lactuca*)	Soil and foliar application	The soil application of NPK fertilizer @ 75% of the recommended dose, combined with 15% concentration of seaweed foliar spray was observed most efficient treatment for the best sweet potato growth, yield and tuber root chemical compositions.	Helaly (2021)
	Potassium, zinc	Foliar application	Foliar spray with zinc was found most effective treatment on vegetative growth and chemical composition, and foliar applied potassium silicate significantly increased tuber yield and quality of sweet potato.	Mohamed El Nagy et al. (2020)
Lettuce	Proline	Exogenous application	Exogenous application of proline in certain concentrations increased the yield and quality of lettuce grown in hydroponic culture under low potassium conditions	Zhang et al. (2020)
	Ca (calcium lactate)	Foliar application	Application of CaL 1.5 g L^{1} increased antioxidant activity and nutrients (N and ca). N led to increase growth and fresh yield of lettuce under normal irrigation and water stress conditions.	Khani et al. (2020)

with high micronutrient concentrations. Commercial varieties can directly be acquainted with new traits via mutagenesis to improve the quality of plants. This can be an effective approach but it is assumed that genetically modified crop production is not ethical to interfere with nature. It may exhibit toxicity which may affect human health. The pollinating insects can also be affected by the toxic pollen produced by GM plants. Furthermore, GM crop seeds are expensive and beyond the approach of developing countries. Because of these major limitations and restrictions on the use of genetically modified crops in many countries the focus is to develop new plants by traditional plant breeding methods as it is found to be more appropriate approach for developing countries. Many international organizations have initiated programs to address micronutrient deficiencies and improvement in the nutritional content of crops through conventional plant breeding.

HarvestPlus program is one among them being launched in 2003 by the Consultative Group on International Agricultural Research (CGIAR) along with the International Center for Tropical Agriculture (CIAT) and the International Food Policy Research Institute. Harvest Plus through an interdisciplinary and global alliance of scientific institutions and implementing agencies invest heavily to enrich major staple food crop like rice, common beans, cassava, maize, sweet potatoes, pearl millet, and wheat in Asia and Africa, with three major nutrients, Fe, Zn, and vitamin A (Jha and Warkentin 2020). Conventional plant breeding approach requires a one-time investment and farmers can grow and multiply plant for years effectively at very low cost. Due to its better acceptability, feasible implementation and availability of healthy productive seeds, it benefits not only a large population but also poor people of remote areas who cannot access commercially marketed fortified food. Major food crops targeted through breeding biofortification are tabulated in Table 2.2.

2.3.3 Transgenic Approach

Transgenic approach can be an effective substitute for the development of biofortified crops when genetic diversity is unavailable in nutrient content among plant cultivars. Transgenic approaches provide an access to the unlimited genetic pool to identify the desirable genes, and its expression involved in specific plant processes. These genes then are utilized to engineer plant metabolism for the development of transgenic crops. Once a useful gene is discovered such as phytoene synthase, carotene desaturase, nicotinamide synthase, and ferritin, it can be utilized for multiple events in multiple crops. Crops can also be acquainted with bacteria and other organisms to engineer metabolism. Transgenic breeding is the only feasible option to fortify crop with the desired micronutrients which does not exist naturally in crops. Incorporation of genes enhances micronutrients concentrations, their bioavailability and reduces concentration of antinutrients Garg et al. (2018).

Biofortification of crops through transgenic approach initially need time, efforts and investment but in long run it is more cost effective and sustainable approach of breeding than nutrition-based organizational and agronomic biofortification

Table 2.2 Summary of studies on different crops quality and productivity through breeding biofortification

Crops	Nutrients	Breeding biofortification	Varieties / results	References
Cereals				
Maize				
Quality protein maize (QPM) Orange maize	β-Carotene, lysine, tryptophan	Marker-assisted gene pyramiding (MAGP)	The pyramided inbred lines (UMI 1200 and UMI 1230) recorded a higher level of β-carotene, lysine, tryptophan, and better agronomic performance on par to donor parent and recurrent parents respectively.	Chandrasekharan et al. (2022)
	Amio acids (lysine and tryptophan)	Selection	Several national and international maize breeding programs led to development of an array of improved nutritional quality of QPM cultivars, CML176, CML176 × CML186, HQPM-1, HQPM4, HQPM-5, HQPM-7, VivekQPM-9, FQH-4567, CML140, CML194, P70 CML161 × CML165 etc., having approximately twice the content of tryptophan (0.07–0.08% in flour) and lysine (0.25–0.40% in flour) as compare to conventional maize cultivars (tryptophan: 0.03–0.04% in flour; lysine: 0.15–0.20% in flour).	CIMMYT Prasanna et al. (2020)
		Marker-assisted selection (MAS)	Improved quality cultivars Pusa HM4 (0.91% tryptophan and 3.62% lysine) Pusa HM8 (1.06% tryptophan and 4.18% lysine), Pusa HM9 (0.68% tryptophan and 2.97% lysine), IQMH 201 (LQMH 1), IQMH 202 (LQMH 2), IQMH 203 (LQMH 3) were released with improved productivity and yield.	CIMMYT; VPKAS, IARI
	Provitamin A, lysine and tryptophan	Marker-assisted backcross breeding (MABB)	Improved varieties of Pusa Vivek QPM9, Pusa HQPM 5, Pusa HQPM 7contain 8.15 ppm provitamin A, 2.67% lysine, 0.74% tryptophan	IARI
	Provitamin A	(MABB)	Improved cultivar Pusa Vivek hybrid 27 possesses 5.49 ppm of provitamin-A as compared to normal maize hybrids that possess very low provitamin-A (<1 ppm) with better yield performance	IARI

(continued)

Table 2.2 (continued)

Crops	Nutrients	Breeding biofortification	Varieties / results	References
		Conventional	Ife maizehyb-3, Ife maizehyb-4 have high yield potential i.e., 6.65 t/ha with good seed quality and high pro-vitamin A but nitrogen use efficiency of Ife maizehyb-4 is more than Ife maizehyb-3. Sammaz 38 (PVA SYN2) have intermediate level of pro- vitamin A content (5.7'g/g), with high yield potential. (6.4 t/ha) whereas Sammaz 39 (PVA SYN8) contain 6.4'g/g pro- vitamin A with high yield potential i.e., 6.8 t/ha.	CIIMYT
	Provitamin A carotenoids	Conventional	Several nutritious maize cultivars like GV662A, GV664A, GV665A, maizehyb-4, Sammaz 38 (OPV), and 39 (OPV) were released without compromising grain yield levels or other important agronomic and adaptive traits. Accepted by consumer and grown by farmers on large scale.	Manjeru et al. (2019)
	Zinc	Conventional	Zinc enriched maize hybrid ICTA HB-18 contains 15% more zinc (34 ppm) compared to commercial maize and ICTA B-15 (36 ppm) zinc maize contains up to 60% more zinc than regular tortillas. BIO-MZN01 has 36% more zinc than other maize varieties with enhanced grain yield.	CIMMYT, HarvestPlus science and technology institute, ICTA
Rice	Zinc	Backcross and pedigree selection	Among DRR Dhan 49, DRR Dhan 48, DRR Dhan 45, DRR Dhan 49 is superior and high yielding variety with high zinc content (25.2 ppm). Its desirable grain type of medium slender is mostly accepted and consumed in different states of India.	IIRR
		Pure line selection	Zinco Rice MS contain 27.4 ppm zinc with better agronomic performance	IGKV

	Iron and zinc	Conventional	First zinc enriched (20–22 ppm) rice varieties BRRI dhan62, BRRI dhan72, BRRI dhan64 were developed by the breeding of old varieties with high concentration of iron and zinc in their grains. Released in 2013 with highly improved agronomic traits by the Bangladesh Rice Research Institute.	CIAT, HarvestPlus
		Marker-assisted back-cross breeding (MABB)	Binadhan-20, developed by Bangladesh Institute of Nuclear Agriculture (BINA) is an iron enriched rice variety contained 20–31 mg/L Fe.	Al Noor et al. (2019)
		Selection	A cross breeding between a high yielding rice variety (IR72) and a tall, old variety (Zawa Bonday) with a high concentration of grain iron (21 ppm) resulted qualitatively and quantatively improved line (IR68144-3B-2-2-3). Further a traditional variety Jalmagna has been identified with double the iron concentration and 40% more zinc concentration than of common rice variety for further breeding programs.	Gregorio et al. (2000) Majumder et al. (2019)
	Phytic acid (PA)	Conventional	Rice mutant varieties Os-lpa-XS-110-1, Os-lpa-XS-110-2, Os-lpa-XQZ-1 were developed by reducing the concentration of PA, an anti-nutrient which reduces the bioavailability. Conventional breeding by using mutants often results in high bioavailability but poor agronomic performance and low seed germination.	Perera et al. (2018)
	Protein	Backcross followed by pedigree selection	CR Dhan 311 (Mukul), CR Dhan 315are high protein variety (10.1%-10.2%) with high Zn content (20 ppm). These high protein varieties had significantly higher glutelin content than Naveen (rice cultivar) with better yield potential.	NRRI, India

(continued)

Table 2.2 (continued)

Crops	Nutrients	Breeding biofortification	Varieties / results	References
Sorghum	Iron	Breeding	12KNICSV 22 (Zabuwa) variety has iron content three times higher at 128.99 ppm compared to 40 ppm. The improved varieties 12KNICSV-188 (Deko) 12KNICSV 22(Zabuwa) yield 2.4–2.8 tons per ha, compared to yields of less than one ton per ha from local varieties. Both varieties are highly acceptable by farmers because of its yields that are double that of the local varieties and short duration.	ICRISAT, HarvestPlus Jha and Warkentin (2020)
	Iron and zinc	Breeding	Five sorghum hybrids enriched with iron and zinc, ICSH 14001 (Fe 49 ppm and Zn 38 ppm), ICSH 14002 (Fe 46 ppm and Zn 32 ppm), ICSA 661 × ICSR 196 (Fe 45 ppm and Zn 36 ppm), ICSA 318 × ICSR 94 (Fe 45 ppm and Zn 34 ppm), ICSA 336 × IS 3760 (Fe 45 ppm and Zn 40 ppm), and a R-line/variety ICSR 14001 (Fe 42 ppm and Zn 35 ppm) have been released with higher yield potential	ICRISAT, HarvestPlus Sheoran et al. (2022)
	Iron and zinc	Breeding	Sorghum variety Parbhani Shakti produced higher grain yield (up to 3.5 t/ ha) with higher grain Fe concentration of up to 45 ppm and zinc up to 32 ppm as compared to the baseline of 30 ppm Fe and 20 ppm Zn in sorghum.	All India coordinated sorghum improvement project (AICSIP)
Wheat	Zinc	Conventional	PBW1Zn variety contains 40.6 ppm zinc in grains released by Punjab agricultural university, India with better nutritional quality and improved grain yield.	PAU, India
			Zincol, 2016, NR 419 and 421 have been released in Pakistan with high concentration of zinc i.e., 33.9 ppm with better agronomic performance.	CIMMYT
			Zn-biofortified wheat cultivar, BARI Gom 33, has 32 and 50–55 mg kg^{-1} Zn without and with soil application of $ZnSO_4$, respectively.	Das et al. (2019)

			Zinc Gahun 1, Zinc Gahun 2, Borlaug 2020, varieties are high in zinc and well known for being high yielding, drought- and heat-resilient, and resistant to wheat blast.	CIMMYT
		Hybridization and selection	HD 3171 variety has 47.1 ppm Zn concentration in grains with high yield potential i.e., 46.33 q/ha. PBW 757 enriched with 42.3 ppm Zinc is high yielding bread wheat variety with average grain yield of 36.7q/ha and yield potential of 44.9q/ha. It is highly resistant to diseases.	Yadav and Kiran (2019) Gupta e al. (2019)
		Conventional	BHU 1, BHU 3, BHU 5, BHU 6, BHU 17, BHU 18, zinc Shakti (Chitra) are wheat varieties released in India in 2013 with high concentration of zinc (40–45 ppm). Two varieties BHU 1 and BHU 6 have high yield potential, disease resistance in addition to high zinc.	CIAT, CIMMYT, harvest plus
	Iron and zinc	Pure line selection	WB2 with high Fe (40 ppm) and Zn (42 ppm) has been developed by Indian Institute of Wheat and Barley Research, India	IIWBR
		Pure line selection	HPBW 01 is a pure line variety contains high Fe (40.0 ppm) and Zn (40.6 ppm) with improved agronomic traits.	Ambati et al. (2020)
		Conventional	HI 8777 (durum) is high yielding nutritive pasta product variety contain 48.7 ppm Fe and 43.6 ppm Zn in grains. Average yield is 19 q/hec and yield potential is 29 q/hec. Higly resistant to diseases and provide good level of yellow pigments i.e., more than 6 ppm	IARI, India
	Fe, Zn, and protein	Pure line selection	Pusa Tejas (HI 8759) is released with high protein (12%), iron (42.1 ppm) and zinc (42.8 ppm). Very suitable for making bread, pasta and other traditional food products. MACS 4028 is developed with high protein (14.7%) iron (46.1 ppm) and zinc (40.3 ppm). Both the durum varieties are with better agronomic traits.	Ambati et al. (2020) Ambati et al. (2019)

(continued)

Table 2.2 (continued)

Crops	Nutrients	Breeding biofortification	Varieties / results	References
	Carotene (provitamin A), anthocyanin	Conventional	The durum wheat variety HI 8627 with high provitamin A (6-9 ppm) was released by IARI, India. The yellow pigment is due to lutein. Lutein and anthocyanin are antioxidants in nature and enhanced quality trait of wheat grains. Black grained wheat and other colored wheat cultivars are rich in protein and selenium. The colored wheat cultivars (purple, blue, and black) have higher antioxidant potential and are the major sources of fibers, proteins, and phytochemicals.	IARI, India Kumari et al. (2020) Havrlentova et al. (2014)
Finger millet	Fe	Pedigree selection	VR 929 (Vegavathi) is developed having 131.8 mg/kg Fe, 33.2 mg/kg Zn, 371 mg/100 g ca, 9.8% protein and 12.6% dietary fiber content with recorded high grain yield i.e., 36.1 q/ha and fodder yield (7.2 t/ha).	ANGRAU, India Raju et al. (2020)
	Ca, Fe, Zn	Pedigree selection	CFMV 1 (Indravathi) variety has the potential of 35–40 q/hec grain yield. Seed quality is maintained by high concentration of Fe (58.0 ppm), Zn (44.0 ppm), ca (428 mg/100 g).	ANGRAU, India; NAU, India
Little millet	Fe, Zn	Pedigree selection	CLMV 1 (JAICAR Sama 1) variety has the grain yield potential of 18–20 q/hec. This variety contains high protein (14.4%) and high grain iron (59 ppm) and zinc (35 ppm).	IIMR, India
Legumes/pulses				
Lentils	Fe	Conventional	Pusa Ageti masoor (L4717) a pure line variety is 13.7% superior over the best performing variety in central zone and matures in only 100 days. It contains 65.0 ppm iron.	IARI, India

	Fe, Zn	Conventional	IPL 220 is high yield and high quality variety. Average yield is recorded 14–18 q/hec and quality is maintained by the presence of 73.0 ppm Fe and 51.0 ppm Zn.	ARI India, ICARDA, HarvestPlus
		Conventional	BARI Masur-8 has high yield potential (1973 kg ha − 1) and high concentrations of Fe (74 mg kg − 1), Zn (61 mg kg − 1), and se (325μgkg − 1) in seed which make it highly acceptable by consumers.	Shahin Uz Zaman et al. (2022)
Cow pea	Fe	Conventional	Pant Lobia-1 contain 82 ppm Fe and 40 ppm Zn; pant Lobia-2 contain 100 ppm Fe, and 37 ppm Zn; pant Lobia-3 contain 67 ppm Fe, and 38 ppm Zn and pant Lobia-4 contain 51 ppm Fe and 36 ppm Zn. All these are also high yielding and multiple disease resistant varieties.	GBPAUT, HarvestPlus Boukar et al. (2011)
Beans	High iron	–	RWR 2245; RWR 2154; MAC 42; MAC 44; CAB 2; RWV 1129; RWV 3006; RWV 3316; RWV 3317; RWV 2887 are iron rich beans varieties being developed by HarvestPlus with high grain yield and yield potential t/hec.	HarvestPlus (Rwanda)
Oil seed crop				
Groundnut	Oleic acid	Marker-assisted breeding	Girnar 4, Girnar 5 are developed with 78.4–78.5% oleic acid.	DGR, India
Linseed	Linoleic acid	Mutagenesis	TL 99 is reported with 58.9% linoleic acid.	BARC, India
Mustard	Erucic acid	Pedigree selection	Erucic acid of Pusa mustard 30 is 1.20% and Pusa mustard 32, 1.32%.	IARI, India
	Erucic acid and Glucosinolates	Pedigree selection	Pusa double zero mustard 31 variety is recorded with 0.76% erucic acid and 29.41 ppm Glucosinolates	IARI, India
Soybean	Kunitz trypsininhibitor free	Marker-assisted backcrossing	NRC 127 is first Kunitz trypsin inhibitor (anti-nutrition) free soybean variety being developed by ICAR-IISR.	ICAR-IISR, India
	Lipoxigenase-2 free	Modified marker-assistedbackcrossing	NRC 142 variety has been developed free from Kunitz trypsin inhibitors as well as Lipoxigenase-2. These are	IISR, India

(continued)

Table 2.2 (continued)

Crops	Nutrients	Breeding biofortification	Varieties / results	References
			two major constraints for behind utilization of soy food uses.	
	Oleic acid	Pedigree selection	NRC 147 soybean variety is having high content of oleic acid i.e., 42.00%	IISR, India
Vegetables				
Potato	Anthocyanin		Kufri Manik potato variety is rich in micronutrients, carotenoids and anthocyanin (0.68 ppm) with yield potential is 23–25 t/ha.	CPRI, India
		Hybridization and selection	Kufri Neelkanth potato cultivar is rich in antioxidants (anthocyanins >100 μg/100 g fresh wt. and carotenoids~200 μg/100 g fresh wt.) and the average yield of the hybrid is estimated to be 35–38 t/ha.	CPRI, India
	Zinc, iron	–	Variability in Fe and Zn concentration has been observed among the evaluated genotypes of potato. It ranges from 34.67 to 76.67 mg $kg^{-}1$ and 12.88 to 66.1 mg kg^{-1} for iron and zinc respectively. Mnandi, Hertha, Buffelspoort cultivars and breeding lines-N105–1, 00-S100–33 and 03–627-50 are reported to be the best in quality and quantity.	Managa (2015)
Sweet potato				
Orange sweet potato	Vitamin A	Clonal selection	Ejumula, Kakamega, Vita, Kabode, Naspot 12O, Naspot 13O, Olympia, Twatasha, Kokota, Chiwoko, Zambezi cultivars have been developed by CIP. Naspot 12 O′ and Naspot 13 O′ have higher storage root, beta-carotene in storage roots and biomass yield.	HarvestPlus, international potato Centre (CIP)

	Provitamin A (β-carotene)	Pure line selection	Bhu Sona contains high β-carotene (14.0 mg/100 g) content as compared to 2.0–3.0 mg/100 g β- carotene in other popular varieties. Tuber yield is 19.8 t/ha, dry matter 27.0–29.0%, starch 20.0% and Total sugar 2.0–2.4%. .	CIP, HarvestPlus Yadava et al. (2020)
		Conventional	Beauregard, Resisto, W-119 are improved varieties in yield coupled with good taste and dry matter content. Resisto is with high beta-carotene content (17–25 mg/ 100 g, fresh weight)	Laurie et al. (2015)
Cassava	Vitamin A	–	HarvestPlus in collaboration with International Institute of Tropical Agriculture have released six vitamin A fortified varieties in Nigeria (2011; TMS 01/1368—MUCASS 36, TMS 01/1412— UMUCASS 37 and 2014; TMS 01/1371—UMUCASS 38 and NR 07/0220—UMUCASS 44, TMS 07/0593—UMUCASS 45, and TMS 07/539—UMUCASS 46) and one in DRC-Democratic Republic of Congo [Kindisa (TMS 2001/1661)] Efforts have been made to improve nutritional quality with high yield.	IITA, HarvestPlus
Cauliflower	Provitamin A (β-carotene)	Pure line selection	Pusa Beta Kesari 1 first bio-fortified cauliflower contain beta carotene (8.0–10.0 ppm), marketable curd weight is about 1.2 kg and average yield is 40.0–45.0 t/ha. Orange Cheddar cauliflower is an excellent source of beta-carotene, the orange pigment found within the cauliflower head that can be converted into vitamin A in the body. It also contains vitamin C, lower amounts of potassium, fiber, iron, calcium, and magnesium. Purple graffiti, cauliflower is an excellent source of vitamin C and is rich in anthocyanins. The pigmented heads contain fiber, some calcium, folate, potassium, and vitamin A.	IARI, India

(continued)

Table 2.2 (continued)

Crops	Nutrients	Breeding biofortification	Varieties / results	References
Fruits				
Tomato	Anthocyanin	Conventional	Anthocyanin biofortified tomato cultivar is sun black. It is high yielding cultivar (7 q / hec) and its deep purple fruit pigmentation is due to high anthocyanin content in the peel i.e., 7.1 mg/100 FW.	Mazzucato et al. (2008)
		Conventional	Another anthocyanin biofortified tomato cultivar is black galaxy generated by similar approach has been reported from Israel. Its high quality is maintained by the presence of high concentration of antioxidants and vitamin C,	Garg et al. (2018)
Banana	Vitamin A	–	Apantu, Lahi, Bira, Lai, and Pelipita are the most promising pro-vitamin A carotenoids (pVACs)-rich banana cultivars with improved nutrient quality and yield.	Bioversity International—Uganda, HarvestPlus Muller et al. (2021)
Mango	Beta-carotenevitamin C	–	Amarpali, Pusa Arunima, Pusa Surya, Pusa Pratibha, Pusa Peetamber, Pusa Lalima, Pusa Shreshth, Mexican Ataulfo have been introduced with enhanced nutritional and agronomical important characters, contain sufficient amount of Beta-carotene vitamin C with high yield/tree. The Mexican grown Ataulfo variety contains the highest level of both vitamin C (ascorbic acid) and beta-carotene.	IARI, India USDA agricultural research service
Pomegranate	Fe, Zn, vitamin C	Conventional	Biofortified Solapur Lal contain 5.6–6.1 mg/100 g Fe, 0.64–0.69 mg/ 100 g Zn, and 19.4–19.8 mg/ 100 g vit C/. no. of fruits/tree is 130–140. Potential yield is 35–39 kg/tree	NRCP, India
Grapes	Antioxidants	–	Pusa Navrang is an improved variety, released by Indian agricultural institute. It contains higher amount of total soluble solids (carbohydrates, organic acids, proteins, fats, and minerals) and antioxidants.	IARI, India

program. It has no taxonomic limitation and even synthetic genes of specific traits can be constructed and transfer to target crops with accuracy in a short time period compare to conventional breeding. Successful experiments have been done on crops by the use of genes from different sources to increase micronutrients for the improvement of food crop nutritional levels i.e. vitamins, essential amino acid, fatty acids and mineral. Transgenic means of breeding enhanced crop quality and quantity in developing countries but safety issues on human and the environment are major obstruct for their wide acceptance. Some successful examples of transgenic method of biofortification on different staple food crop are tabulated in Table 2.3.

2.4 Conclusion

Biofortification approaches through agronomy, breeding or transgenic manipulation hold great potential to improve not only the content and bioavailability of micronutrients such as Fe, Zn, Se, and pro-vitamin plant but also the productivity or yield of staple food crops like cereals, legumes, vegetable and fruits. Being promising and cost effective it is implemented though out the world and international initiatives are working and investing to achieve the set targets. Among different crops, biofortified by different approaches largest number of cereals, legumes and vegetables have been biofortified by all three biofortification approaches in almost equal percentage. Transgenic approach covers highest number of crops specially oilseed crops, due to their limited genetic variability (Garg et al. 2018). Researches have demonstrated that agronomy and traditional plant breeding approaches of biofortifications are finding widespread and easy acceptance and work to enhance quality and quantity of crops. Transgenically fortified crops face hurdles of acceptance by consumers and different expensive and time consuming regulatory approval processes, implemented by different countries. Beside these challenges biofortification approaches have an immense potential to enhance crop yield, biomass production, increase the nutrient contents of crops to remove micronutrients malnutrition among poor people of developing countries, suppress crop diseases, abiotic stressors, and ultimately increase farming income. All these approaches are in need to be incorporated with ongoing national agriculture, food, nutrition and health education programs for its effective implementation and sustainability.

Table 2.3 Summary of studies on different crops quality and productivity through transgenic biofortification

Crops	Nutrients	Donor Organism	Gene involved	Results	References
Cereals					
Rice	Fe	Arabidopsis, Common bean	*AtIRT1*, *AtNAS1*, *Pvferritine*	Iron increased upto 9.67 μg/g DW in the polished grains that is 2.2-fold higher than NFP rice cultivar	Boonyaves et al. (2016, 2017)
			OsNAC5	Overexpression of OsNAC5 enlarges roots significantly and enhances drought tolerance and grain yield (i.e., 22%–63%) under field conditions.	Jeong et al. (2013)
		Rice cultivar	*OsNAS2*	An effective increase of iron i.e., 19 μg/g DW was observed in rice endosperm	Johnson et al. (2011)
		Common bean	*Phaseolus ferritin (PHAVU)*	An elevation in Fe concentration is recorded i.e. 22.07 μg Fe/g DW.	Lucca et al. (2001)
		Soybean, *aspergillus fumigatus*, rice	*Ferritin*, phytase, *AtNAS1*	More than six-fold increase in iron content was found in transgenic rice endosperm. Phytase may reduce the iron antinutrient phytate.	Wirth et al. (2009)
		Soybean	*Ferritin* gene *Soyfer H-1*	38.1 μg Fe/g DW The high-iron rice did not face yield penalty or significant changes in trait characters	Goto et al. (1999)
	Zinc	Barley	*HvNAS 1*	More than three-fold Fe and two-fold Zn concentrations increased in both polished and brown seeds of transgenic rice	Malik and Maqbool (2020);

				plants. Both nutritional and agronomic traits combined to produce high quality-high yield crops.	Masuda et al. (2009)
	Fe, Zn, β-Carotene	Arabidopsis, common bean, maize	*AtNAS1, PvFERRITIN, CRTI, ZmPSY*	All transgenic rice lines have found with increased β-carotene, iron, and zinc content without compromising grains yield and its quality.	Singh et al. (2017)
	Vitamin E	Arabidopsis thaliana	*AtTC, AtHP*	Vitamin E biofortification produced quality rice grain with immense benefits for human nutrition	Sundararajan et al. (2021)
	Vitamin B1	Arabidopsis thaliana	*THIC, THI1, TH1*	Variants with high levels of vitamin B1could potentially provide a yield benefit	Strobbe et al. (2021)
	Vitamin B6	Arabidopsis thaliana	*AtPDX1.1, AtPDX02*	Vitamin B_6 in leaves was increased up to 28.3-fold, in roots up to 12-fold and in rice seeds up to 3.1-fold with insignificant impact on growth. Though, seed yield was affected in some cultivars with enhanced vitamin B_6.	Mangel et al. (2019)
	Vitamin B9	Arabidopsis thaliana	*ADCS, AtGTP cyclohydrolase 1*	Folate (vitamin B9) biofortification obtained up to 150 fold higher folate concentrations in rice grains than those of wild-type rice	Blancquaert et al. (2015)
	Provitamin A (β-carotene)	Daffodil, *Erwinia uredovora, maize*	*Carotene desaturase (crtI), daffodil PSY*	Total carotenoids were increased upto 37 μg/g as compared to the original Golden Rice and observed a better accumulation of β-carotene.	Paine et al. (2005)

(continued)

Table 2.3 (continued)

Crops	Nutrients	Donor Organism	Gene involved	Results	References
	Vitamin A	Daffodil, *Erwinia uredovora, maiz*	*Phytoene synthase (PSY), phytoene desaturase* (*CrtI*) gene; psy and lycopene β-cyclase (β-lcy)	Nutritional quality of grains were enhanced by increased level of vitamin A up to 1.6 μg/g DW	Kumar (2019)
	Methionine and cysteine	Sesame	*Sulfur-rich protein, S2SA*	Methionine contents of transgenic rice were elevated by 29–76% and cysteine 31–75% as compared to wild-type rice grains. Growth rate and grain yield of the transgenic plants, did not differ from those of the wild-type control plants	Lee et al. (2003)
	Lysine	Bacteria	*lysC, dapA*	Free lysin levels in seeds of transgenic line (high free lysin; HFL) were found increased upto 25-folds than that of wild type. HFL transgenic rice growth was normal with a slight difference in height and grain color.	Yang et al. (2016)
	Flavonoid, linoleic acid	Rice, *Arabidopsis thaliana*, soybean, parsley	*GmFAD3, ZmC1, chalcone synthase, phenylalanine ammonia lyase*	Flavonoids and linoleic acid accumulation in rice effectively improve the nutritional value of rice grains.	Zhu et al. (2021) Ogo et al. (2013)
Maize	Iron	Soybean, *aspergillus*	*GmFER, aspergillus phytase, aspergillus phy2*	Bioavailability of iron has been increased in transgenic lines of maize.	Garg et al. (2018)

		Maize and soybean	*lpa1–1, ferritin*	More than two-fold improvement was observed in iron bioavailability. High-iron *lpa1–1* seeds have higher germination rates and seedling vigor as compared to transgenic seeds.	Aluru, et al. (2011)
	Low-phytate	*phyA2*	*Aspergillus Niger phyA2*	Phytase activity in transgenic maize seeds increased about a 50-fold as compare to non-transgenic maize seeds. It leads to the development of new maize hybrids with improved phosphorus availability.	Chen et al. (2008)
	Vitamin A or multivitamin	Bacteria	*crtB* and *crtI*,	Transgenic line has been reported with a provitamin A content of 9.8 μg g^{-1} seed dry weight. Carotenoid was increased upto 34-fold with β carotene accumulation in the maize endosperm.	Aluru et al. (2008)
		Maize	Phytoene synthase (*psy1)*	Kernels of transgenic lines contain 59.32 μg/g DW vitamins and 169 fold the normal amount of β carotene, six fold the normal amount of ascorbic and double the normal amount of folic acid.	Naqvi et al. (2009)
	Vitamin E	Barley	*HGGT*	An increase in tocotrienol and tocopherol content as much as six-fold was found in corn seeds. It results in an increase of antioxidant content.	Cahoon et al. (2003)

(continued)

Table 2.3 (continued)

Crops	Nutrients	Donor Organism	Gene involved	Results	References
	Carotenoid	Bacteria	*crtI*	Carotenoid biosynthesis was enhanced by more than 20-fold in seed endosperm of transgenic lines.	Decourcelle et al. (2015)
	Lysin and total protein	*Solanum berthaultii*	*sb401*	Nutritional value of transgenic maize was improved by an increased in lysine content i.e. 16.1% to 54.8% and total protein content i.e. 11.6% to 39.0% as compare to non-transgenic maize control.	Yu et al. (2004)
	Lipid, protein (lysin) and starch	Arabidopsis, rice, maize	*AtGIF1*, *OstGIF1*, *ZmGIF1*	Increased in activity of invertase and sugar content were observed in the transgenic plants leaves which led to increased grain yield and improved grain nutrients.	Guo et al. (2018)
Wheat	Iron	Soybean	*Ferritin*	The grain iron content increased, ranging from 4.93 to 64.03%.	Xiaoyan et al. (2012)
		Wheat	*TaFer1* and *TaFer2*	50–85% higher iron content was observed in the grain.	Borg et al. (2012)
	Provitamin A	Maize, bacteria	*psy1*, *crtI*, *CrtB*+ *Crtl*	The provitamin A content was significantly increased in all transgenic wheat lines, which ranges from 0.18 μg g^{-1} to 3.86 μg g^{-1} of seed dry weight.	Wang et al. (2014)
	Amino acid composition	*Amaranthus hypochondriacus*	*Ama1*	The expression of the AmA1 protein increased essential amino acid content and all those parameters which are associated with functional quality.	Tamas et al. (2009)

	Low Phytat	*Aspergillus Niger*	*phyA*	Overexpression of the phytochrome (*PhyA*) gene enhanced production of phytase and anti-nutrient activity, whereas silencing of *phyA* and *TaABC13* transporter gene decreased phytic acid production upto 18–19% and increased iron and zinc bioavailability.	Brinch-Pedersen et al. (2000)
	Amylose	Wheat	*SBEIIa*	The silencing of the *SBEIIA* gene increased concentration of less digestible amylose starch. No significant differences were found in grain weight between transgenic plants and wild type controls.	Sestili et al. (2010)
	Anthocyanin	Barley	*Dhn12*, *Itr1*, and *Ltp1*	The antioxidant activity of wheat was improved by expressing maize regulatory genes (*C1*, *B-Peru*) involved in anthocyanin production	Doshi et al. (2006)
Sorghum	Carotenoids	–	–	Carotenoid bioaccessibility was increased with the increase in the amount of coformulated lipid from 5% w/w to 10% w/w. transgenic sorghum contained 4–eight fold increased bioaccessible β-carotene as compare to non-transgenic sorghum.	Lipkie et al. (2013)
	Lysin, vitamin A, Fe and Zn	Maize, *Pantoea ananatis*, Arabidopsis, barley	*PSY1*, *CRTI*, *at-DXS HGGT*	Nutritionally enhanced sorghum lines contained enhanced levels of pro-vitamin A, lines with	Zhao et al. (2019)

(continued)

Table 2.3 (continued)

Crops	Nutrients	Donor Organism	Gene involved	Results	References
				reduced phytate (90% reduction) increased 40–80% iron and zinc bioavailability.	
Barley	Zinc	Arabidopsis	Zinc transporter	Zinc content has been improved by overexpression of zinc transporters which lead to increased yields and quality of grains.	Ramesh et al. (2004)
	β-D-glucan	Oat, cauliflower mosaic virus	*HvCslF*	Over-expression of cellulose synthase-like gene *HvCslF* result in significantly elevated amounts of β-D-glucan in the grain and vegetative tissues of transgenic barley lines.	Burton et al. (2011)
	Phytase	Barley	*HvPAPhy*_a	To increase the bioavailability of iron and zinc in barley seeds, phytase activity has been increased by expression of phytase gene *HvPAPhy*_a	Holme et al. (2012)
	Lysine	*Escherichia coli*	*dapA*	Essential amino acid lysine content in leaves of transgenic barley was 50% higher than in wild type plants.	Ohnoutkova et al. (2012)
Legumes/pulses					
Soybean	Sulfur amino acids	Maize	*Zein*	Amino acid analysis of the transgenic lines confirmed that there was a 12–20% increase in methionine, and 15–35% increase in cysteine content as compared to the control.	Dinkins et al. (2001)

	Amino acid	Soybean	*MB-16*	Amino acid analysis of mature seed of transgenic line, showed a significant increase of 16.2 and 65.9% in methionine and cysteine, respectively, as compared to the control. Improvements in the essential amino acids profile of seeds maintain nutritional quality of soybean.	Zhang et al. (2014)
	Vitamin E	Arabidopsis	*At-VTE3*	*At-VTE3 coexpressed with at-VTE4 result 08-fold increase of* α-tocopherol and upto five-fold increase in seed vitamin E activity as compare to normal.	Van Eenennaam et al. (2003)
	β-Carotenoid	*Capsicum* and *Pantoea ananatis*	*PAC*	*β-PAC* transgenic seeds accumulated 146 μg/g of total carotenoids about 62-fold higher than non-transgenic seeds of which 112 μg/g (77%) was β-carotene.	Kim et al. (2012)
	α-Linolenic acids	Soybean	Gm*FAD3*	A significant reduction of this fatty acid increased the stability of the seed oil and enhanced the seed agronomical value.	Flores et al. (2008)
	γ-Linolenic acid + stearidonicacid (STA)	*Borago officinalis*	Δ^6 desaturase	Average levels of γ-linolenic acid (GLA) ranged from 3.4 up to 28.7%, and stearidonic acid (STA) from 0.6 to 4.2% in transgenic soybean (T_1 generation).	Sato et al. (2004)
Common bean	Methionine and cysteine		*uidA* and *be2s2*	In two transgenic lines of beans the methionine content was significantly increased i.e., 14% and 23% over the values found in untransformed plants.	Aragao et al. (1999)

(continued)

Table 2.3 (continued)

Crops	Nutrients	Donor Organism	Gene involved	Results	References
Vegetables					
Potato	Amino acid composition	Amaranthus	*AmA1*	Transgenic tubers showed a significant increase in concentration of essential amino acids and 60% increase in total protein content. Total biomass and tuber yield was also increased with the increase in photosynthetic activity.	Chakraborty et al. (2010)
	Anthocyanins + phenolic acids	Barley and *Petunia hybrida*	*CHS*, *CHI* and *DFR*	A significant increase was recorded in phenolic acids and anthocyanins.	Lukaszewicz et al. (2004)
	Vitamin C	Strawberry	*GalUR*	Transgenic line with enhanced vitamin C performed better under different abiotic stresses as compared to control.	Upadhyay et al. (2009)
	Beta carotene	Cauliflower	*Or*	The *or* transgenic tubers showed accumulation of increased levels of carotenoids and three additional metabolites intermediates of phytoene, phytofluene and ζ-carotene	Lopez et al. (2008)
	Methionine	*Solanum tuberosum*	*StMGL1*	Elevated accumulation of free methionine in some transgenic lines.	Huang et al. (2014)
Sweet potato	Beta-carotene	Sweet potato	Lycopene ϵ-cyclase (*LCY-ϵ*)	In the transgenic calli, the β-carotene content was found 21-fold higher than in control calli. The transgenic calli also	Kim et al. (2013)

				showed enhanced salt stress tolerance (200 mM).	
	Antioxidants	Sweet potato	*IbMYB1*	Increased anthocyanin levels promoted the elevation of proanthocyanidin and total phenolic levels in fresh storage roots of transgenic plant. Transgenic plants displayed much higher antioxidant activities than control. Yield of storage root varied between the transgenic lines.	Park et al. (2015)
Cassava	Beta caroteneprovitamin A	*Pantoea ananatis*	*nptII*, *crtB* and *DXS* *PSY*, *CrtI*	Roots of all the transgenic cassava were positive for the DXS, *crtB* and *npt*II genes. Cassava plants over expressing a *PSY* transgene produce yellow-fleshed, high-carotenoid roots.	Telengech et al. (2015); Welsch et al. (2010)
	Iron	Arabidopsis	*Vascular iron transporter VIT1*, *iron transporter IRT1*, *ferritin*(*FER1*)	In transgenic plants levels of iron accumulation was 7–18 times higher and zinc 3–10 times higher than in nontransgenic controls whereas growth parameters and storage-root yields were unaffected.	Narayanan et al. (2019)
Oilseeds					
Linseed/flax	Carotenoid	*Pantoea ananatis*	*crtB*	Total carotenoid amounts in seeds of the transgenic flax plants were 65.4–156.3 μg/g fresh weight, which is parallel to 7.8- to 18.6-fold increase, as compared to non-transgenic controls.	Fujisawa et al. (2008)

(continued)

Table 2.3 (continued)

Crops	Nutrients	Donor Organism	Gene involved	Results	References
	Flavonoid	*Petunia hybrida*	*CHS, CHI, DFR*	In transgenic flax plants the simultaneous expression of genes resulted an increase in flavanones, flavones, flavonols and anthocyanins levels.	Lorenc-Kukuła et al. (2007)
	Essential amino acids	*Arabidopsis thaliana, Escherichia coli*	*Als, nos, nptII, Bla, spc*	Transgenic linseed rich in essential amino acids {(cv CDC triffid flax (FP967)} has been released by University of Saskatchewan, in Colombia, USA, and Canada	University of Saskatchewan, Canada
Canola	Carotenoid	*Pantoea ananatis* and *Brevundimonas* sp.	*crtB*, *crtE*, *crtZ*, *crtY*, *crtI*, *crtW*, and *idi*	The total amount of carotenoids in seeds of three transgenic lines was 412–657 μg g^{-1} fresh weight, parallel to 19- to 30-fold increase as compare to nontransgenic control. The total amount of ketocarotenoids was 60–190 μg g^{-1} fresh weight. β-carotene was detected with significant amount of α-carotene, echinenone, phytoene, lutein, and canthaxanthin also detected in the transgenic seeds	Fujisawa et al. (2009)
	Lysine	*Corynebacterium and Escherichia coli*	*AK* and *DHDPS*	Expression of *DHDPS* plus lysine-insensitive AK caused several hundred-fold increases in free lysine and total seed lysine by as much as five-fold.	Falco et al. (1995)

	Fatty acids	*Cuphea hookeriana*	*Ch FatB2*	Canola normally does not accumulate caprylate (8:0) and caprate (10:0) in seed oil, but in seeds of transgenic canola a dramatic increase in the levels of these two fatty acids were observed accompanied by a decrease in the levels of linoleate (18:2) and linolenate (18:3).	Dehesh et al. (1996)
	Phytate degradation (increase inavailable P)	–	–	Phytaseed™ canola (MPS 961–965) engineered for phytase degradation to increase the availability of phosphorus in canola has been produced and released by BASF in USA	BASF
Mustard	γ-Linolenic acid	*Pythium irregulare*	*Pid6*	Expression of the desaturase in *Brassica juncea* resulted in production of three Δ6 unsaturated fatty acids (18:2–6, 9; 18:3–6, 9, 12; and 18:4–6, 9, 12, 15) in seeds. γ-linolenic acid (GLA; 18:3–6, 9, 12) was found in abundance and accounts for up to 40% of total seed fatty acids. GLA is almost completely incorporated into triacylglycerol (98.5%), and within the triacylglycerol, the GLA is found in abundance at the *sn*-2 position.	Hong et al. (2002)

(continued)

Table 2.3 (continued)

Crops	Nutrients	Donor Organism	Gene involved	Results	References
Camelinasativa	Low polyunsaturated fatty acids	Targeted mutagenesis by CRISPR/Cas9	*FAD2*	Camelina transgenic lines with 10% to 62% oleic acid accumulation in the oil led to reduction in polyunsaturated fatty acids and improvement in agronomical and biotechnological potential of crop.	Morineau et al. (2017)
Fruits					
Tomato	*b*-carotene	Arabidopsis	*b-Lcy*	The overexpression of gene increases the levels of beta-carotene up to sevenfold higher than non-transgenic plant.	Rosati et al. (2000)
		Erwinia herbicola, narcissuspseudonarcissus	*b-cyclase*	Plastid expression of a plant lycopene β-cyclase does not only trigger lycopene conversion to β-carotene, but also results >50% increase in total carotenoid accumulation.	Apel and Bock (2009)
	Astaxanthin	*Chlamydomonas reinhardtii* and *Haematococcus pluvialis*	–	A massive accumulation of free astaxanthin was recorded in leaves (3.12 mg/g), esterified astaxanthin in fruits (16.1 mg/g) and a 16-fold increase of total carotenoid without affecting the normal growth and development.	Huang et al. (2013)
	Xanthophyll	Arabidopsis and pepper	*b-Lcy*, *b-Chy*	A significant increase in β-carotene, β-cryptoxanthin and zeaxanthin were observed in fruits of the transformants.	Dharmapuri et al. (2002)

	Iodine	*Solanum lycopersicum* L	*HMT, S3H,* and *SAMT*	Iodosalicylates, iodobenzoates, KI and KIO_3, good source of iodine for tomato biofortificatio, did not have a negative effect on growth and development of tomato plants and led to an increase of the iodine content in all the parts of tomato plants.	Halka et al. (2019)
	Folate	*Arabidopsis*	*AtADCS*	Transgenic fruit accumulated up to 25-fold more folate (25 nmol/g fresh weights) than controls.	De la Graza et al. (2007)
Apple	Resistance against fungal diseases	Grapevine (*Vitis vinifera* L.) Kiwi (*Actinidia deliciosa*)	*Vst1,* PGIP	Results revealed the accumulation of a resveratrol-derivate, a glycoside, in transgenic plants.	Szankowsk et al. (2003)
Banana	Vitamin A	Asupina banana	*PSY2a*	The gene construct can increase beta carotene level, a precursor to vitamin A in bananas to 20 mg per gram dry weight.	Waltz (2014)
Tobacco	Protein	CRISPR/Cas9	*XylT, FucT*	In knock-out lines no trace of β (1,2)-xylose or α (1,3)-fucose could be detected which shows that *N. tabacum* BY-2 cells can be engineered to humanize pharmacological glycoproteins.	Mercx et al. (2017)
Fodder					
Alfalfa	Methionine	*Arabidopsis*	AtCGS	In transgenic lines, the levels of free cysteine (the Sulphur donor for methionine synthesis), glutathione (the cysteine storage and transport form), and protein-bound cysteine increased up to 2.6-fold, 5.5-fold, and 2.3-fold, respectively, comparative to that in wild-type plants.	Avaram et al. (2005)

References

Aditi C, Tripathi S, Singh N, Saini L, Govind (2019) Effect of fertilizer levels, biocompost and biofertilizer on growth and yield attributes of fodder sorghum (*Sorghum bicolor* L.). J Pharm Phytochem 8(6):617–620

Afsahi K, Nazari M, Omidi H, Shekari F, Bostani AA (2020) The effects of different methods of zinc application on canola seed yield and oil content. J Plant Nut 43(8):1070–1079. https://doi.org/10.1080/01904167.2020.1724299

Ahmad M, Adil Z, Hussain A, Mumtaz M, Nafees M, Ahmad I, Jamil M (2019) Potential of phosphate solubilizing bacillus strains for improving growth and nutrient uptake in mungbean and maize crops. Pak J Agri Sc 56:283–289

Ait-El-Mokhtar M, Labbilta T, Anli M, Boutasknit A, Abouliatim Y, Khouloud M, Mesnaoui M, Meddich A (2022) Phosphate glass-based controlled-release fertilizers improve wheat growth, yield and grain nutritional quality under field conditions. Gesunde Pflanzen 74(3):715–727. https://doi.org/10.1007/s10343-022-00647-w

Al Noor M, Quadir QF, Naher J, Jewel ZA, Rashid HO, Chakrobarty T, Razia S, Nath UK (2019) Inducing variability in rice for enriched iron and zinc content through in vitro culture. Plant Tissue Cult Biotech 29(2):161–174. https://doi.org/10.3329/ptcb.v29i2.44505

Al-Bayati HJM, Allela WBAM, Salim NS, Ibraheem FFR (2021) IOP Confer Earth Environ Sci 910:012115

Ali AM, Awad MYM, Hegab SA, Gawad AMAE, Eissa MA (2021) Effect of potassium solubilizing bacteria (*Bacillus cereus*) on growth and yield of potato. J Plant Nutr 4(3):411–420

Alsaady MHM, Salim HA, Abdulrazzaq AK, Saleh UN, Jassim NH, Hamad AR, Attia JA, Jamal Jumaa Darwish JJ, Hassan AF (2020) Response of cabbage plants to foliar application of yeast suspension and nitrogen fertilizer. Ecol Environ Cons 26(2):832–836

Aluru M, Xu Y, Guo R, Wang Z, Li S, White W, Wang K, Rodermel S (2008) Generation of transgenic maize with enhanced provitamin a content. J Exp Bot 59:3551–3562

Aluru MR, Rodermel SR, Reddy MB (2011) Genetic modification of low phytic acid 1-1 maize to enhance iron content and bioavailability. J Agric Food Chem 59:12954–12962

Ambati D, Prasad SS, Singh JB, Verma DK, Mishra AN, Prakasha TL, Phuke RM, Sharma KC, Singh AK, Singh GP et al (2019) High yielding durum wheat variety HI 8759 PusaTejas—A new rust and kernel bunt resistant. Ind Farm 69:20–22

Ambati DM, Sontakke D, Raut GS (2020) Biofortification of crop for nutritional security just agriculture (Multidisciplinary newsletter) 1(4):049

Anwar Z, Basharat Z, Hafeez MB, Khan S, Zahra N, Rafique Z, Maqsood M (2022) Biofortification of maize with zinc and iron not only enhances crop growth but also improves grain quality. Asian J Agric Biol. https://doi.org/10.35495/ajab.2021.02.079

Apel W, Bock R (2009) Enhancement of carotenoid biosynthesis in transplastomic tomatoes by induced lycopene-to-provitamin a conversion. Plant Physiol 151:59–66

Aragao FJL, Barros LMG, De Sousa MV, Grossi de Sá MF, Almeida ERP, Gander ES, Rech EL (1999) Expression of a methionine-rich storage albumin from the Brazil nut (*Bertholletia excels* HBK, Lecythidaceae) in transgenic bean plants (*Phaseolus vulgaris* L., Fabaceae). Genet Mol Biol 22:445–449

Asif M, Abbas B, Aziz A, Adnan M, Safdar M, Ali A, Ali R, Hanif M (2020) Bio-fortification of calcium, zinc and iron improves yield and quality of forage sorghum (*Sorghum bicolor* L.). J Pure Appl Agri 5:71–81

Avaram T, Badani H, Galili S, Aamir R (2005) Enhanced levels of methionine and cysteine in transgenic alfalfa (*Medicago sativa* L.) plants over-expressing the *Arabidopsis* cystathionine γ-synthase gene. Plant Biotech J 3:71–79. https://doi.org/10.1111/j.1467-7652.2004.00102.x

Bailey RL, West KP Jr, Black RE (2015) The epidemiology of global micronutrient deficiencies. Ann Nutr Metab 66:2233

Barczak B, Jastrzębska M, Kostrzewska MK (2019) Biofortification of spring barley grain with microelements through sulfur fertilization. J Chem. https://doi.org/10.1155/2019/8214298

Bekbayeva L, Zakaria Z, Karpenyuk T, Goncharova A, Negim ES, Kaldanay K (2021) Effects of NPK fertilizers on the growth, yield and chemical content of tomato (*Lycopersicon esculentum* L.). In: Abdul Karim SA, Abd Shukur MF, Fai Kait C, Soleimani H, Sakidin H (eds) Proceedings of the 6th international conference on fundamental and applied sciences. Springer, Singapore, pp 37–50. https://doi.org/10.1007/978-981-16-4513-6_4

Blancquaert D, Van Daele J, Strobbe S, Kiekens F, Storozhenko S, De Steur H et al (2015) Improving folate (vitamin B9) stability in biofortified rice through metabolic. Nat Biotech 33: 1076–1078. https://doi.org/10.1038/nbt.3358

Boldrin PF, Faquin V, Ramos SJ, Boldrin KVF, Ávila FW, Guilherme LRG (2013) Soil and foliar application of selenium in rice biofortification. J Food Comp Anal 31(2):238–244. https://doi.org/10.1016/J.JFCA.2013.06.002

Boonyaves K, Gruissem W, Bhullar NK (2016) NOD promoter-controlled AtIRT1 expression functions synergistically with NAS and FERRITIN genes to increase iron in rice grains. Plant Mol Biol 90:207–215

Boonyaves K, Wu TY, Gruissem W, Bhullar NK (2017) Enhanced grain iron levels in rice expressing an iron-regulated metal transporter, nicotianamine synthase, and ferritin gene cassette. Front Plant Sci 8:130

Borg S, Brinch-Pedersen H, Tauris B, Madsen LH, Darbani B, Noeparvar S, Holm PB (2012) Wheat ferritins: improving the iron content of the wheat grain. J Cereal Sci 56:204–213

Bouis HE (2000) Enrichment of food staples through plant breeding: a new strategy for fighting micronutrient malnutrition. Nutrition 16:701–704. https://doi.org/10.1016/S0899-9007(00)00266-5

Boukar O, Massawe F, Muranaka S, Franco J, Maziya-Dixon B, Singh B, Fatokun C (2011) Evaluation of cowpea germplasm lines for protein and mineral concentrations in grains. Plant Genet Res 9(4):515–522

Brinch-Pedersen H, Olesen A, Rasmussen SK, Holm PB (2000) Generation of transgenic wheat (*Triticum aestivum* L.) for constitutive accumulation of an *aspergillus phytase*. Mol Breed 6: 195–206

Bueno AM, Flores RA, de Brito Ferreira EP (2022) Effects of foliar Silicon application, seed inoculation and splitting of N fertilization on yield, physiological quality, and economic viability of the common bean. SILICON 14:4169–4418. https://doi.org/10.1007/s12633-021-01208-y

Buriro KA, Soomro NS, Siddiqui MA, Chang MS, Sootaher JK, Nangraj GM, Siyal AL, Wadho M, Buriro KH, Majeedano NA (2022) Effects of foliar application of gibberellic–Acid3 on the growth and yield of canola (*Brassica napus* L.) genotypes. BioSci Rev 4(1):14. https://doi.org/10.32350/BSR.0401.01

Burton RA, Collins HM, Kibble NA, Smith JA, Shirley NJ, Jobling SA et al (2011) Over-expression of specific *HvCslF* cellulose synthase-like genes in transgenic barley increases the levels of cell wall (1,3;1,4)-β-d-glucans and alters their fine structure. Plant Biotech J 9(2): 117–135. https://doi.org/10.1111/j.1467-7652.2010.00532.x

Cahoon EB, Hall SE, Ripp KG, Ganzke TS, Hitz WD, Coughlan SJ (2003) Metabolic redesign of vitamin E biosynthesis in plants for tocotrienol production and increased antioxidant content. Nat Biotech 21:1082–1087

Cakmak I (2008) Enrichment of cereal grains with zinc: agronomic or genetic biofortification. Plant Soil 302:1–17. https://doi.org/10.1007/s11104-008-9584-6

Cakmak I, Chanakan P, Luiz G, Abdul R, Katja H, Atilla Y, Erdinc S, Kalayci M, Yusuf T, Piyawan P, Rizwan M, Martins FAD, Guilherme D, Levent O (2017) Iodine biofortification of wheat, rice and maize through fertilizer strategy. Plant Soil 418:1–17

Cakmak I, Kutman UB (2017) Agronomic biofortification of cereals with zinc: a review. Eur J Soil Sci 69:172–180. https://doi.org/10.1111/ejss.12437

Chakraborty S, Chakraborty N, Agrawal L, Ghosh S, Narula K, Shekhar S (2010) Next- generation protein-rich potato expressing the seed protein gene *AmA1* is a result of proteome rebalancing in transgenic tuber. Proc Natl Acad Sci 107(41):17533–17538. https://doi.org/10.1073/pnas.1006265107

Chandrasekharan N, Ramanathan N, Pukalenthy B, Chandran S, Manickam D, Adhimoolam K (2022) Development of β-carotene, lysine, and tryptophan-rich maize (*Zea mays*) inbreds through marker-assisted gene pyramiding. Sci Rep 12:8551. https://doi.org/10.1038/s41598-022-11585-y

Chaudhary M, Mandal A, Muduli S, Deepasree A, Abshiba A (2022) Agronomic biofortification of food crops: a sustainable way to boost nutritional security. IntechOpen. https://doi.org/10.5772/intechopen.103750

Chen R, Xue G, Chen P, Yao B, Yang W, Ma Q, Fan Y, Zhao Z, Tarczynski MC, Shi J (2008) Transgenic maize plants expressing a fungal phytase gene. Transgenic Res 17:633–643

Dar JS, Cheema MA, Rehmani MIA, Khuhro S, Rajput S, Virk AL (2021) Potassium fertilization improves growth, yield and seed quality of sunflower (*Helianthus annuus* L.) under drought stress at different growth stages. PLoS One 16(9):e0256075. https://doi.org/10.1371/journal.pone.0256075

Das S, Chaki AK, Hossain A (2019) Breeding and agronomic approaches for the biofortification of zinc in wheat (*Triticum aestivum* L.) to combat zinc deficiency in millions of a population: a Bangladesh perspective. Acta Agrobot 72:1770

De la Graza RD, Gregory JF III, Hanson AD (2007) Folate biofortification of tomato fruit. Proc Natl Acad Sci 104(10):4218–4222. https://doi.org/10.1073/pnas.0700409104

Decourcelle M, Perez-Fons L, Baulande S, Steiger S, Couvelard L, Hem S, Zhu C, Capell T, Christou P, Fraser P (2015) Combined transcript, proteome, and metabolite analysis of transgenic maize seeds engineered for enhanced carotenoid synthesis reveals pleotropic effects in core metabolism. J Exp Bot 66:3141–3150

Dehesh K, Jones A, Knutzon DS, Voelker TA (1996) Production of high levels of 8: 0 and 10: 0 fatty acids in transgenic canola by overexpression of *Ch FatB2*, a thioesterase cDNA from *Cuphea hookeriana*. Plant J 9:167–172

Dhaliwal SS, Sharma V, Shukla AK, Kaur J, Verma V, Kaur M, Singh P, Kaur L, Verma G, Singh J (2022) Biofortification of soybean (*Glycine max* L.) through $FeSO_4{\cdot}7H_2O$ to enhance yield, iron nutrition and economic outcomes in sandy loam soils of India. Agriculture 12:586. https://doi.org/10.3390/agriculture12050586

Dhaliwal SS, Sharma V, Shukla AK, Verma V, Behera SK, Singh P, Alotaibi SS, Gaber A, Hossain A (2021a) Comparative efficiency of mineral, chelated and nano forms of zinc and iron for improvement of zinc and iron in chickpea (*Cicer arietinum* L.) through biofortification. Agronomy 11:2436. https://doi.org/10.3390/agronomy11122436

Dhaliwal SS, Sharma V, Shukla AK, Verma V, Sandhu PS, Behera SK, Singh P, Kaur J, Singh H, Abdel-Hafez SH, Gaber A, Sayed S, Hossain A (2021b) Interactive effects of foliar application of zinc, iron and nitrogen on productivity and nutritional quality of indian mustard (*Brassica juncea* L.). Agronomy 11(11):2333. https://doi.org/10.3390/agronomy11112333

Dharmapuri S, Rosati C, Pallara P, Aquilani R, Bouvier F, Camara B, Giuliano G (2002) Metabolic engineering of xanthophyll content in tomato fruits. FEBS Lett 519:30–34

Dimkpa CO, Bindarban PS (2016) Fortification of micronutrients for efficient agronomic production: a review. Agron Sustain Dev 36:7. https://doi.org/10.1007/s13593-015-0346-6

Dinkins RD, Reddy MSS, Meurer CA, Yan B, Trick H, Thibaud-Nissen F, Finer JJ, Parrott WA, Collins GB (2001) Increased sulfur amino acids in soybean plants overexpressing the maize 15 kDa zein protein. Vitr Cell Dev Biol Plant 37:742–747

Doshi KM, Eudes F, Laroche A, Gaudet D (2006) Transient embryo specific expression of anthocyanin in wheat. In Vitro Cell Dev Biol Plant 42:432–438. https://doi.org/10.1079/IVP2006778

Eata AEM, El-Sherbini MAA, Mahmoud AR, Ali AH (2020) Effect of fertilizer management on growth, yield and quality of pea (*Pisum sativum* L.). Sing J Sci Res 10:327–335. https://doi.org/10.17311/sjsres.2020.327.335

Ewais MA, Abd El-Rahman LA, Sayed DA (2020) Effect of foliar application of boron and potassium sources on yield and quality of potato (*Solanum tuberosum* L.). mid east. J Appl Sci 10(1):120–137. https://doi.org/10.36632/mejas/2020.10.1.15

Falco SC, Guida T, Locke M, Mauvais J, Sanders C, Ward RT, Webber P (1995) Transgenic canola and soybean seeds with increased lysine. Nat Biotech 13:577–582

Flores T, Karpova O, Su X, Zeng P, Bilyeu K, Sleper DA et al (2008) Silencing of *GmFAD3* gene by siRNA leads to low alpha-linolenic acids (18:3) of fad3-mutant phenotype in soybean [*Glycine max* (Merr.)]. Transgenic Res 17(5):839–850. https://doi.org/10.1007/s11248-008-9167-6

Foroughbakhch R, Benavides-Mendoza A, Rocio CS, Cruz LS, Heya MN (2020) Agronomic biofortification with selenium in tomato crops (*solanum lycopersicon* L. mill). Agriculture 10: 11. https://doi.org/10.3390/agriculture10100486

Fuge R, Johnson CC (2015) Iodine and human health, the role of environmental geochemistry and diet, a review. Appl Geochem 63:282–302

Fujisawa M, Takita E, Harada H, Sakurai N, Suzuki H, Ohyama K, Shibata D, Misawa N (2009) Pathway engineering of Brassica napus seeds using multiple key enzyme genes involved in ketocarotenoid formation. J Exp Bot 60:1319–1332

Fujisawa M, Watanabe M, Choi SK, Teramoto M, Ohyama K, Misawa N (2008) Enrichment of carotenoids in flaxseed (*Linum usitatissimum*) by metabolic engineering with introduction of bacterial phytoene synthase gene *crtB*. J Biosci Bioeng 105:636–641

Garg M, Sharma N, Sharma S, Kapoor P, Kumar A, Chunduri V, Arora P (2018) Biofortified crops generated by breeding, agronomy, and transgenic approaches are improving lives of millions of people around the world. Front Nutr 5:12. https://doi.org/10.3389/fnut.2018.00012

Golubkina N, Gomez LD, Kekina H, Cozzolino E, Simister R, Tallarita A, Torino V, Koshevarov A, Cuciniello A, Maiello R, Cenvinzo V, Caruso G (2020) Joint selenium–iodine supply and arbuscular mycorrhizal fungi inoculation affect yield and quality of chickpea seeds and residual biomass. Plan Theory 9(7):804. https://doi.org/10.3390/plants9070804

Goto F, Yoshihara T, Shigemoto N, Toki S, Takaiwa F (1999) Iron fortification of rice seed by the soybean ferritin gene. Nat Biotechnol 17:282–286

Gould J (2017) Nutrition: a world of insecurity. Nat Outlook 544:S7. https://doi.org/10.1038/544S6

Graham RD, Welch RM, Saunders DA, Ortiz-Monasterio I, Bouis HE, Bonierbale M (2007) Nutritious subsistence food systems. Adv Agron 92:1–74. https://doi.org/10.2134/agronj2005.0222

Gregorio GB, Senadhira D, Htut H, Graham RD (2000) Breeding for trace mineral density in rice. Food Nutr Bul 21:382–386. https://doi.org/10.1177/156482650002100407

Gupta A, Singh C, Kumar V, Singh G, Singh GP (2019) Special features of newly released wheat and barley varieties for cultivation in India. J Cer Res 11(2):168–171

Halka M, Smolen S, Czernicka M, Klimek-Chodacka M, Pitala J, Tutaj K (2019) Iodine biofortification through expression of HMT, SAMT and S3H genes in *Solanum lycopersicum* L. Plant Physiol Biochem 144:35–48

Havrlentova M, Psenakova I, Zofajova A, Ruckschloss L, Kraic J (2014) Anthocyanins in wheat seed–a mini review. Nova Biotech Chim 13(1):1–12. https://doi.org/10.2478/nbec-2014-0001

Helaly AA (2021) Green seaweed extract: a complementary bio-fertilizer and bio-stimulator for growth and yield of sweet potato plants. Sci J Agric Sci 3(1):1–14. https://doi.org/10.21608/sjas.2021.60170.1067

Hodge J (2016) Hidden hunger: approaches to tackling micronutrient deficiencies. In: Gillespie S, Hodge J, Yosef S, Pandya-Lorch R (eds) Nourishing millions: stories of change in nutrition. Int Food Policy Res Ins (IFPRI), pp 35–43

Holme IB, Dionisio G, Brinch-Pedersen H, Wendt T, Madsen CK, Vincze E (2012) Cisgenic barley with improved phytase activity. Plant Biotech J 10(2):237–247. https://doi.org/10.1111/j.1467-7652.2011.00660.x

Hong H, Datla N, Reed DW, Covello PS, MacKenzie SL, Qiu X (2002) High-level production of γ-linolenic acid in *Brassica juncea* using a Δ6 desaturase from pythium irregular. Plant Physiol 129(1):354–362. https://doi.org/10.1104/pp.001495

Hotz C, McCafeerty B, Hawkes C, Ruel M, Babu S (2007) From harvest to health: challenges for developing biofortified staple foods and determining their impact on micronutrient status. Food and Nutr Bull 28:271–279

Huang JC, Zhong YJ, Liu J, Sandmann G, Chen F (2013) Metabolic engineering of tomato for high-yield production of astaxanthin. Metab Eng 17:59–67

Huang S, Wang P, Yamaji N, Ma JF (2020) Plant nutrition for human nutrition: hints from rice research and future perspectives. Mol Plant 13:825–835. https://doi.org/10.1016/j.molp.2020.05.007

Huang T, Joshi V, Jander G (2014) The catabolic enzyme methionine gamma-lyase limits methionine accumulation in potato tubers. Plant Biotech J 12:883–893

Jalal A, Shah S, Filho CMT, Khan MA, Shah T, Ilyas M, Aparecida Leonel Rosa PAL (2020) Agro-biofortification of zinc and iron in wheat grains. Gesunde Pflanzen 72:227–236. https://doi.org/10.1007/s10343-020-00505-7

Jeong JS, Kim YS, Redillas MC, Jang G, Jung H, Bang SW, Choi YD, Ha SH, Reuzeau C, Kim JK (2013) OsNAC5 overexpression enlarges root diameter in rice plants leading to enhanced drought tolerance and increased grain yield in the field. Plant Biotech J 11(1):101–114. https://doi.org/10.1111/pbi.12011

Jha AB, Warkentin TD (2020) Biofortification of pulse crops: status and future perspectives. Plants 9:73. https://doi.org/10.3390/plants9010073

Johnson AA, Kyriacou B, Callahan DL, Carruthers L, Stangoulis J, Lombi E, Tester M (2011) Constitutive overexpression of the *OsNAS* gene family reveals single-gene strategies for effective iron- and zinc-biofortification of rice endosperm. PLoS One 6:e24476. https://doi.org/10.1371/journal.pone.0024476

Keerio RA, Soomro NS, Soomro AA, Siddiqui MA, Khan MT, Nizamani GS, Kandhro MN, Siddiqui M, Khan H, Soomro FD (2020) Effect of foliar spray of zinc on growth and yield of sunflower (*Helianthus annuus* L.). Pak J Agric Res 33(2):264–269

Khani A, Barzegar T, Nikbakht J, Ghahremani Z (2020) Effect of foliar spray of calcium lactate on the growth, yield and biochemical attribute of lettuce (*Lactuca sativa* L.). under water deficit stress Adv Hortic Sci 34(1):11–24

Kim MJ, Kim JK, Kim HJ, Pak JH, Lee JH, Kim DH, Choi HK, Jung HW, Lee JD, Chung YS (2012) Genetic modification of the soybean to enhance the β-carotene content through seed-specific expression. PLoS One 7:e48287

Kim SH, Kim YH, Ahn YO, Ahn MJ, Jeong JC, Lee HS (2013) Downregulation of the lycopene ε-cyclase gene increases carotenoid synthesis via the β-branch-specific pathway and enhances salt-stress tolerance in sweet potato transgenic calli. Physiol Plant 147(4):432–442. https://doi.org/10.1111/j.1399-3054.2012.01688.x

Kumar S, Palve A, Joshi C, Srivastava RK (2019) Crop biofortification for iron (Fe), zinc (Zn) and vitamin a with transgenic approaches. Heliyon 5:e01914. https://doi.org/10.1016/j.heliyon.2019.e01914

Kumari A, Sharma S, Sharma N, Chunduri V, Kapoor P, Kaur S, Goyal A, Garg M (2020) Influence of biofortified colored wheats (purple, blue, black) on physicochemical, antioxidant and sensory characteristics of chapatti (Indian flatbread). Molecules 25:5071

Laurie S, Faber M, Adebola P, Belete A (2015) Biofortification of sweet potato for food and nutrition security in South Africa. Food Res Int 76:962–970

Lee TTT, Wang MMC, Hou RCW, Chen LJ, Su RC, Wang CS, Tzen JTC (2003) Enhanced methionine and cysteine levels in transgenic rice seeds by the accumulation of sesame 2S albumin. Biosci Biotech Biochem 67:1699–1705

Lipkie TE, De Moura FF, Zhao ZY, Albertsen MC, Che P, Glassman K, Ferruzzi M (2013) Bioaccessibility of carotenoids from transgenic provitamin a biofortified sorghum. J Agric Food Chem 61:5764–5771

Liu W, Muzolf-Panek M, Kleiber T (2022) The effect of various foliar treatments and nitrogen nutrition levels on the yield and physicochemical parameters of flowering Chinese cabbage. Agronomy 12:737. https://doi.org/10.3390/agronomy12030737

Lopez AB, Van Eck J, Conlin BJ, Paolillo DJ, O'Neill J, Li L (2008) Effect of the cauliflower or transgene on carotenoid accumulation and chromoplast formation in transgenic potato tubers. J Exp Bot 59:213–223

Lorenc-Kukuła K, Wróbel-Kwiatkowska M, Starzycki M, Szopa J (2007) Engineering flax with increased flavonoid content and thus fusarium resistance. Physiol Mol Plant Pathol 70:38–48

Lucca P, Hurrell R, Potrykus I (2001) Genetic engineering approaches to improve the bioavailability and the level of iron in rice grains. Theor Appl Genet 102:392–397

Lukaszewicz M, Matysiak-Kata I, Skala J, Fecka I, Cisowski W, Szopa J (2004) Antioxidant capacity manipulation in transgenic potato tuber by changes in phenolic compounds content. J Agric Food Chem 52:1526–1533

Luo H, He L, Du B, Pan S, Mo Z, Duan M, Tian H, Tang X (2020) Biofortification with chelating selenium in fragrant rice: effects on photosynthetic rates, aroma, grain quality and yield formation. Field Crops Res 255. https://doi.org/10.1016/j.fcr.2020.107909

Ma BL, Biswas DK, Herath AW, Whalen JK, Ruan SQ, Caldwell C, Earl H, Vanasse A, Scott P, Smith DL (2015) Growth, yield, and yield components of canola as affected by nitrogen, sulfur, and boron application. J Plant Nutr Soil Sci 178:658–670

Maertens M, Velde KV (2017) Contract-farming in staple food chains: the case of rice in Benin. World Dev 95:73–87

Majumder S, Datta K, Datta SK (2019) Rice biofortification: high iron, zinc, and vitamin-a to fight against "hidden hunger". Agronomy 9:803

Maleki FS, Chaichi MR, Mazaheri D, Tavakkol AR, Savaghebi G (2011) Barley grain mineral analysis as affected by different fertilizing systems and by drought stress. J Agric Sci Technol 13:315–326

Malik KA, Maqbool A (2020) Transgenic crops for biofortification. Front Sustain Food Syst 4:182. https://doi.org/10.3389/fsufs.2020.571402

Managa LR (2015) Evaluation of iron (Fe) and zinc (Zn) concentration among selected potato (*Solanum Tuberosum*) genotypes in South Africa. http://hdl.handle.net/10500/19792

Mangel N, Fudge JB, Li KT, Wu TY, Tohge T, Fernie AR (2019) Enhancement of vitamin B6 levels in rice expressing Arabidopsis vitamin B6 biosynthesis *de novo* genes. Plant J 99:1047–1065. https://doi.org/10.1111/tpj.14379

Manjeru P, Van Biljon A, Labuschagne M (2019) The development and release of maize fortified with provitamin a carotenoids in developing countries. Crit Rev Food Sci Nutr 59:1284–1293. https://doi.org/10.1080/10408398.2017.1402751

Marcelle R (1995) Mineral nutrition and fruit quality. Acta Horti 383:219–226

Marschner P (2012) Marschner's mineral nutrition of higher plants, 3rd edn. Elsevier, Oxford, Amsterdam, Netherlands

Manojlović MS, Lončarić Z, Cabilovski RR, Popović B, Karalić K, Ivezić V, Ademi A, Singh BR (2019) Biofortification of wheat cultivars with selenium. Soil Plant Sci 69(8):715–724. https://doi.org/10.1080/09064710.2019.1645204

Martins dos Santos AC, Marques KR, Rodrigues LU, Gomes de Faria AJ, Nascimento VL, Fidélis RR (2019) Biofortification of soybean grains with foliar application of Li sources. J Plant Nutr 42(19):2522–2531. https://doi.org/10.1080/01904167.2019.1659339

Masuda H, Usuda K, Kobayashi T, Ishimaru Y, Kakei Y, Takahashi M, Higuchi K, Nakanishi H, Mori S, Nishizawa NK (2009) Overexpression of the barley Nicotianamine synthase gene *HvNAS1* increases iron and zinc concentrations in rice grains. Rice 2:155–166

Maxime Merlin DT, Edith Marius FK, Bertrand KE, Mariette A, François ZN et al (2020) Effect of fertilizers' types on yield parameters, sweetness and nutritional quality of carrot (Daucus carota L.) genotypes. Int J Agric Sc Food Technol 6(1):079–087. https://doi.org/10.17352/2455815X.000059

Mazzucato A, Papa R, Bitocchi E, Mosconi P, Nanni R, Negri V et al (2008) Genetic diversity, structure and marker-trait associations in a collection of Italian tomato (*Solanum lycopersicum* L.) landraces. Theor Appl Genet 116(5):657–669. https://doi.org/10.1007/s00122-007-0699-6

Mercx S, Smargiasso N, Chaumont F, De Pauw E, Boutry M, Navarre C (2017) Inactivation of the β (1,2)-xylosyltransferase and the α(1,3)-fucosyltransferase genes in *Nicotiana tabacum* BY-2 cells by a multiplex CRISPR/Cas9 strategy results in glycoproteins without plant-specific glycans. Front Plant Sci 8:403

Mohamed El Nagy MM, Abou El- Salehein EH, Fekry WA, Wahdan HM (2020) Effect of foliar spray with different potassium sources and zinc rates on growth and yield of sweet potato (*Ipomoea batatas* L.). J Prod Dev 25(2):231–246

Mohamed MHM, Ali M, Eid RSM, El-Desouky HS, Petropoulos SA, Sami R, Al-Mushhin AAM, Ismail KA, Zewail RMY (2021) Phosphorus and biofertilizer application effects on growth parameters, yield and chemical constituents of broccoli. Agronomy 11(11):2210. https://doi.org/10.3390/agronomy11112210

Mohtashami R, Mohsen Movahhedi Dehnavi MM, Balouchi H, Faraji H (2020) Improving yield, oil content and water productivity of dryland canola by supplementary irrigation and selenium spraying. Agric Water Manag 232. https://doi.org/10.1016/j.agwat.2020.106046

Morineau C, Bellec Y, Tellier F, Gissot L, Kelemen Z, Nogué F, Faure JD (2017) Selective gene dosage by CRISPR-Cas9 genome editing in hexaploid *Camelina sativa*. Plant Biotechnol J 15: 729–739

Mubarak MU, Kiran A, Shahzad AN, Qayyum MF, Ishfaq M, Mahmood K (2022) Mineral biofortification of vegetables through soil-applied poultry mortality compost. PLoS One 17(2):e0262812. https://doi.org/10.1371/journal.pone.0262812

Muller K, Alice S, Charles S, Shimwela M, Deborah N, Beatrice E (2021) Agronomic performance of provitamin A-rich banana cultivars in Burundi and the Democratic Republic of Congo. Afr J Agric Res 17(9):1209–1220

Mulualem T (2015) Application of bio-fortification through plant breeding to improve the value of staple crops. Biomed Biotechnol 3(1):11–19

Nadeem F, Farooq M, Aman U, Rehman A, Nawaz A, Naveed M (2020) Influence of Zn nutrition on the productivity, grain quality and grain biofortification of wheat under conventional and conservation rice–wheat cropping systems. Agron Soil Sci 66(8):1042–1057. https://doi.org/10.1080/03650340.2019.1652273

Naqvi S, Zhu C, Farre G, Ramessar K, Bassie L, Breitenbach J, Dario Perez Conesa DP, Ros G, Sandmann G, Capell T, Christou P (2009) Transgenic multivitamin corn through biofortification of endosperm with three vitamins representing three distinct metabolic pathways. Proc Natl Acad Sci 106(19):7762–7767. https://doi.org/10.1073/pnas.0901412106

Narayanan N, Beyene G, Chauhan RD, Gaitán-Solís E, Gehan J, Butts P, Siritunga D, Okwuonu I, Woll A, JiménezAguilar DM (2019) Biofortification of field-grown cassava by engineering expression of an iron transporter and ferritin. Nat Biotechnol 37:144–151

Naseer I, Ahmad M, Hussain A, Jamil M (2020) Potential of zinc solubilizing bacillus strains to improve rice growth under axenic conditions. Pak J Agri Sci 57(4):1057–1071

Nikhil K, Salakinkop S (2018) Agronomic biofortification of maize with zinc and iron micronutrients. Mod Con Dev Agron 1(4):5

Ogo Y, Ozawa K, Ishimaru T, Murayama T, Takaiwa F (2013) Transgenic rice seed synthesizing diverse flavonoids at high levels: a new platform for flavonoid production with associated health benefits. Plant Biotechnol J 11:734–746. https://doi.org/10.1111/pbi.12064

Ohnoutkova L, Zitka O, Mrizova K, Vaskova J, Galuszka P, Cernei N (2012) Electrophoretic and chromatographic evaluation of transgenic barley expressing a bacterial dihydrodipicolinate synthase. Electrophoresis 33(15):2365–2373. https://doi.org/10.1002/elps.201200033

Paine JA, Shipton CA, Chaggar S, Howells R, Kennedy MJ, Vernon G, Wright SY, Hinchliffe E, Adams JL, Silverstone AL (2005) Improving the nutritional value of Golden Rice through increased pro-vitamin a content. Nat Biotechnol 23:482–487. https://doi.org/10.1038/nbt1082

Palacio-Márquez A, Ramírez-Estrada CA, Gutiérrez-Ruelas NJ, Sánchez E, Barrios DLO, Chávez-Mendoza C, Sida-Arreola JP (2021) Efficiency of foliar application of zinc oxide nanoparticles versus zinc nitrate complexed with chitosan on nitrogen assimilation, photosynthetic activity, and production of green beans (*Phaseolus vulgaris* L.). Sci Hortic 288:110297

Pandey H, Kumar A, Pathak R, Pandey R, Tiwari U, Darshan D, Prakash V, Dubey S (2022) Biofortification effect of vermicompost, zinc and iron on yield and quality of hybrid maize (*Zea mays* L.). AMA 53(4):7195–7212

Park SC, Kim YH, Kim SH, Jeong YJ, Kim CY, Lee JS (2015) Overexpression of the *IbMYB1* gene in an orange-fleshed sweet potato cultivar produces a dual-pigmented transgenic sweet potato with improved antioxidant activity. Physiol Plant 153(4):525–537. https://doi.org/10.1111/ppl.12281

Perera I, Seneweera S, Hirotsu N (2018) Manipulating the phytic acid content of rice grain toward improving micronutrient bioavailability. Rice 11:4. https://doi.org/10.1186/s12284-018-0200-y

Prasanna BM, Palacios-Rojas N, Hossain F, Muthusamy V, Menkir A, Dhliwayo T, Ndhlela T, San Vicente F, Nair SK, Vivek BS (2020) Molecular breeding for nutritionally enriched maize: status and prospects. Front Genet 10:1392

Qureshi MT, Ahmad MF, Iqbal N, Waheed H, Hussain S, Brestič M, Anjum A, Noorka IR (2021) Agronomic bio-fortification of iron, zinc and selenium enhance growth, quality and uptake of different sorghum accessions. Plant Soil Environ 67:549–557

Raju PRK, Sarma NDRK, Anuradha N, Patro TSSK, Jamuna PP, Prabhakar TVA (2020) Notification of crop varieties and registration of germplasm. Indian J Genet 80(2):226–230

Ramesh SA, Choimes S, Schachtman DP (2004) Over-expression of an *Arabidopsis* zinc transporter in *Hordeum vulgare* increases short-term zinc uptake after zinc deprivation and seed zinc content. Plant Mol Biol 54(3):373–385. https://doi.org/10.1023/B:PLAN.0000036370.70912.34

Ramzan Y, Hafeez MB, Khan S, Nadeem M, Rahman S, Batool S, Ahmad J (2020) Biofortification with zinc and iron improves the grain quality and yield of wheat crop. Int J Plant Prod 14:501–510. https://doi.org/10.1007/s42106-020-00100-w

Rengel Z, Batten GD, Crowley DE (1999) Agronomic approaches for improving the micronutrient density in edible portions of field crops. Field Crops Res 60:27–40. https://doi.org/10.1016/S0378-4290(98)00131-2

Rodrigo S, Santamaría O, López-Bellido FJ, Poblaciones MJ (2013) Agronomic selenium biofortification of two-rowed barley under Mediterranean conditions. Plant Soil Environ 59(3):115–120

Rosati C, Aquilani R, Dharmapuri S, Pallara P, Marusic C, Tavazza R, Bouvier F, Camara B, Giuliano G (2000) Metabolic engineering of beta-carotene and lycopene content in tomato fruit. Plant J 24:413–420

Safina N, Muhammad L, Sakeena H, Sajid A, Hasan S, Riaz A, Akbar AM (2021) Potential of foliar application of micronutrients on growth, yield, quality and nutraceutical properties of pea (*Pisum sativum* L.) using multivariate approach. Pak J Bot 54. https://doi.org/10.30848/PJB2022-3(9)

Sakellariou M, Mylona PV (2020) New uses for traditional crops: the case of barley biofortification. A review Agronomy 10:1964. https://doi.org/10.3390/agronomy10121964

Saleem M, Elahi E, Gandahi AW, Bhatti SM, Hajra I, Shaikh MA (2019) Effect of Sulphur application on growth, oil content and yield of sunflower. Sar J Agric 35(4):1198–1203

Sarhan IH, Mahmood SA (2021) The effect of foliar spraying with licorice extract and some nutrients on the growth and yield of the red cabbage. IOP Conf Ser Earth Environ Sci 910: 01205. https://doi.org/10.1088/1755-1315/910/1/012055)

Sato S, Xing A, Ye X, Schweiger B, Kinney A, Graef G, Clemente T (2004) Production of γ-linolenic acid and stearidonic acid in seeds of marker-free transgenic soybean. Crop Sci 44(2):646–652. https://doi.org/10.2135/cropsci2004.0646

Sayed EG, Ouis MA (2022) Improvement of pea plants growth, yield, and seed quality using glass fertilizers and biofertilizers. Environ Technol Innov 26:102356. https://doi.org/10.1016/j.eti.2022.102356

Sestili F, Janni M, Doherty A, Botticella E, D'Ovidio R, Masci S, Lafiandra D (2010) Increasing the amylose content of durum wheat through silencing of the *SBEIIa* genes. BMC Plant Biol 10:144

Shahin Uz Zaman M, Altaf Hossain M, Jahangir Alam M, Mahbubul Alam AKM, Omar Ali M, Sarker D, Sarker A (2022) 'BARI Masur-8': a high-yielding and biofortified lentil cultivar in Bangladesh. J Plant Regul 1–7. https://doi.org/10.1002/plr2.20210

Sharma S, Malhotra H, Borah P, Meena MK, Bindraban P, Chandra S, Pande V, Pandey R (2019) Foliar application of organic and inorganic iron formulation induces differential detoxification response to improve growth and biofortification in soybean. Plant Physiol Rep 24:119–128. https://doi.org/10.1007/s40502-018-0412-6

Sheoran S, Kumar S, Ramtekey V, Kar P, Meena RS, Jangir CK (2022) Current status and potential of biofortification to enhance crop nutritional quality: an overview. Sustainability 14:3301. https://doi.org/10.3390/su14063301

Sher A, Sami UA, Abdul S, Muhammad I, Abdul Q, Abdul M, Muhammad S (2022) Zinc sulfate application to grass forages (oat, barley, annual ryegrass and triticale) for increasing their yield, quality and profitability. Crop Pasture Sci 73:473–483. https://doi.org/10.1071/CP21476

Singh SP, Gruissem W, Bhullar NK (2017) Single genetic locus improvement of iron, zinc and β-carotene content in rice grains. Sci Rep 7:147

Smith SE, Read DJ (2007) Mycorrhizal Symbiosis, 3rd edn. Elsevier, London, UK

Solanki RL, Sharma M, Indoria D (2018) Effect of phosphorus, Sulphur and psb on yield of indian mustard (*Brassica juncea* L.) and available macronutrients in soil. J Ind Soc Soil Sci 66(4): 415–419. https://doi.org/10.5958/0974-0228.2018.00052.X

Stangoulis JCR, Knez M (2022) Biofortification of major crop plants with iron and zinc achievements and future directions. Plant Soil 474:57–76. https://doi.org/10.1007/s11104022-05330-7

Strobbe S, Verstraete J, Stove C, Van Der Straeten D (2021) Metabolic engineering of rice endosperm towards higher vitamin B1 accumulation. Plant Biotech J 19:1253–1267. https://doi.org/10.1111/pbi.13545

Subbarao, Dawson J (2022) Effect of phosphorous and zinc on growth and yield of field pea (*Pisum sativum* L.). Pharma Inno J 11(5):254–256

Sundararajan S, Rajendran V, Sivakumar HP, Nayeem S, Chandra HM, Sharma A (2021) Enhanced vitamin E content in an indica rice cultivar harbouring two transgenes from *Arabidopsis thaliana* involved in tocopherol biosynthesis pathway. Plant Biotechnol J 19:1083. https://doi.org/10.1111/pbi.13578

Szankowski I, Briviba K, Fleschhut J, Schönherr J, Jacobsen HJ, Kiesecker H (2003) Transformation of apple (*Malus domestica* Borkh.) with the stilbene synthase gene from grapevine (*Vitis vinifera* L.) and a *PGIP* gene from kiwi (*Actinidia deliciosa*). Plant Cell Rep 22:141–149. https://doi.org/10.1007/s00299-003-0668-8

Tamas C, Kisgyorgy BN, Rakszegi M, Wilkinson MD, Yang MS, Lang L (2009) Transgenic approach to improve wheat (*Triticum aestivum* L.) nutritional quality. Plant Cell Rep 28(7): 1085–1094. https://doi.org/10.1007/s00299-009-0716-0

Telengech PK, Maling'a JN, Nyende AB, Gichuki ST, Wanjala BW (2015) Gene expression of beta carotene genes in transgenic biofortified cassava. Biotechnology 5:465–472

Upadhyay CPH, Ko EY, Nookaraju A, Kim HS, Heung JJ, Oh MO, Reddy AC, Chun SC, Kim DH, Park SW (2009) Over-expression of strawberry D-galacturonic acid reductase in potato leads to accumulation of vitamin C with enhanced abiotic stress tolerance. Plant Sci 177:659–667

Vajinder K, Singh G, Dhaliwal S (2021) A new approach in agronomic biofortification for improving zinc and iron content in chickpea (*Cicer arietinum* L.) grain with simultaneous foliar application of zinc sulphate, ferrous sulphate and urea. J Soil Sci Plant Nutr 21. https://doi.org/10.1007/s42729-021-00408-0

Van Eenennaam A, Lincoln K, Durrett TP, Valentin HE, Shewmaker CK, Thorne GM, Jiang J, Baszis SR, Levering CK, Aasen ED (2003) Engineering vitamin E content: from Arabidopsis mutant to soy oil. Plant Cell 15:3007–3019

Waltz E (2014) Vitamin A super banana in human trials. Nat Biotechnol 32:857. https://doi.org/10.1038/nbt0914-857

Wang C, Zeng J, Li Y, Hu W, Chen L, Miao Y et al (2014) Enrichment of provitamin a content in wheat (*Triticum aestivum* L.) by introduction of the bacterial carotenoid biosynthetic genes *CrtB* and *CrtI*. J Exp Bot 65:2545–2556. https://doi.org/10.1093/jxb/eru138

Wang X, Zhao F, Zhang G, Zhang Y, Yang L (2017) Vermicompost improves tomato yield and quality and the biochemical properties of soils with different tomato planting history in a greenhouse study. Front Plant Sci 8. https://doi.org/10.3389/fpls.2017.01978

Waqar M, Rahman MH, Hasnain MU, Iqbal S, Ghaffar A, Iqbal R, Husain MI, Sabagh A (2022) Effect of slow release nitrogenous fertilizers and biochar on growth, physiology, yield, and nitrogen use efficiency of sunflower under arid climate. Environ Sci Pollut Res. https://doi.org/10.1007/s11356-022-19289-6

Welsch R, Arango J, Bär C, Salazar B, Al-Babili S, Beltran J, Chavarriaga P, Ceballos H, Tohme J, Beyer P (2010) Provitamin a accumulation in cassava (*Manihot esculenta*) roots driven by a single nucleotide polymorphism in a phytoene synthase gene. Plant Cell 22:3348–3356

Wiltgren AR, Booth AO, Kaur G, Cicerale S, Lacy KE, Thorpe MG, Keast RS, Riddell LJ (2015) Micronutrient supplement use and diet quality in university students. Nutrients 7(1094):1107

Wirth J, Poletti S, Aeschlimann B, Yakandawala N, Drosse B, Osorio S, Tohge T, Fernie AR, Günther D, Gruissem W (2009) Rice endosperm iron biofortification by targeted and synergistic action of nicotianamine synthase and ferritin. Plant Biotechnol J 7:631–644

Xiaotong G, Duan X, Wu Y, Cheng J, Zhang J, Zhang H, Bei Li J (2018) Genetic engineering of maize (*Zea mays* l.) with improved grain nutrients. J Agric Food Chem 66(7):1670–1677. https://doi.org/10.1021/acs.jafc.7b05390

Xiaoyan S, Yan Z, Shubin W (2012) Improvement Fe content of wheat (*Triticum aestivum*) grain by soybean ferritin expression cassette without vector backbone sequence. J Agric Biotechnol 20: 766–773

Yadav R, Kiran G (2019) Wheat variety HD 3171 for rainfed timely sown conditions of north eastern plain zone comprising of India. https://doi.org/10.13140/RG.2.2.14502.57921

Yadava DK, Choudhury PR, Hossain F, Kumar D, Mohapatra T (2020) Bio-fortified varieties: sustainable way to alleviate malnutrition. Ind Coun Agric Res 86

Yang QQ, Zhang CQ, Chan ML, Zhao DS, Chen JZ, Wang Q, Li QF, Yu HX, Gu MH, Sun SSM et al (2016) Biofortification of rice with the essential amino acid lysine: molecular characterization, nutritional evaluation, and field performance. J Exp Bot 67:4285–4296

Yanthan MR, Singh R (2021) Effect and boron and zinc level on growth yield and yield parameters of mustard (*Brassica campestris* L.). Pharma Inno 10(12):474–476. https://doi.org/10.22271/tpi.2021.v10.i12g.9340

Yeboah S, Asibuo J, Oteng-Darko P, Adjei EA, Lamptey M, Danquah EO, Waswa B, Butare L (2021) Impact of foliar application of zinc and magnesium aminochelate on bean physiology and productivity in Ghana. Int J Agron 9766709:9. https://doi.org/10.1155/2021/9766709

Yu J, Peng P, Zhang X, Zhao Q, Zhy D, Sun X, Liu J, Ao G (2004) Seed-specific expression of a lysine rich protein sb401 gene significantly increases both lysine and total protein content in maize seeds. Mol Breed 14:1–7

Zhang G, Yan Z, Wang Y, Feng Y, Yuan Q (2020) Exogenous proline improve the growth and yield of lettuce with low potassium content. Sci Hortic 271. https://doi.org/10.1016/j.scienta.2020.109469

Zhang Y, Schernthaner J, Labbé N, Hefford MA, Zhao J, Simmonds DH (2014) Improved protein quality in transgenic soybean expressing a de novo synthetic protein, MB-16. Trans Res 23: 455–467

Zhao ZY, Che P, Glassman K, Albertsen M (2019) Nutritionally enhanced sorghum for the arid and semiarid tropical areas of Africa. In: Clifton NJ (ed) Methods in molecular biology, vol 1931, pp 197–207. https://doi.org/10.1007/978-1-4939-9039-9_14

Zhu C, Naqvi S, Gomez-Galera S, Pelacho AM, Capell T, Christou P (2007) Transgenic strategies for the nutritional enhancement of plants. Trends Plant Sci 12:548–555

Zhu Q, Tan J, Wang B, Liu YG (2021) Genetic engineering for increasing antioxidant content in rice: recent progress and future perspectives. In: Hossain MA, Hassan L, Iftekharuddaula K, Kumar A, Henry R (Eds) Molecular breeding for rice abiotic stress tolerance and nutritional quality Wiley Blackwell Hoboken, NJ. pp. 358–381. doi:https://doi.org/10.1002/9781119633174.ch19

Chapter 3
Improving Zinc Biofortification in Plants

Qudrat Ullah Khan, Muhammad Safdar Baloch, Asghar Ali Khan, Muhammad Amjad Nadim, and Umar Khitab

Abstract Zinc deficiency in plants and human beings is gaining importance due to its key role in various functions in both plants and human beings. Zinc deficiency has been linked with malnutrition in human beings due to its inadequate concentration in the daily diet especially in the developing countries where the people are mostly dependent on cereal crops for their staple diet. As most of the farmer are focused on macro—nutrients for attaining higher yield, Zn being a micro—nutrient is rarely applied. Also, the frequent use of phosphatic fertilizer has aggravated Zn deficiency due to their antagonistic relation with each other. The antagonism is found in both soil and plant. In soil Zinc and phosphorus react to form an insoluble compound Zinc phosphate, while in plant presence of phytate decreases the bioavailability of Zn. In this regard an effort has been made to enhance the availability of Zn in food grains by different means, amongst these biofortification is widely used method for fortification of minerals in food. The biofortification techniques used throughout the world includes Zn enrichment through breeding methods, which comprises of conventional and genetical engineering methods. There are different pros and cons relates to these breeding and genetics methods. The second most common methods are the agronomic technique of biofortification it involves the fertilizer application both as soil applied and foliar applied, biofertilizers, seeding priming etc. It has been concluded at the end that the agronomics methods are more user friendly and cost effective as compared to the breeding methods which are time consuming. Amongst the agronomic methods priming with Zn solution has resulted to be promising when compared with other techniques.

Q. U. Khan (✉)
Department of Soil Science, Faculty of Agriculture, Gomal University, Dera Ismail Khan, Pakistan
e-mail: qudrat@gu.edu.pk

M. S. Baloch · A. A. Khan · M. A. Nadim · U. Khitab
Department of Agronomy, Faculty of Agriculture, Gomal University, Dera Ismail Khan, Pakistan
e-mail: drasghar@gu.edu.pk; dramjad@gu.edu.pk

M. Hasanuzzaman et al. (eds.), *Mineral Biofortification in Crop Plants for Ensuring Food Security*, https://doi.org/10.1007/978-981-99-4090-5_3

Keywords Agronomic · Biofortification · Breeding method · Cereal · Zinc

3.1 Introduction

Zinc (Zn) is an important micronutrient required for normal plant growth and development. The key role of Zn in plants is activation of enzymes, metabolism of carbohydrate and auxin (growth regulator), synthesis of protein, expression of genes, guard against the heat stress and photo-oxidative damage and protect against certain pathogens (Alloway 2008). Plant deficient in Zn usually have retarded photosynthesis and nitrogen metabolism, reduction in development of flowers and fruits, delay maturity due to prolong growth and yield and quality of fruits is decreased. Symptoms most observed in plant having Zn deficiency are older leaves turn light green or yellow and interveinal areas have bleached spots; younger leaves are often left small in size and are termed as "little leaf", also may have resetting i.e., leaves appear in bunch and come out from the same point. Like plants, Zn is also essential for human health. It is a vital constituent of the human body the adult body comprises of 2–3 grams of Zn. This may be present in different parts including cells, fluids, bones, tissue, and organs. Zinc is crucial for activation of more than 300 enzymes which have a key role in the human body including growth stimulation, cell division, immune system, fertility, taste, smell and appetite, skin, hair and nails and vision. Hussain et al. (2022) reported that in human being Zn deficiency has become the health issue globally and approximately two billion people are affected across the globe. This is mainly due to Zn deficiency in the grains, fruits, and vegetables. It has been further reported that the Zn deficiency in humans cannot be easily diagnosed as the symptoms do not appear until the damage is caused to the vital organs.

In the developing countries the Zn deficient cereal are used as a staple food. Cakmak (2010a) reported that optimum Zn required for humans is 40–50 mg kg^{-1}. As there is small concentration and low bioavailability of Zn in the cereal crops which cannot meet the human requirement of Zn (Cakmak and Kutman 2018). Some of the plants like nuts, beans and whole grains contain Zn, but their bioavailability is lower than the animal foods due to presence of phytate or phytic acid. Phytate is the storage form of phosphorus in plant tissues, which binds Zn and forms complex in the gastrointestinal tract. Negative correlation between the ratio of Phytate: Zn and bioavailable Zn was recorded (Fig. 3.1) in the wheat data of some of the high yielding varieties of Pakistan as reported by Hussain et al. (2012). Thus, its absorption is reduced (King et al. 2015). Also, the protein is required for absorption of Zn usually people with low protein intake face the deficiency of Zn. Most of the improved cultivars have inherently low capacity to uptake Zn. This has aggravated the problem by increasing the gap between Zn concentration in crops and the amount crucial for good health.

Most of the farmers seldom supplement the soil with Zn fertilizer for most of the crops, so the soil is becoming deficient in Zn. Zinc deficiency has been reported to be most common in alkaline calcareous soil, saline and waterlogged soil and heavy clay textured soils (Alloway 2008).

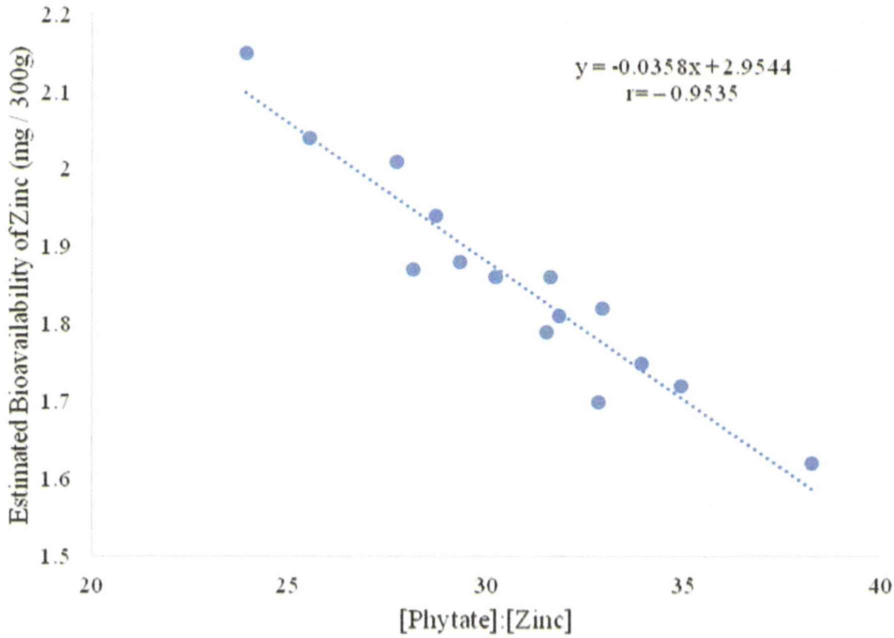

Fig. 3.1 Correlation between Phytate: Zn and Zn bioavailability

Zinc is recognized for its interaction with other plant nutrients in soil and plants. The interaction of Zn is synergistic with nitrogen and potassium, while antagonistic interaction of Zn and phosphorus, calcium and iron has been reported (Prasad et al. 2016). The negative interaction with phosphorus is due to increasing the rate of Zn adsorption by application of phosphatic fertilizers (Cakmak 2004). Also, the Phosphorus may react with Zn to form insoluble zinc phosphate [$Zn_3(PO_4)_2.4H_2O$] (Lambert et al. 2007). Mostly calcium is found in the soil as lime (calcium carbonate) which reacts with Zn to form calcium zincate [$CaZn_2$ $(OH)_6.2H_2O$], which is unavailable to the plants due to its least solubility (Zeigler and Johnson 2001). Agib and Jarcass (2008) reported that Zn also negatively interacts with Fe in soil to form oxide Franklinite [$ZnFe_2O_4$] which precipitate to become unavailable to plants. Keeping in view the problems related to the Zn bioavailability in plants, it has become inevitable to enhance the Zn content of the plants, especially grains. Biofortification has emerged as an important agricultural strategy to enhance the concentration of micronutrients in grains.

Bio-fortification is the enhancement of nutrients in crops by breeding and agronomic measures. Biofortification is the cheapest and the easiest methods for enhancing the nutrient concentration in food.

3.2 Biofortification Techniques

3.2.1 Breeding Methods of Zn Biofortification

Through the breeding method of biofortification the grains having the potential to uptake higher Zn content may be selected as a trait that may be transferred in the improved varieties by selective breeding or through genetical modification and biotechnological techniques. In this way the Zn bioavailability in the food may be enhanced. Breeding methods may be further divided in to two major categories:

3.2.1.1 Conventional Breeding for Zn Fortification

In the conventional or traditional breeding method, the genetic variation is produced for achieving the desire traits i.e., higher Zn content. Further, it involves the crossing of traits for many generations to achieve all the favorable traits. This is the most common method used in different parts of the world for fortification of Zn. Swamy et al. (2016) reported that the rice varieties having higher grain Zn were effectively used in rice breeding for alleviating the Zn content of rice varieties. They considered this method to be more targeted, sustainable, cost effective and food-based in biofortification of Zn.

In Wheat several studies have been conducted to enhance the Zn concentration wheat genotypes. *Cakmak et al. (*2010b*)* reported enhancement of Zn trait in the modern wheat genotype from its wild relative. Ibrahim et al. (2021) have reported that through the conventional breeding method some of the high Zn content varities have been developed. These include Zincol-2016, BHU-1, BHU-6. The have the potential of 35–42 μg Zn g^{-1}. The relationship between the Zn content and grain yield has been reported to be inverse. Mostly the higher Zn concentration in the grains have lower grain yield (Mc Donald et al. 2008; Fan et al. 2008).

The conventional breeding approach aiming to develop genotype with high content of Zn initially require the valuable genetic variation amongst the germplasms selected for the proper accumulation of Zn in the grains (Rehman et al. 2018; Ludwig and Slamet-Loedin 2019). Also in another study carried out by Ullah et al. (2020) on characterization of 16 genotypes of gram (*Cicer aritenum*) for enhancing the Zn concentration in the grain by breeding biofortification method. They found very low to moderate genetical variation amongst the different genotypes tested during the study. Concluded that Zn bioavailability (3.72–4.42 mg day^{-1}) and concentration (37.5–48.6 mg kg^{-1}) are at a very narrow range of diversity. However, in lentil genotype the genetic variation was recorded higher when compared with the Phytate content in the different genotypes of lentil (Rasheed et al. 2020).

The breeding method, despite of having several constraints may be used as a tool of biofortification of Zn in different crops. The goal of the breeder is currently focused on the developing countries is on enhancing productivity rather than the nutrient content. There is great potential to explore genetic resources for improving the qualitative traits of the crop by breeding biofortification method.

3.2.1.2 Genetic Engineering/Modification

Genetic engineering/modifications in plants refers to the improved conventional breeding, transgenic breeding, molecular breeding, mutation breeding, and somatic hybridization, (Yan and Kerr 2002; Bouis 2003). The desirable traits can be produced in the new genotypes by the modern genetic engineering techniques. In this technique the infinite gene pool is used for transferring the desirable traits from one organism to another, which is evolutionary and distinct from one another. Likewise, if certain micronutrients is not naturally found is a specific crop, than it is only the transgenic method which is the most appropriate for fortifying the nutrient in crop (Perez-Massot et al. 2013). Transgenic crops are produced by the introduction of some unique genes, which dominate and mask the gene already present by overexpressing itself. In addition, it controls the expression of some genes, or inhibits them by interrupting their pathway. Malik and Maqbool (2020) revealed that Zn content in staple food may be enhanced by specific coding for Zn that sequences the protein which binds Zn. Also, by dominance and overexpressing of Zn storage protein and increasing the expression of protein which are responsible for Zn uptake by plants. The transgenic approach has been used for enhancement of vitamins in different crops i.e., wheat, maize, potato and rice. Cong et al. (2009) reported 10.8 folds higher carotenoid in wheat by transgenic method. Borrill et al. (2014) reported Zn content in the grain can be enhanced by Zn mobilization and translocation through the over-expression of genes without decreasing the yield of the crop is the actual goal of transgenic approach.

Scientists have discovered metal transporter proteins in many crop species which can be used as multiple metal substrate for Zn, Fe and for Cd. These may be extracted from the plant roots and soil. Like for Fe ferritin is the iron storage protein, which may assist the accumulation of Fe in plant tissues. Lately, Zn transporter family (ZIP) has been identified by Li et al. (2019) for carrying the divalent cations in plant. Genetically modified crops should be analyzed socio-economically for their adaptability and future impacts on the consumers. The researcher should study its cost effectiveness for end users and orient the farmers and consumers about the importance of genetically modified crops specially for nutrients like Zn De-steur et al. (2014).

3.2.2 Agronomic Methods of Zn Biofortification

The process of enriching micronutrients in the crop edible portion by application of inorganic fertilizers and other cultural practices is termed as Agronomic biofortification. This type of biofortification is not only necessary for cultivars obtained by conventional breeding or genetic modification but also for genetically inefficient cultivars. Agronomic biofortification is the cheapest method and effective in eradicating micronutrient deficiency in plants (Velu et al. 2014). Agronomic biofortification includes the following methods.

3.2.2.1 Application of Zinc Fertilizers

Inorganic Zn fertilizers may be used for enhancing the Zn content of the plants. The common Zn fertilizer available includes zinc sulphate ($ZnSO_4$), zinc oxide (ZnO) and Zn-chelates (Fageria 2009). Soil application of Zn has been commonly used in the soil deficient in Zn. But Zn interaction with other nutrients usually suppresses the Zn availability to plants. In this regard the foliar application of Zn solely and in combination with soil application have proved effective. Chattha et al. (2017) reported that soil Zn application may increase the yield of crop but does not increase the Zn grain content due to its low efficiency.

Foliar application of Zn has showed better result in terms of Zn grain content as compared with soil application, as in soil application the losses are more (Johnson et al. 2005). The growth stage also influences the rate of uptake of Zn by plants. Zou et al. (2012) revealed that foliar application was more effective when applied at later stage of plant growth as compared with early vegetative growth (Yilmaz et al., 1997). Often the high dosage of foliar $ZnSO_4$ fertilizer causes burning of leaves which may result in decline in growth and yield of crop.

Due to different problems pertaining to the Zn availability to plants through soil and foliar application. Chelated and Nano Zn fertilizer have been investigated in terms of crop response and Zn enrichment by the plant. Dhaliwal et al. (2021) found that the nano fertilizer for Zn has responded better both for crop yield and Zn enrichment by chickpea plants. It was attributed to the altered structure of the nano-fertilizer which was responsible for higher translocation of Zn than the Chelated and other Zn fertilizers.

3.2.2.2 Biofertilizers

Biofertilizers are soil microorganisms that help in promotion of growth and yield of plants. They are referred to as plant growth promoting microorganisms. Biofertilizers are cost effective, simple, accessible, and sustainable option for enhancing the nutrient content in plants. Ramesh et al. (2014) recorded that Zn biofortification in grains of different crops may be enhanced by microorganisms, viz., cyanobacteria (*Azotobacter* sp. and *Anabaena* sp.) and *Bacillus aryabhattai* facilitate Zn in corn, wheat and soyabean respectively (Prasanna et al. 2015). Dotaniya et al. (2016) reported that Zn deficiency from the plant may be eliminated using microbial interventions.

3.2.2.3 Seed Priming

Seed treated with Zn solution has been effective in enhancing the seed grain content. It has been reported that seed priming with Zn increased the concentration of Zn in seed of chickpea, maize, and wheat by 29, 19, and 12% respectively (Harris et al. 2008). Also, seed priming was considered most cost effective for the three crops. The grain Zn content is not enhanced as much as it has been reported for the foliar application of fertilizer, however it is effective under stress environment i.e., moisture stress. Choukri et al. (2022) studied the Zn priming techniques on corn and described that this method enhanced the yield of the crop and enriched Zn in the grain. The seed was treated for 24 hrs with 0.5% zinc sulphate which enhanced the Zn content by 15% and the yield by 47%. They concluded that Zn priming was more effective than the soil application of zinc sulphate fertilizer.

3.3 Conclusion

The Biofortification technique for enhancing the Zn content of the plant has been found to be more applicable and cost effective. Zinc is the most inevitable nutrient both for human beings and plants. Different methods of biofortification have shown their tendency to accumulate Zn and make it bioavailable. Comparing the methods of biofortification it may be concluded that conventional breeding methods though had several constraints but were effective in producing wheat varieties Zincol-2016, BHU-1 and BHU-6. Genetic modification methods have been employed worldwide and have shown to enhance the Zn content of plants, but its social and economic aspects needs to be determined for future studies. Finally, agronomic practices have been extensively used for biofortification worldwide and have been reported as the most effective, easily adaptable, and economically feasible method amongst the biofortification techniques. However, there is still need to further research regarding the biofortification methods for different crops including cereals and legumes for improving the biofortification.

References

Agib A, Jarcass F (2008) Prediction of zinc precipitation accompanying sorption process in calcareous and basaltic soils. Tishreen Univ J Res Sci Stud–Bio Sci Series 30(5):235–251

Alloway BJ (2008) Zinc in soils and crop nutrition. Paris, France: IFA; and Brussels, Belgium: IZA

Bouis HE (2003) Micronutrient fortification of plants through plant breeding: can it improve nutrition in man at low cost? Proc Nut Soc 62:403–411

Borrill P, Connorton JM, Balk J, Miller AJ, Sanders D, Uauy C (2014) Biofortification of wheat grain with iron and zinc: integrating novel genomic resources and knowledge from model crops. Front Plant Sci 21:53–60

Cakmak I (2004) Identification and correction of widespread zinc deficiency in Turkey–a success story. Internat Fertiliser Soc

Cakmak I, Pfeiffer WH, Mc Clafferty B (2010a) Biofortification of durum wheat with zinc and iron. Cereal Chem 87:10–20

Cakmak I, Kalayci M, Kaya Y, Torun AA, Aydin N, Wang Y, Arisoy Z, Erdem H, Yazici A, Gokmen O, Ozturk L, Horst WJ (2010b) Biofortification and localization of zinc in wheat grain. J Agric Food Chem 58:9092–9102. https://doi.org/10.1021/jf101197h

Cakmak I, Kutman UB (2018) Agronomic biofortification of cereals with zinc: a review. Europ J Soil Sci 69:172–180. https://doi.org/10.1111/ejss.12437

Chattha MU, Hassan MU, Khan I, Chattha MB, Mahmood A, Chattha MU, Nawaz M, Subhani MN, Kharal M, Khan S (2017) Biofortification of wheat cultivars to combat zinc deficiency. Front Plant Sci 8. https://doi.org/10.3389/fpls.2017.00281

Choukri M, Abouabdillah A, Bouabid R, Abd-Elkader OH, Pacioglu O, Boufahja F, Bourioug M (2022) Zn application through seed priming improves productivity and grain nutritional quality of silage corn. Saudi J Bio Sci 29(12):103456

Cong L, Wang C, Chen L, Liu H, Yang G, He G (2009) Expression of phytoene synthase1 and carotene desaturase *crtI* genes result in an increase in the total carotenoids content in transgenic elite wheat (*Triticum aestivum* L.). J Agric Food Chem 57:8652–8660. https://doi.org/10.1021/jf9012218

Dhaliwal SS, Sharma V, Shukla AK, Verma V, Behera SK, Singh P, Alotaibi SS, Gaber A, Hossain A (2021) Comparative efficiency of mineral, chelated and nano forms of zinc and iron for improvement of zinc and iron in chickpea (*Cicer arietinum* L.) through biofortification. Agronomy 11:2436. https://doi.org/10.3390/agronomy11122436

De Steur H, Feng S, Xiaoping S, Gellynck X (2014) Consumer preferences for micronutrient strategies in China: a comparison between folic acid supplementation and folate biofortification. Pub Health Nutr 17:1410–1420

Dotaniya ML, Datta SC, Biswas DR, Dotaniya CK, Meena BL, Rajendiran S, Regar KL, Lata M (2016) Use of sugarcane industrial by-products for improving sugarcane productivity and soil health. Int J Rec Org Waste Agric 5(3):185–194

Fageria NK (2009) The use of nutrients in crop plants. Boca Raton,Fl: CRC press

Fan M, Zhao F, Fairweathertait S, Poulton P, Durham S, McGrath S (2008) Evidence of decreasing mineral density in wheat grain over the last 160 years. J Trace Elem Med Bio 22:315–324. https://doi.org/10.1016/j.jtemb.2008.07.002

Harris D, Rashid A, Miraj G, Arif M, Yunas M (2008) On-farm' seed priming with zinc in chickpea and wheat in Pakistan. Plant Soil 306:3–10. https://doi.org/10.1007/s11104-007-9465-4

Hussain A, Jiang W, Wang X, Shahid S, Saba N, Ahmad M, Masood SU, Imran M, Mustafa A (2022) Mechanistic impact of zinc deficiency in human development. Front Nutr Sec Nut Food Sci Tech 9. https://doi.org/10.3389/fnut.2022.717064

Hussain S, Maqsood MA, Miller LV (2012) Bioavailable zinc in grains of bread wheat varieties of Pakistan. Cereal Res Commun 40(1):62–73

Ibrahim S, Saleem B, Naeem MK, Arain SM, Khan MR (2021) Next-generation technologies for iron and zinc biofortification and bioavailability in cereal grains. Crop Pasture Sci 73(2):77–92. https://doi.org/10.1071/CP20498

Johnson S, Lauren J, Welch R, Duxbury J (2005) A comparison of the effects of micronutrient seed priming and soil fertilization on the mineral nutrition of chickpea (*Cicer arietinum*), lentil (*Lens culinaris*), rice (*Oryza sativa*) and wheat (*Triticum aestivum*) in Nepal. Exp Agric 41(4): 427–448. https://doi.org/10.1017/S0014479705002851

King JC, Brown KH, Gibson RS, Krebs NF, Lowe NM, Siekmann JH, Raiten DJ (2015) Biomarkers of nutrition for development (bond)-zinc review. J Nutr 146:858S–885S

Lambert R, Grant C, Sauve S (2007) Cadmium and zinc in soil solution extracts following the application of phosphate fertilizers. Sci Total Env 378:298–305

Li S, Liu X, Zhou X, Li Y, Yang W, Chen R (2019) Improving zinc and iron accumulation in maize grains using the zinc and iron transporter ZmZIP5. Plant Cell Physiol 60(9):2077–2085. https://doi.org/10.1093/pcp/pcz104

Ludwig Y, Slamet-Loedin IH (2019) Genetic biofortification to enrich rice and wheat grain iron: from genes to product. Front Plant Sci 10:833. https://doi.org/10.3389/fpls.2019.00833

Malik KA, Maqbool A (2020) Transgenic crops for biofortification. Fron Sust Food Sys 4:571402. https://doi.org/10.3389/fsufs.2020.571402

Perez-Massot E, Banakar R, Gómez-Galera S, Zorrilla-López U, Sanahuja G, Arjó G, Miralpeix B, Vamvaka E, Farré G, Rivera SM, Dashevskaya S, Berman J, Sabalza M, Yuan D, Bai C, Bassie L, Twyman RM, Capell T, Christou P, Zhu C (2013) The contribution of transgenic plants to better health through improved nutrition: opportunities and constraints. Genes Nut 8: 29–41. https://doi.org/10.1007/s12263-012-0315-5

Prasanna R, Bidyarani N, Babu S, Hossain F, Shivay YS, Nain L (2015) Cyanobacterial inoculation elicits plant defence response and enhanced Zn mobilization in maize hybrids. Cogent Food Agric 1(1):998507. https://doi.org/10.1080/23311932.2014.998507

Prasad R, Shivay YS, Kumar D (2016) Interactions of zinc with other nutrients in soils and plants- a review. Indian J Fert 12(5):16–26

Ramesh A, Sharma SK, Sharma MP, Yadav N, Joshi OP (2014) Inoculation of zinc solubilizing *bacillus aryabhattai* strains for improved growth, mobilization and biofortification of zinc in soybean and wheat cultivated in Vertisols of Central India. Appl Soil Ecol 73:87–96. https://doi.org/10.1016/j.apsoil.2013.08.009

Rasheed N, Maqsood MA, Aziz T, Jabbar A (2020) Characterizing lentil germplasm for zinc biofortification and high grain output. J Soil Sci Plant Nutr 24:1–4. https://doi.org/10.1007/s42729-020-00216-y

Rehman A, Farooq M, Ozturk L, Asif M, Siddique KHM (2018) Zinc nutrition in wheat-based cropping systems. Plant Soil 422:283–315. https://doi.org/10.1007/s11104-017-3507-3

Swamy PBM, Rahman MA, Inabangan-Asilo MA, Amparado A, Manito C, Chadha-Mohanty P, Reinke R, Slamet-Loedin IH (2016) Advances in breeding for high grain zinc in Rice. Rice 9:49. https://doi.org/10.1186/s12284-016-0122-5

Ullah A, Al-Sadi AM, Al-Subhi AM, Farooq M (2020) Characterization of chickpea genotypes of Pakistani origin for genetic diversity and zinc grain biofortification. J Sci Food Agric 100:4139–4149. https://doi.org/10.1002/jsfa.10453

Velu G, Ortiz-Monasterio I, Cakmak I, Hao Y, Singh RPP (2014) Biofortification strategies to increase grain zinc and iron concentrations in wheat. J Cereal Sci 59:365–372

Yan L, Kerr PS (2002) Genetically engineered crops: their potential use for improvement of human nutrition. Nutr Rev 60:135–141

Yilmaz A, Ekiz H, Torun BI, Gultekin S, Bagci SA, Cakmak I (1997) Effect of different zinc application methods on grain yield and zinc concentration in wheat cultivars grown on zinc-deficient calcareous soils. J Plant Nutr 20(4–5):461–471

Zeigler F, Johnson CA (2001) The solubility of calcium zincate [$CaZn_2(OH)_6.2H_2O$]. Cement Con 31:1327–1332

Zou CQ, Zhang YQ, Rashid A, Ram H, Savasli E, Arisoy RZ, Ortiz-Monasterio I, Simunji S, Wang ZH, Sohu V, Hassan M, Kaya Y, Onder O, Lungu O, Mujahid MY, Joshi AK, Zelenskiy Y, Zhang FS, Cakmak I (2012) Biofortification of wheat with zinc through zinc fertilization in seven countries. Plant Soil 361(1):119–130

Chapter 4
Zinc Biofortification: Role of ZIP Family Transporters in the Uptake of Zinc from the Soil up to the Grains

Shyam Narain Pandey and Murtaza Abid

Abstract It is a known fact that plants and animals need micronutrients like zinc (Zn) for their proper growth and development. Zinc plays a significant role as activator of many enzymes, in biosynthetic pathway of several biomolecules and regulative and protective functions in plants. Its poor availability in soils causes low crop yield and low Zn content in food grains which often promotes adverse effects on human health. Therefore, this overview describes the role of transporters in the plant physiological processes that maintain the Zn homeostasis. It includes absorption of Zn from the soil via roots, control of Zn transport from roots to aerial plant parts. Soil condition play significant role in availability of Zn to the plant roots for absorption, thereafter transporters facilitate their translocation up to the grains. Zinc homeostasis is highly regulated in a complex process. The families of Zn-regulated transporter (ZIP)-like proteins are involved in the cellular uptake of Zn, as well as its intracellular trafficking and detoxification in plants. Very little information is available on the structural features and Zn transport mechanisms of plant ZIP family transporters (ZRT-IRT-like proteins). In this overview, we elucidate a comprehensive structure, functions, and regulations of ZIP carriers. We also described the structure of plant ZIPs through homology modeling and multiple sequence alignment with *Bordetella bronchiseptica* ZIP (*BbZIP*) protein whose crystal structure has been solved recently. The details on ZIP transporter genes identified and characterized in some plants till date may play crucial role in biofortification of Zn in food grains.

Keywords Zinc availability · Biofortification · ZIP transporters · Homology modeling · *Bordetella bronchiseptica*

S. N. Pandey (✉) · M. Abid
Department of Botany, University of Lucknow, Lucknow, India

M. Hasanuzzaman et al. (eds.), *Mineral Biofortification in Crop Plants for Ensuring Food Security*, https://doi.org/10.1007/978-981-99-4090-5_4

4.1 Introduction

The element zinc (Zn) is vital for all living beings that are absorbed from the soil and used by crops for their structural and functional needs (Sharma 2006; Pandey 2020). Increase the Zn concentration in food crops, their biofortification in grains, is a natural solution to improve the status of human nutritional requirement on a large scale. On an average of the whole world's countries, 25.1 mg Zn kg^{-1} field-grown wheat grains show a big gap to the biofortification target for human health, which needs upto the 40 mg zinc kg^{-1} in grains of wheat. Likewise, a big gap in Zn contents in other edible crops sources also exists (Pandey and Verma 2020). The optimal supply of micronutrients in soils is vital for normal crop production, but low bioavailability of nutrients is responsible for reduction in crop production. Zincis the most vital micronutrient required growth of the plant and its related metabolism (Sharma 2006; Pandey 2020). The deficiency or surplus of zinc cripples and challenges the physiological processes which negatively affect the biochemical functioning (Cakmak 2000). Shortfall of Zinc is widely spread global complication which reduces productivity of plants up to great extent (Ruel and Bouis 1998; Sadeghzadeh 2013; Pandey and Verma 2020). Zinc deficiency in plants causes yield loss as well as the reduction in content of Zn in food (Krithika and Balachandar 2016; Pandey 2018).

About 50% of agricultural soils are poor in Zn in the world. Zinc deficiencies have mainly reported in a diversified soil conditions like soils with greater alkaline pH and calcareous soils, coarse-textured sandy soils, and fertilized soils with high phosphorus (P) content (Marschner 1995; Sharma 2006). Major limiting factor that affects the plant growth can be attributed to Zn deficiency as its direct role in many cellular biochemical activities (Pandey 2018). Zinc scarcity is the major phenomenon degrading crops yield because of its direct role in many cellular biochemical activities (Pandey 2018). Zinc deficiency in crops is caused by reduced bioavailability to plants determined by the soil conditions (Texture, pH, organic matter content, interactions of chemical constituents, etc.). In addition, other environmental factors and cultural practices also influence the availability of plant micronutrients (Sharma 2006). Above soil factors contribute Zn deficiency in rice plants has been described (Pandey 2020). The significance of different soil factors affecting micronutrients uptake in crops have also been reported earlier (Sharma 2006; Pandey 2020).Zinc However, at concentrations greater than 0.2 mgg^{-1} of plant dry matter, potential phytotoxic effects on development of leaf tissue (Bonnet et al. 2000), mainly due to the low photosynthetic process (Sharma 2006) being affected the photochemical reactions, carbonic anhydrase activity (Lopez-Millan 2005) and chlorophyll synthesis as well as integrity of cell membrane. Thus, to improve the yield and varieties of crops, Zn uptake efficiency is helpful in overcoming Zn deficiency issues to optimize agricultural production (Pandey 2018). To overcome the low availability of Zn, inter-twined set of tightly regulated adaptive mechanisms through their cellular mechanisms. ZIP family have wide prevalence in prokaryotes, eukaryotes and archaeotes; have a basic function of absorption and transport of metals, that includes zinc (Grotz and Guerinot 2006; Kavitha et al. 2015). The central role of zinc transport homeostatis in food crops has been established by the prevalance of ZIP transporters in a wide variety of phytogroups (Kavitha et al. 2015; Pandey 2018). Apart from Zn^{2+} ZIP transporters are engaged in the transport of other trace metals like iron (Fe^{2+}), manganese (Mn^{2+}), copper (Cu^{2+}), cadmium (Cd^{2+})and cobalt (Co^{2+}) (Sharma 2006; Pedas and Husted 2009). IRTs (Iron-regulated transporters) are the part of ZIP family transporters and are key transporters of iron in plants (Eide et al. 1996; Conte and Walker 2011; Lilay et al. 2020). In *Arabidopsis* AtIRT1 is the chief transporter of Fe^{2+} reported. Likewise the transporters AtIRT1 (Eide et al. 1996), AtIRT2 (Vert et al. 2001) and AtIRT3 (Lin et al. 2009) are also reported to function in absorption and transport of Fe^{2+} in *Arabidopsis*. Many divalent cations such as Zn^{2+}, Mn^{2+} and Cd^{2+} have been reported to be transported by IRTs in *Arabidopsis* (Chiang et al. 2006; Lilay et al. 2018). Recently in bacteria *B. bronchiseptica* the crystal structure of the ZIP protein was inferred and the mechanism by which it is transported has also been described (Zhang et al. 2017). Thus, in this overview, we have discussed about the structure, mechanisms and phylogenesis relatives of ZIP transporters of plants and had developed perception into the structure of plant ZIPs through homology modeling and multiple sequence alignment using ZIP protein (*BbZIP*) as a template of *B. bronchiseptica*. We also throw some light on ZIP transporter genes that have been identified and characterized and this overview will constitutes precious understandings for upcoming studies on zinc transporters in cultures under stress conditions.

4.2 Role of zinc in Plants

Zinc is an essential element functions as constituent of many enzymes (alcohol dehydrogenase, carbonic anhydrase and carboxypeptidase etc.) and works as activator in plant defense system (Pandey 2020). Initially zinc is available in soil and is taken up by the plants through root system. Zinc deficiency is now a days is considered as the most wide spread micronutrient deficiency disorders in several crop plants (Fig. 4.1) (Wissuwa et al. 2006). Zinc is an essential element for plants its disorders can cause adverse effect on various metabolic activities and the qualitative production of food grains (low protein, sugar and mineral contents etc.) (Pandey et al. 2002). Zinc is also involved in defense system of plants as being constituents of enzymes and functions as anti-oxidant against reactive oxygen species (ROS) during various stress conditions (Alscher et al. 2002).

In crops, Zn deficiency was reported primarily in rice on calcareous soils in India (Nene 1966; Yoshida and Tanaka 1969). Plant cells require a certain prescribed level of Zn for normal growth and physiological functions (Sharma 2006; Pandey 2020). Zinc play role in the maintaining the structural and functional integrity of cell membranes and facilitates the protein synthesis, gene expression and, regulations of various metabolic reactions (Cakmak 2000; Andreini et al. 2006; Sadeghzadeh 2013; Pandey 2018). Zinc is required in minimal amounts (<1 ppm) in the plant and

Fig. 4.1 Zinc deficiency symptoms shown in different crops

is barely available from alkaline soil. Thus, becomes unfavorable for proper growth and functioning of plants (Krishna et al. 2017; Pandey 2020), as important for many physiological, metabolic functions and biosynthetic pathways (Sharma 2006). A Zn content that falls below 15–20 μg per gram of dry plant leaf tissue is indicative of Zn deficiency (Sharma 2006; Mitra 2015). Zinc is important for execution of some specific metabolic pathway reactions, Zn plays an important role in the biosynthesis of tryptophan which is an active precursor ofauxin (Horak and Trcka 1976; Marschner 2011). The plant hormones experience a downfall in the absence of Zn (Kumar et al. 2016), which is significantly affects plant cell division and differentiation (Eloy Navarro-León et al. 2016). Cakmak et al. (1989) stunted plant growth, yield, and quality of crops is indicative of Zn deficient soil (Chen et al. 2008; Pandey and Verma 2020), mung bean (Samreen et al. 2017) and maize (Wang and Ji 2005). A decrease of Zn concentration can negatively affect the activity of the main photosynthetic enzymes, namely carbonic anhydrase (Pandey 2020), which is crucial for agricultural production (Ali et al. 2008). Male sterility in maize and wheat are suggestive of Zn deficiency (Sharma et al. 1987) and wheat (Sharma et al. 1987). Zinc deficiency is global hazard, especially in the alkaline soil pH range in arid and semi-arid regions (Pandey 2018; Lilay et al. 2020). Most Zn deficiency problems are due to unavailability influenced by soil factors and their translocations to aerial plant parts (Sharma 2006; Pandey 2020). Therefore, extensive research is needed for the effective elimination of Zn deficiency in plants for optimum growth and production of plants growing in zinc deficient soil.

4.3 Function of ZIP transporters

The bioavailable zinc is absorbed from the soil and translocated in the form of divalentcationZn^{2+}. Zinc cations can also be fixed in roots seepage like low molecular weight oxalic acid, malic acid and citric acid which facilitates movement to the surface of roots (Suzuki et al. 2006). Zn^{2+} cannot permeate the lipid bilayer membrane easily (Eide 2005). It first diffuses the free space of the root cell wall by the process of diffusion from soil concoction (Hacisalihoglu and Kochian 2003). The Zn^{2+}cations are transferred by apoplastic or symplastic route in the cortex (Kumar et al. 2016). The transportation of Zn^{2+} is done through transfer proteins into and out of cellular compartments. They include low affinity (Km = 2–5 μM) and high affinity (Km = 0.6 − ++2 nM) cellular membrane transport systems (Hacisalihoglu et al. 2001). Under low Zn concentration in the soil solution the high affinity transport system is mainly active (Hacisalihoglu et al. 2001). Zn^{2+} ions are transported across plasma membrane into cytosol through ZIP and IRT transporters (Eide 2005; Krishna et al. 2017). Then the Zn^{2+} ions crosses through Casparian strip to the endoderm and cells of parenchyma of the xylem and then unladed into xylem. The P-type ATPase family heavy metal transporters ATPase 2 (HMA2) and HMA4 are involved in loading of Zn^{2+}into the xylem (Hussain et al. 2004; Hanikenne et al. 2008). In plants ZIP family transporters are involved mainly facilitates the uptake, translocation, and distribution of Zn micronutrients (Figs. 4.2 and 4.3) (Pandey 2020). The importance of the ZIP transporters for zinc uptake and homeostasis cannot be undermined (Van de Mortel et al. 2006). Therefore, elucidation of their genes functions, location is very crucial. Biotechnological approaches and use of functional genomics to design crops that are tolerant to zinc deficiency can be employed in the near future. Globally Zn deficiency hazards can be improved by endowment of Zn constituent in grains. The group of zinc transporters can be genetically modified to make crops efficient for zinc (Fu et al. 2017).

4.4 Locations of ZIP Family Transporters in Plant Cell

The identification and functioning of ZIP transporter genes were performed in model plants like *Arabidopsis thaliana* and *Oryza sativa*, but their characterization and functions are still unclear. ZIP family genes expression levels displayed patterns in cultures. Under Zn depleted situations only few ZIP transporters shows expression, hence their articulation declines in less than 2 h as soon as Zn is added to the medium (Van de Mortel et al. 2006). At cellular levels there are evidences that ZIP transporters are important in uptake of Zn when experiments are performed in yeast complementation assays (Fu et al. 2017). ZIP transporters are restricted to cell membranes and membranes of many intracellular organelles (Fig. 4.4). In model plant *Arabidopsis* sub-families of ZIP transporters such as *AtZIP1*, *AtZIP2*, *AtZIP3* and *AtZIP4* have been classified functionally as being the transporters with distinct affinities (Grotz et al. 1998). For enhancing Zn accumulation in edible parts of plants

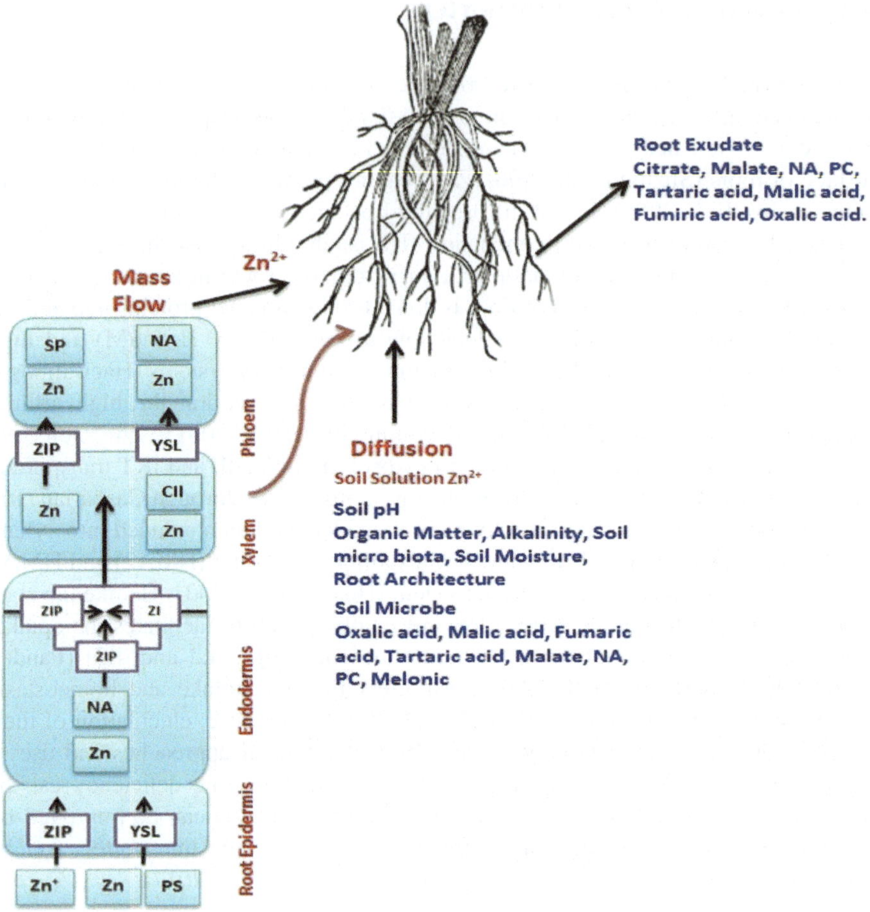

Fig. 4.2 Mechanisms of uptake and translocation of Zn^{2+} by roots in higher plants

the direct role ZIP transporters has been reported (Ramegowda et al. 2013). As compared to other crops *Oryza sativa*is the only model whose ZIP family transporters have been investigated to a great extent. Thus, in *Oryza sativa* model ZIP transporters are discussed in detailthat may play an important role to understand the mechanisms of ZIP and IRT transporters in other crops and cereals (Claus and Chavarría-Krauser 2012).

4.5 ZIP Transporters Regulation in Plants

Every biotic organism must sustain optimal amount of nutrients, encompassing Zn, for normal cell growth and function (Shaul et al. 1999; Sharma 2006; Pandey 2020). The plants are equipped with homeostatic systems that maintain the zinc levels

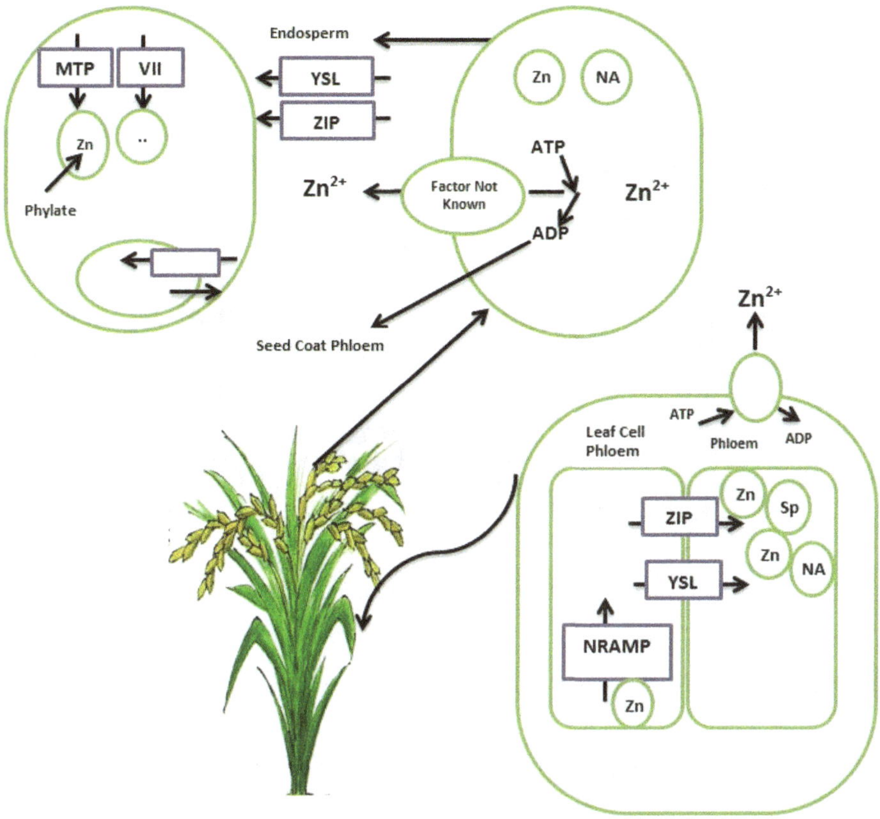

Fig. 4.3 Mechanism of translocation of Zn^{2+} in shoot and seeds incrops

within detectable concentration range that mitigate toxicity or scarcity through their well-organized mechanism (Pandey 2018). Zinc homeostasis requires a complex web of cellular functions like Zn uptake, translocation, assembly, trafficking, sequestration, and remobilization and detoxification processes (Chavarría-Krauser 2012). ZIP transporters are regulated to supply an appropriate amount of Zn in all different types of cells and at all periods of development (Clemens 2001). The balance of Zn homeostasis has been developed for both internal and external Zn requirements (Claus and Chavarría-Krauser 2012). However, very scarce documentation is accessible in regard to the regulatory framework of ZIP transporters and the modes involved in homeostasis of Zn. The deployment of different transporters to transfer divalent cations—Zn^{2+} can be challenging. It may be difficult to uncover the role of plant ZIPs for zinc. In zinc deficient soils, to make up for the zinc scarcity overexpression of genes related to ZIP transporters can be observed. In barley up-regulated, *HvZIP3*, *HvZIP5* and *HvZIP8* genes are strongly expressed in the roots and actively guide absorptionof Zn in Zn scars conditions (Pedas and Husted

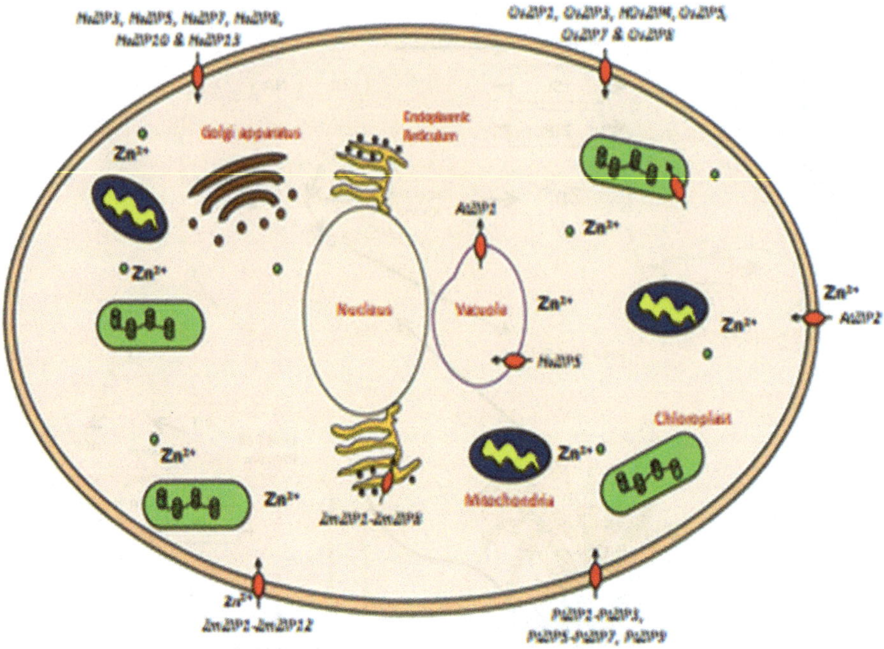

Fig. 4.4 Family of ZIP transporter proteins in the plant cells

2009). Plant detection and transfer of Zn deficiency still requires more elucidation. However, based on the current knowledge, certain transcription factors (TFs) are considered pinpointed for the expression of target genes and the maintenance of Zn homeostasis. Two TFs pinpointed in the model plant *Arabidopsis* have been shown to be pivotal in the adaptive response to Zn deficiency (Sharma 2006).

The basic leucine zipper (*bZIP*) is a DNA binding motif and is a member of F group bZIP family contains histidine rich repeats at the basic N-terminal region it helps in the regulation of ZIP transporters in *Arabidopsis* through the entire time of scarcity (Corrêa et al. 2008; Nijhawan et al. 2008; Assunção et al. 2013). In plants, *bZIP19* and *bZIP23*proteins are present in the form of monomers in enough Zn situations but during Zn scarcity conditions *bZIP19* and *bZIP23* forms dimer that encourage the ZIP gene expression. The dimer bZIP19 and bzip23 binds together at 10 bp, Zn deficiency responsive elements (RTGTCGACAY) available at target genes of the promoter region (Lilay et al. 2018). F-group bZIP TFs contains histidine rich motifs which are conserved and it is regarded to assume responsibility of a zinc sensor (Assunção et al. 2013). So far 33 *bZIP19* genes have been recognized in cereal crops with the help of in silicotechniques and in most of bZIP19 TFs histidine rich motifs are present (Henriques et al. 2017). In *Arabidopsis* under Zn scarcity the TFs *AtbZIP19* and *AtZIP23* influence the up regulation of the *AtZIP4* gene (Lilay et al. 2018). These TFs also promotes the stimulation and expression of genes like ATZIP1, ATZIP2, *ATZIP4*, *ATZIP5*, *ATZIP9*, *ATZIP10* and *ATZIP12* (Lilay et al.

2020). Similar to *Arabidopsis*TFs*bZIP19*, *bZIP23* and *bZIP24*, 3*bZIP* TFs are *bZIP1*, *bZIP2* and *bZIP3* have been recognized in common beans (Astudillo et al. 2013).

Likewise, 7bZIP TFs are recognized in the wheat genome out of these 4TabZIP TFs, viz., TabZIPF1-7DL, TabZIPF3b-7BL, TabZIPF4-7AL and TabZIPF4-7DL are implicated in the functional complementation test (Evens et al. 2017) and the function of these 4TabZIP TFs in the Zn homeostatis mechanism was established. In *Arabidopsis* bZIP19–4bZIP23–2 double mutant line under the Zn deficient conditions. *TabZIPF1-7DL* and *TabZIPF4-7AL* can only feeble complement the double mutant line (*bZIP19–4*, *bZIP23–2*) during Zn scarcity conditions. Slight increments were registered in the development of the double mutant line in zinc scarce environment for TabZIPF4-7DL. But TabZIPF4-7DL didn't complement the double mutant line (Evens et al. 2017), and likewise, in barley 7 *bZIP TFs HvbZIP1*, *HvbZIP10*, *HvbZIP55*, *HvbZIP56*, *HvbZIP57*, *HvbZIP58* and *HvbZIP62* are recognized. Among these TFs *HvbZIP56* and HvbZIP62 can reinitiate the growth of *Arabidopsis* bZIP19 and bZIP23 double mutants during the Zn scarcity conditions (Nazri et al. 2017).

In rice 3 *bZIP* TFs have been recognized and characterized such as *OsbZIP48*, *OsbZIP49* and *OsbZIP50*. *OsbZIP48* and *OsbZIP50* complemented the *Arabidopsis* double mutant (*bZIP19/bZIP23*) under Zn scarcity conditions (Lilay et al. 2020). In plants, the amount of micronutrients and macronutrients affect one another through molecular interactions for example phosphorous interacts with Fe and Zn (Khan et al. 2014).

A ZIP transport protein helps in the absorption, transport and homeostasis of Zn^{2+} in cells of plants (Li et al. 2013). Many ZIP transporters (*ZIP1-ZIP12*) are functional under low Zn^{2+} conditions and are localized on the membranes of different intracellular organelles (Ramesh et al. 2003). Minimal work has been done on the classification of Zn sensitive and many studies are limited only to plants like *Arabidopsis*. Therefore, there is still a scope of research to identify and characterized Zn-responsive TFs in cultures. This may be helpful in elucidating the genetic mode of Zn scarcity tolerance and could help in developing Zn deficiency tolerant crops (Tables 4.1 and 4.2).

4.6 Conclusion

Zinc deficiencies hamper the growth and development of crops, their low availability in plants and their low content in food grains affect human health due to malnutrition problems worldwide. The scarcity of these micronutrients in food crops and grains mainly causes health problems in poor, children and women who depend on food crops. Experimenters must pay attentiveness for Zn biofortificationin crops through soil management and the use of efficient crop varieties capable of transporting Zn and iron from the root to the grains. The ZIP transporters can exert its positive influence in biofortification of zinc. In the soil, a fertilizer consisting of Zn usually

Table 4.1 Genes of ZIP family transporter identified in some crops

Name of the plant	ZIP genes name	Characterization Method	References
Arabidopsis	*AtZIP1-AtZIP4*	Yeast complementation assay	Grotz et al. (1998)
Rice	*OsZIP1- OsZIP16*	qRT-PCR and complementation assay	Chen et al. (2008); Yang et al. (2009)
Maize	*ZmZIP1- ZmZIP12*	qRT-PCR and yeast complementation assay	Li et al. (2013)
Wheat	*TaZIP1-TaZIP3, TaZIP5-TaZIP7, TaZIP9-TaZIP11, TaZIP13, TaZIP14 and TaZIP16*	qRT-PCR and yeast complementation assay	Evens et al. (2017)
Barley	*HvZIP1-HvZIP3, HvZIP5- HvZIP8, HvZIP10, HvZIP11, HvZIP13, HvZIP14 and HvZIP16*	qRT-PCR and yeast complementation assay	Pedas and Husted (2009); Tiong et al. (2015)

solves the Zn scarcity problems. But this is not the only answer because soil factors greatly influence their availability to plant roots. Plant factors also interfere with their translocation to food grains. Genes are identified linked to ZIP transporters, their characterization and the exploitation of this knowledge to develop plants via transgenic or marker-assisted selection may guide in the development of Zn efficient crops cultures with ZIP genes is also useful for enhancing cultures related to Zn utilization efficacy. So, having knowledge of structure and function of the plant ZIP transporter are important.

The in silico studies was helpful to decipher the different overlapping centers for plant ZIP transporters and *BbZIPs*. It is well known that due to greater variation in ZIP transporters, there are different sites for metal binding and transport which are not conserved. For future investigations these results would constitute key theoretical knowledge for understanding the genes, proteins and functional covers of Zn transporters in crops and could help in overcoming the hazards associated with Zn deficiency. In many model plants ZIP family genes which regulate Zn transport have been characterized. Therefore, this review focuses on the location of ZIP transport in different parts of the plants as well as the function and characterization of ZIP transporter for a fluid knowledge of Zn uptake by plant roots. We also provide insight into the knowledge of the effect of physiological matter on the uptake of micronutrients from soil. We hope that this review has provided a pathway to investigate the mechanism of different transporters for maximum absorption of Zn and to improve food grains quality and reduce deficiencies of nutrients.

Table 4.2 Zn Transporters Names, family and functions in crops

Name of the Plant (Family of Zn Transporter)	Name of Transporter	Role	References
Barley (ZIP)	HvZIP7	Uptake of Zn from root to shoot	Tiong et al. (2014)
Barley (ZIP)	*HvZIP3, HvZIP5, HvZIP7, HvZIP8, HvZIP13*	Translocation of Zn	Tiong et al. (2015)
Barley (ZIP)	*HvZIP3, HvZIP5, HvZIP8*	Uptake of Zn	Pedas and Husted (2009)
Barley (CDF)	*HvMTP1*	Translocation of Zn	Menguer et al. (2018)
Barley (P-type ATPase)	*HvHMA2*	Translocation of Zn from root to shoot	Mills et al. (2012)
Rice (P-type ATPase)	*OsHMA3, OsHMA7, OsHMA9*	Homeostasis of Zn	Yamaji et al. (2013); Yamaji and Ma (2019); Satoh-Nagasawa et al. (2011)
Rice (ZIP)	*OsZIP1, OsZIP6, OsZIP7*	Homeostasis of Zn	Gao et al. (2019); Liu et al. (2019)
Rice (ZIP)	*OsHMA3, OsHMA4, OsHMA5*	Translocation of Zn	Lee et al. (2010a); Tan et al. (2020)
Rice (ZIP)	*OsZIP8, OsZIP9,*	Uptake of Zn and translocation	Yang et al. (2009); Lee et al. (2010b)
Rice (ZIP)	*OsMTP1*	Homeostasis of Zn	Lan et al. (2013)
Wheat (ZIP)	*TaZIP3, TaZIP6, TaZIP7, TaZIP9, TaZIP13*	Translocation of Zn	Evens et al. (2017)
Wheat (P-type ATPase)	*TaHMA2*	Translocation of Zn from root to shoot	Qiao et al. (2018)
Emmer wheat (ZIP)	*TdZIP1*	Translocation of Zn	Emel et al. (2011)
Red wild einkorn wheat (CDF)	*TuMTP1*	Zn translocation and homoestatsis	Wang et al. (2018)
Maize (ZIP)	*ZmZIP1-ZmZIP8, ZmIRT1*	Translocation of Zn	Li et al. (2013)
Maize (ZIP)	*ZmIRT1, ZmZIP3*	Uptake of Zn and translocation	Li et al. (2015)

Acknowledgements We are highly grateful to Prof. Y.K. Sharma, Ex-Head, Department of Botany, University of Lucknow, for his precious recommendations and effective conversations in gaining knowledge on various aspects of plant nutritional dynamics.

References

Ali S, Khan AR, Mairaj G, Arif M, Fida M, Bibi S (2008) Assessment of different crop nutrient management practices for yield improvement. Aust J Crop Sci 2:150–157

Andreini C, Banci L, Bertini I, Rosato A (2006) Zinc through the three domains of life. J Proteome Res 5:3173–3178. https://doi.org/10.1021/pr0603699

Alscher RG, Erturk N, Heath LS (2002) Role of superoxide dismutase (SoDs) in controlling oxidative stress. J Exp Bot 153:1331–1341

Astudillo C, Fernandez A, Blair MW, Cichy KA (2013) The *Phaseolus vulgaris* ZIP gene family: identification, characterization, mapping, and gene expression. Front Plant Sci 4:286. https://doi.org/10.3389/fpls.2013.00286

Assunção A, Persson D, Huste S, Schjørring J, Alexander R, Aarts M (2013) Model of how plants sense zinc deficiency. Metallomics 5:1110–1116. https://doi.org/10.1039/c3mt00070b

Bonnet M, Camares O, Veisseire P (2000) Effects of zinc and influence of Acremonium lolii on growth parameters, chlorophyll a fluorescence and antioxidant enzyme activities of ryegrass (Lolium perenne L. cv Apollo). J Exp Bot 51(346):945–953

Cakmak I (2000) Tansley review no. 111 possible roles of zinc in protecting plant cells from damage by reactive oxygen species. New Phytol 146:185–205. https://doi.org/10.1046/j.1469-8137.2000.00630.x

Cakmak I, Marschner H, Bangerth F (1989) Effect of zinc nutritional status on growth, protein metabolism and levels of Indole-3-acetic acid and other phytohormones in Bean (Phaseolus vulgaris L.). J Exp Bot 40(3):405–412

Chen W, Feng Y, Cha Y (2008) Genomic analysis and expression pattern of *OsZIP1*, *OsZIP3*, and *OsZIP4* in two rice (*Oryza sativa* L.) genotypes with different zinc efficiency. Russ J Plant Physiol 55:400–409. https://doi.org/10.1134/S1021443708030175

Chiang HC, Lo JC, Yeh KC (2006) Genes associated with heavy metal tolerance and accumulation in Zn/cd hyperaccumulator *Arabidopsis halleri*: a genomic survey with cDNA microarray. Environ Sci Technol 40:6792–6798. https://doi.org/10.1021/es061432y

Claus J, Chavarría-Krauser A (2012) Modeling regulation of zinc uptake via ZIP transporters in yeast and plant roots. PLoS One 8:e37193. https://doi.org/10.1371/journal.pone.0037193

Clemens S (2001) Molecular mechanisms of plant metal tolerance and homeostasis. Planta 212:475–486

Conte SS, Walker EL (2011) Transporters contributing to iron trafficking in plants. Mol Plant 4: 464–476. https://doi.org/10.1093/mp/ssr015

Corrêa LGG, Riaño-Pachón DM, Schrago CG, Santos RV, Mueller Roeber B, Vincentz M (2008) The role of bZIP transcription factors in green plant evolution: adaptive features emerging from four founder genes. PLoS One 3:e2944. https://doi.org/10.1371/journal.pone.0002944

Eide D, Broderius M, Fett J, Guerinot ML (1996) A novel iron-regulated metal transporter from plants identified by functional expression in yeast. Proc Natl AcadSciUSA 93:5624–5628. https://doi.org/10.1073/pnas.93.11.5624

Eide DJ (2005) "The zip family of Zn transporters," in Zinc Finger Proteins. In: Iuchi S, Kuldell N (eds) Molecular biology intelligence unit. Springer, Boston, MA, pp 261–264. https://doi.org/10.1007/0-387-27421-9_35

Emel D, Coruh C, Dinler G, Grusak MA, Peleg Z, Saranga Y, Fahima T, Yazici A, Ozturk L, Cakmak I, Budak H (2011) Expression and cellular localization of ZIP1 transporter under zinc deficiency in wild emmer wheat. Plant Mol Biol Report 29:582–596

Navarro-León E, Albacete A, de la Torre-González A, Ruiz JM, Blasco B (2016) Phytohormone profile in *Lactuca sativa* and *Brassica oleracea* plants grown under Zn deficiency. Phytochemistry 130:85–89

Evens NP, Buchner P, Williams LE, Hawkesford MJ (2017) The role of ZIP transporters and group F bZIP transcription factors in the Zn−deficiency response of wheat (*Triticumaestivum*). Plant J 92:291–304. https://doi.org/10.1111/tpj.13655

Fu XZ, Zhou X, Xing F, Ling LL, Chun CP, Cao L (2017) Genomewide identification, cloning and functional analysis of the zinc/iron-regulated transporter-like protein (ZIP) gene family in trifoliate orange (*Poncirustrifoliata* L. Raf.). Front. Plant Sci 8:588. https://doi.org/10.3389/fpls.2017.00588

Gao S, Xiao Y, Xu F, Gao X, Cao S, Zhang F, Wang G, Sanders D, Chu C (2019) Cytokinin-dependent regulatory module underlies the maintenance of zinc nutrition in rice. New Phytol 224(1):202–215. https://doi.org/10.1111/nph.15962

Grotz N, Guerinot ML (2006) Molecular aspects of cu, Fe and Zn homeostasis in plants. BBA Mol Cell Res 1763:595–608. https://doi.org/10.1016/j.bbamcr.2006.05.014

Grotz N, Fox T, Connolly E, Park W, Guerinot ML, Eide D (1998) Identification of a family of zinc transporter genes from *Arabidopsis* that respond to zinc deficiency. Proc Natl Acad Sci U S A 95:7220–7224. https://doi.org/10.1073/pnas.95.12.7220

Hacisalihoglu G, Kochian LV (2003) How do some plants tolerate low levels of soil zinc? Mechanisms of zinc efficiency in crop plants. New Phytol 159:341–350. https://doi.org/10.1046/j.1469-8137.2003.00826.x

Hacisalihoglu G, Hart JJ, Kochian LV (2001) High-and low-affinity zinc transport systems and their possible role in zinc efficiency in bread wheat. Plant Physiol 125:456–463. https://doi.org/10.1104/pp.125.1.456

Hanikenne M, Talke IN, Haydon MJ, Lanz C, Nolte A, Motte P (2008) Evolution of metal hyperaccumulation required *cis*-regulatory changes and triplication of HMA4. Nature 453:391–395. https://doi.org/10.1038/nature06877

Henriques A, Farias D, de Oliveira Costa A (2017) Identification and characterization of the bZIP transcription factor involved in zinc homeostasis in cereals. Genet Mol Res 16(2):gmr16029558

Horak V, Trcka I (1976) The influence of Zn^{2+} ions on the tryptophan biosynthesis in plants. Biol Plant 18:393–396. https://doi.org/10.1007/BF02922471

Hussain D, Haydon MJ, Wang Y, Wong E, Sherson SM, Young J (2004) P-type ATPase heavy metal transporters with roles in essential zinc homeostasis in Arabidopsis. Plant Cell 16:1327–1339. https://doi.org/10.1105/tpc.020487

Kavitha P, Kuruvill S, Mathew M (2015) Functional characterization of a transition metal ion transporter, OsZIP6 from rice (*Oryza sativa* L.). Plant Physiol Biochem 97:165–174. https://doi.org/10.1016/j.plaphy.2015.10.005

Khan GA, Bouraine S, Wege S, Li Y, de Carbonnel M, Berthomieu P (2014) Coordination between zinc and phosphate homeostasis involves the transcription factor PHR1, the phosphate exporter PHO1, and its homologue PHO1; H3 in *Arabidopsis*. J Exp Bot 13:871–884. https://doi.org/10.1093/jxb/ert444

Krishna TPA, Ceasar SA, Maharajan T, Ramakrishnan M, Duraipandiyan V, Al-Dhabi N (2017) Improving the Zinc-use efficiency in plants: A review. SABRAO J Breed Genet 49:221–230

Krithika S, Balachandar D (2016) Expression of zinc transporter genes in rice as influenced by zinc-solubilizing Enterobacter cloacae strain *ZSB14*. Front Plant Sci 7:446. https://doi.org/10.3389/fpls.2016.00446

Kumar L, Meena NL, Singh U, Singh U, Praharaj C, Singh S (2016) Zinc transporter: mechanism for improving Zn availability.In. In: Singh U, Praharaj C, Singh S, Singh N (eds) Biofortification of food crops. Springer, New Delhi, pp 129–146. https://doi.org/10.1007/978-81-322-2716-8_11

Lan H-X, Wang Z-F, Wang Q-H, Wang M-M, Bao Y-M, Huang J, Zhang H-S (2013) Characterization of a vacuolar zinc transporter OZT1 in rice (Oryza sativa L.). Mol Biol Rep 40(2):1201–1210. https://doi.org/10.1007/s11033-012-2162-2

Lee S, Jeong HJ, Kim SA, Lee J, Guerinot ML, An G (2010a) OsZIP5 is a plasma membrane zinc transporter in rice. Plant Mol Biol 73(4):507–517. https://doi.org/10.1007/s11103-010-9637-0

Lee S, Kim SA, Lee J, Guerinot ML, An G (2010b) Zinc defciencyinducible OsZIP8 encodes a plasma membrane-localized zinc transporter in rice. Mol Cells 29(6):551–558. https://doi.org/10.1007/s10059-010-0069-0

Liu XS, Feng SJ, Zhang BQ, Wang MQ, Cao HW, Rono JK, Chen X, Yang ZM (2019) OsZIP1 functions as a metal efux transporter limiting excess zinc, copper and cadmium accumulation in rice. BMC Plant Biol 19(1):283. https://doi.org/10.1186/s12870-019-1899-3

Li S, Zhou X, Huang Y, Zhu L, Zhang S, Zhao Y, Guo J, Chen J, Chen R (2013) Identifcation and characterization of the zinc-regulated transporters, iron-regulated transporter-like protein (ZIP) gene family in maize. BMC Plant Biol 13(1):114. https://doi.org/10.1186/1471-2229-13-114

Li S, Zhou X, Li H, Liu Y, Zhu L, Guo J, Liu X, Fan Y, Chen J, Chen R (2015) Overexpression of ZmIRT1 and ZmZIP3 enhances iron and zinc accumulation in transgenic Arabidopsis. PloS one 10(8):1–21. https://doi.org/10.1371/journal.pone.0136647

Lilay GH, Castro PH, Guedes JG, Almeida DM, Campilho A, Azevedo H (2020) Rice F-bZIP transcription factors regulate the zinc deficiency response. J Exp Bot 71:1–14. https://doi.org/10.1093/jxb/eraa115

Lilay GH, Castro PH, Campilho A, Assunção AG (2018) The Arabidopsis *bZIP19* and *bZIP23* activity requires zinc deficiency–insight on regulation from complementation lines. Front Plant Sci 9:1955. https://doi.org/10.3389/fpls.2018.01955

Lin Y-F, Liang H-M, Yang S-Y, Boch A, Clemens S, Chen C-C, Jing-Fen W, Huang J-L, Yeh K-C (2009) Arabidopsis IRT3 is a zinc-regulated and plasma membrane localized zinc/iron transporter. New Phytol 182:392–404

Lopez-Millan A, Ellis D, Grusak MA (2005) Effect of zinc and manganese supply on the activities of superoxide dismutase and carbonic anhydrase in *Medicagotuncatula* wild type and raz plants. Plant Sci 168:1015–1022

Marschner H (1995) Marschner's mineral nutrition of higher plants. Academic, Cambridge, MA

Marschner (2011) Marschner's mineral nutrition of higher plants, 3rd edn. Academic Press, Elsevier. ISBN 978-0-12-3849005-2

Menguer PK, Vincent T, Miller AJ, Brown JKM, Vincze E, Borg S, Holm PB, Sanders D, Podar D (2018) Improving zinc accumulation in cereal endosperm using HvMTP1, a transition metal transporter. Plant Biotechnol J 16(1):63–71. https://doi.org/10.1111/pbi.12749

Mills RF, Peaston KA, Runions J, Williams LE (2012) HvHMA2, a P1B-ATPase from barley, is highly conserved among cereals and functions in Zn and cd transport. PLoS One 7(8):1–14. https://doi.org/10.1371/journal.pone.0042640

Mitra GN (2015) Zinc (Zn) uptake. In: Regulation of nutrient uptake by plants–a biochemical and molecular approach. Springer, New Delhi, pp 127–133

Nazri AZ, Griffin JH, Peaston KA, Alexander-Webber DG, Williams LE (2017) F−group bZIPs in barley–a role in Zn deficiency. Plant Cell Environ 40:2754–2770. https://doi.org/10.1111/pce.13045

Nene Y (1966) Symptoms, cause and control of Khaira disease of paddy. Bull Indian Phytopathol Soc 3:97–191

Nijhawan A, Jain M, Tyagi AK, Khurana JP (2008) Genomic survey and gene expression analysis of the basic leucine zipper transcription factor family in rice. Plant Physiol 146:333–350. https://doi.org/10.1104/pp.107.112821

Pandey SN, Verma I (2020) Zinc-induced biochemical constituents and reproductive yield of wheat with zinc supply in sand culture conditions. Int J Plant Environ 6(03):178–181. https://doi.org/10.1007/s11099-005-0092-0

Pandey SN (2020) Role of micronutrients in biochemical responses of crops under abiotic stresses. Sust Agric Era Clim Chan 57(2):129–138

Pandey SN (2018) Biomolecular functions of micronutrients towards abiotic stress tolerance in plants. Plant Nutr Abiotic Stress Toler:153–170

Pandey N, Pathak GC, Singh AK, Sharma CP (2002) Enzymic changes in response to zinc nutrition. J Plant Physiol 159:1151–1153

Pedas P, Husted S (2009) Zinc transport mediated by barley ZIP proteins are induced by low pH. Plant Signal Behav 4:842–845. https://doi.org/10.4161/psb.4.9.9375

Qiao K, Gong L, Tian Y, Wang H, Chai T (2018) The metal-binding domain of wheat heavy metal ATPase 2 (TaHMA2) is involved in zinc/cadmium tolerance and translocation in Arabidopsis. Plant Cell Rep 37:1343–1352

Ramesh SA, Shin R, Eide DJ, Schachtman DP (2003) Differential metal selectivity and gene expression of two zinc transporters from rice. Plant Physiol 133:126–134. https://doi.org/10.1104/pp.103.026815

Ramegowda Y, Venkategowda R, Jagadish P, Govind G, Hanumanthareddy RR, Makarla U (2013) Expression of a rice Zn transporter, *OsZIP1*, increases Zn concentration in tobacco and finger millet transgenic plants. Plant Biotechnol Rep 7:309–319. https://doi.org/10.1007/s11816-012-0264-x

Ruel MT, Bouis HE (1998) Plant breeding: a long-term strategy for the control of zinc deficiency in vulnerable populations. Am J Clin Nutr 68:488–494. https://doi.org/10.1093/ajcn/68.2.488S

Satoh-Nagasawa N, Mori M, Nakazawa N, Kawamoto T, Nagato Y, Sakurai K, Takahashi H, Watanabe A, Akagi H (2011) Mutations in rice (Oryza sativa) heavy metal ATPase 2 (OsHMA2) restrict the translocation of zinc and cadmium. Plant Cell Physiol 53(1): 213–224. https://doi.org/10.1093/pcp/pcr166

Sadeghzadeh B (2013) A review of zinc nutrition and plant breeding. J Soil Sci Plant Nutr 13:905–927. https://doi.org/10.4067/S0718-95162013005000072

Samreen T, Shah HU, Ullah S, Javid M (2017) Zinc effect on growth rate, chlorophyll, protein and mineral contents of hydroponically grown mungbeans plant (*Vignaradiata*). Arab J Chem 10: S1802–S1807. https://doi.org/10.1016/j.arabjc.2013.07.005

Sharma PN, Chatterjee C, Sharma CP, Agarwala SC (1987) Zinc deficiency and anther development in maize. Plant Cell Physiol 28:11–18. https://doi.org/10.1093/oxfordjournals.pcp.a077265

Sharma CP (2006) Plant Micronutrient, 1st edn. Science Publisher, New Hampshire, USA

Shaul O, Hilgemann DW, de-Almeida-Engler J, Van Montagu MV, Inze D, Galili G (1999) Cloning and characterization of a novel Mg 21/H1 exchanger. EMBO J 8:3973–3980

Suzuki M, Takahashi M, Tsukamoto T, Watanabe S, Matsuhashi S, Yazaki J (2006) Biosynthesis and secretion of mugineic acid family phytosiderophores in zinc–deficient barley. Plant J 48: 85–97. https://doi.org/10.1111/j.1365-313X.2006.02853.x

Tan L, Qu M, Zhu Y, Peng C, Wang J, Gao D, Chen C (2020) ZINC transporter 5 and zinc transporter9 function synergistically in zinc/cadmium uptake. Plant Physiol 183(3):1235–1249. https://doi.org/10.1104/pp.19.01569

Tiong J, McDonald GK, Genc Y, Pedas P, Hayes JE, Toubia J, Langridge P, Huang CY (2014) HvZIP7 mediates zinc accumulation in barley (H ordeumvulgare) at moderately high zinc supply. New Phytol 201(1):131–143. https://doi.org/10.1111/nph.12468

Tiong J, McDonald G, Genc Y, Shirley N, Langridge P, Huang CY (2015) Increased expression of six ZIP family genes by zinc (Zn) defciency is associated with enhanced uptake and root-to-shoot translocation of Zn in barley (Hordeumvulgare). New Phytol 207(4):1097–1109. https://doi.org/10.1111/nph.13413

Van de Mortel JE, Villanueva LA, Schat H, Kwekkeboom J, Coughlan S, Moerland PD (2006) Large expression differences in genes for iron and zinc homeostasis, stress response, and lignin biosynthesis distinguish roots of *Arabidopsis thaliana* and the related metal hyperaccumulator *Thlaspicaerulescens*. Plant Physiol 142:1127–1147. https://doi.org/10.1104/pp.106.082073

Vert G, Briat JF, Curie C (2001) Arabidopsis *IRT2* gene encodes a root–periphery iron transporter. Plant J 26:181–189. https://doi.org/10.1046/j.1365-313x.2001.01018.x

Wang H, Ji J (2005) Photosynthetic rate, chlorophyll fluorescence parameters, and lipid peroxidation of maize leaves as affected by zinc deficiency. Photosynthetica 43:591–596

Wang F-H, Qiao K, Liang S, Tian S-Q, Tian Y-B, Wang H, Chai T-Y (2018) Triticumurartu MTP1: its ability to maintain Zn2+ and Co2+ homeostasis and metal selectivity determinants. Plant Cell Rep 37(12):1653–1666. https://doi.org/10.1007/s00299-018-2336-z

Wissuwa M, Ismail AM, Yanagihara S (2006) Effects of zinc deficiency on Rice growth and genetic factors contributing to tolerance. Plant Physiol 142:731–741

Yamaji N, Xia J, Mitani-Ueno N, Yokosho K, Feng Ma J (2013) Preferential delivery of zinc to developing tissues in rice is mediated by P-type heavy metal ATPase OsHMA2. Plant Physiol 162(2):927–939. https://doi.org/10.1104/pp.113.216564

Yamaji N, Ma JF (2019) Bioimaging of multiple elements by highresolution LA-ICP-MS reveals altered distribution of mineral elements in the nodes of rice mutants. Plant J 99(6):1254–1263. https://doi.org/10.1111/tpj.14410

Yang X, Huang J, Jiang Y, Zhang H-S (2009) Cloning and functional identifcation of two members of the ZIP (Zrt, Irt-like protein) gene family in rice (Oryza sativa L.). Mol Biol Rep 36(2): 281–287. https://doi.org/10.1007/s11033-007-9177-0

Yoshida S, Tanaka A (1969) Zinc deficiency of the rice plant in calcareous soils. Soil Sci Plant Nutr 15:75–80. https://doi.org/10.1080/00380768.1969.10432783

Zhang T, Liu J, Fellner M, Zhang C, Sui D, Hu J (2017) Crystal structures of a ZIP zinc transporter reveal a binuclear metal center in the transport pathway. Sci Adv 3:e1700344. https://doi.org/10.1126/sciadv.1700344

Chapter 5
Mechanisms of Iron Uptake and Homeostasis in Plants: Implications for Biofortification in Cereal Grains

Usman Zulfiqar, Ghadeer M. Albadrani, and Saddam Hussain

Abstract Iron (Fe) is an essential micronutrient that is indispensable for plant growth, human development, and overall health. Rice and wheat are the main staple cereals worldwide, but their common cultivars contain meager amounts of Fe, which are mostly located in the outer bran layers. The development of substantial Fe-rich rice and wheat germplasms through traditional breeding has been challenging due to the limited availability of significant Fe resources. To survive, plants have evolved two distinct molecular strategies to perceive and interact with their environment and absorb Fe from the soil. Iron is then transported to sink tissues, where it is mainly used to produce enzyme cofactors or electron transport chain components. The uptake, distribution, and metabolism of Fe are tightly regulated by transcriptional and post-transcriptional mechanisms to avoid toxic excess concentrations. Iron is also stored in seeds, making it crucial for human nutrition, as seeds are a critical part of the diet of many crop species. Therefore, increasing Fe levels in seeds and other tissues is a primary objective of biofortification, which is achieved through conventional breeding, biotechnological approaches, and agronomic interventions. The use of transgenic methods provides a precise and efficient means of developing crops with improved nutritional value, which serves as a valuable complement to traditional breeding and mineral fertilization practices in addressing the issue of nutrient deficiency.

Keywords Biofortification · Iron · Xylem · Phloem · Iron translocation · Seed

---

U. Zulfiqar
Department of Agronomy, Faculty of Agriculture and Environment, The Islamia University of Bahawalpur, Bahawalpur, Pakistan

G. M. Albadrani
Department of Biology, Princess Nourah bint Abdulrahman University, Riyadh, Saudi Arabia

S. Hussain (✉)
Department of Biology, Princess Nourah bint Abdulrahman University, Riyadh, Saudi Arabia

Department of Agronomy, University of Agriculture, Faisalabad, Pakistan
e-mail: shussain@uaf.edu.pk

M. Hasanuzzaman et al. (eds.), *Mineral Biofortification in Crop Plants for Ensuring Food Security*, https://doi.org/10.1007/978-981-99-4090-5_5

5.1 Introduction

Iron (Fe) is an essential element that is abundantly available on earth, but its limited bioavailability creates a significant constraint for agriculture and poses a severe problem for human health (Ma and Ling 2009). Although Fe is widely available in the earth's crust, it is usually present in an oxidized form that plants cannot easily absorb (Connorton et al. 2017). Plants are primary producers and play a critical role in the food chain by facilitating the uptake of Fe. Iron-deficiency anemia is a significant global health concern, impacting over two billion people worldwide, leading to severe health complications such as an increased risk of mortality, limited physical and mental development, weakened immunity, anemia, and fatigue (Li et al. 2023). Iron is an essential micronutrient for plants, playing a critical role in their development. In particular, it is a key component in the synthesis of chlorophyll, which is vital for the process of photosynthesis. Moreover, Fe is involved in several other enzymatic and metabolic processes that are essential for the plant lifecycle (Kroh and Pilon 2020). Iron, however, is predominantly found in a ferric form [Fe (III)] in mineral soils. This form of Fe is not very soluble under aerobic conditions in the physiological pH range. This creates a significant abiotic stress in agriculture worldwide, 30% of arable land has a higher soil pH that is too alkaline to support optimal crop production, leading to reduced availability of Fe, resulting in low Fe availability (Kobayashi et al. 2018). To acquire poorly bioavailable Fe from the soil, higher plants have evolved different uptake strategies. Although Fe is vital for plant survival, but it can also be highly toxic and reactive as a result of the Fenton reaction (Morrissey and Guerinot 2009). Therefore, plants tightly regulate Fe homeostasis and respond to both Fe deficiency and toxicity. The availability of Fe in plants has a significant impact on human nutrition, as it affects crop productivity and the Fe content of edible plant parts. Therefore, understanding the mechanisms involved in Fe uptake and transport is vital for developing crops that can thrive in Fe-limited soils and provide better nutrition to humans. Moreover, biofortification is one of the balanced strategies to improve plants' nutritional profile and to combat the occurrence of deficiency of vital nutrients in humans (Zulfiqar et al. 2021a, b). This chapter focuses on Fe transport and homeostasis in plants, covering the uptake of Fe into roots, its long-distance transport to the shoot, and its ultimate loading into seeds. Moreover, this chapter highlights the recent advancements in breeding and agronomic strategies for increasing the Fe content of cereals.

5.2 Iron Uptake and Transport from Root to Seeds

The uptake mechanism of Fe varies among plant species and there are several strategies of Fe uptake by plants i.e., strategy-I, strategy-II, and combined strategy (Fig. 5.1). Dicotyledonous plants uptake Fe by strategy-I, firstly insoluble Fe^{3+} compounds in the soil get acidified by H^+ released from roots (Santi and Schmidt

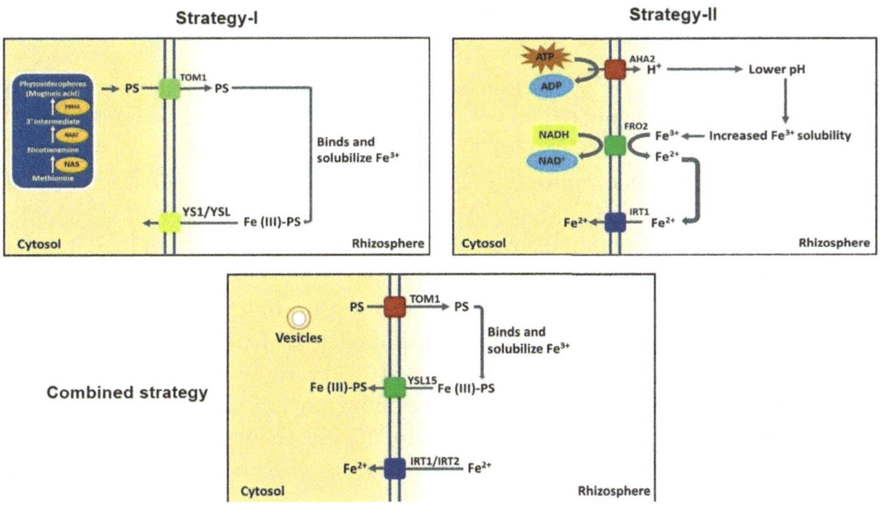

Fig. 5.1 Fe acquisition strategies in higher plants: Strategy I in non-graminaceous plants, Strategy II in graminaceous plants and Combined strategy in rice

2008, 2009), then a membrane-bound ferric chelate reductase (FRO_2) reduces Fe^{3+} to Fe^{2+}, which are consequently taken up into root cells via Fe-regulated transporter 1 (IRT1) (Ivanov et al. 2012). Conversely, cereal plants such as maize and barley use chelation-based Fe uptake strategy-II which consists of a sequence of pathways. The synthesis of mugineic acids (MAs) is primarily achieved through a conserved pathway that commences with the trimerization of three *S*-adenosyl-L-methionine molecules into nicotianamine (NA) via the enzyme nicotianamine synthase (NAS) (Bonneau et al. 2016). Nicotianamine aminotransferase (NAAT) and deoxymugineic acid synthase (DMAS) further convert NA into 2`-deoxymugineic acid and MAs (Bashir et al. 2006). These MAs are secreted into the rhizosphere through the transporter TOM1. MAs chelate insoluble Fe in the soil and the resulting Fe^{3+}-MAs complex is subsequently translocated into root cells by stripe-like family transporters (YS1/YSL) (Curie et al. 2009). Rice despite being a strategy-II plants uptake Fe through combined strategies of reduction-based strategy and chelation-based strategy. It possesses a Fe^{2+} transporter, *OsIRT1* which allows it to absorb Fe^{2+} (Ishimaru et al. 2006). Rice acquires Fe^{3+} through strategy-I and Fe^{2+} via IRT1 and IRT2 directly from the surrounding. However, rice has low Fe^{3+} chelate reductase levels (Kim and Guerinot 2007). The possible reason is the adaptation of rice when grown in anaerobic flooded conditions rich in Fe^{2+} than Fe^{3+} (Sperotto et al. 2012a, b). Like strategy-II, MAs will be secreted to the soil rhizosphere to bind with Fe^{3+} and the complex will be transported with the help of *YSL15* (*YS-like 15*). Among both strategies, rice can uptake Fe more efficiently from surrounding with the help of Fe^{3+}-MAs complexes in comparison with direct uptake of Fe^{2+} (Ishimaru et al. 2006).

5.2.1 Transport of Iron through Plant Roots

After uptake via roots, Fe^{2+} is transported across the root epidermal layer by particular carriers or channels known as ZIP family transporters or carriers (ZRT IRT-like Protein) (Kim and Guerinot 2007). Various types of endodermal as well as epidermal efflux carriers or transporters chiefly transport Fe-phytosiderophores into vascular bundles of roots (Conte and Walker 2011). Iron is transported via apoplast and symplast pathways within the root system of plants. In barley and rice, there are numerous efflux transporters (ETs) (*OsTOM1, OsTOM2, and HvTOM1*) that chelate to deoxymugineic acid (DMA) and export it to soil from the root. Deoxymugineic acid binds the Fe^{3+} in soil which is then transported as DMA-Fe^{3+} complex via YSL (yellow stripe-like) influx transporters located in the roots (Nozoye et al. 2011). Accompanied by TOM proteins, phenolics efflux zero 1 (PEZ1) carry out solubilization of the precipitated apoplasmic Fe by the secretions of caffeic acid and phenolics (Ishimaru et al. 2011).

5.2.2 Long-Distance Transport of Iron Via Xylem and Phloem

Metals can transport via apoplast to senescent leaves via the xylem. For the transportation of metals, there are various ETs existent in the endodermis as well as the pericycle membrane. In barley and rice, TOM1 transports Fe^{3+} into the xylem and phloem as DMA-chelated Fe (Nozoye et al. 2011). Similarly, ferric reductase defective like (FRDL) transporter is reported to transport Fe^{3+} as citrate-chelated Fe in rice, whereas Fe^{2+} is chelated to NA and carried in the xylem cells as NA-Fe^{2+} (Kakei et al. 2009; Yokosho et al. 2009). The nodular region of cereals possesses crucial importance, as nutrients are transferred from xylem to phloem in this region, afterward translocated towards flag leaf as well as spike (Yamaji and Ma 2009). Phloem-mediated transport is the most important, in which Fe is symplastically transported to different plant parts (Wang et al. 2011). Numerous influx as well as efflux transporters like ZIP and YSL play a vital role in Fe transportation in the phloem cells. In rice, *OsZip4* and *OsZip8* are associated with Fe^{2+} transport in vascular tissues, leaf mesophyll cells, and apical meristems of plants (Lee et al. 2010). Iron-regulated transporter 1 (IRT 1) is reported to involve in carrying NA bounded Fe^{2+} into the phloem cells whereas TOM1 efflux channel facilitates the Fe^{3+} transportation in the chelated form (Nozoye et al. 2011).

5.2.3 Iron Uptake and Distribution through Xylem

The understanding of Fe translocation in plants from roots to different tissues is difficult as Fe-ions are highly reactive and poorly soluble. For transport, the ions are carried by making complexes with nicotinamine (NA), and phenolic compounds to prevent contamination and precipitation (Kobayashi and Nishizawa 2012). The xylem sap mainly consists of Fe in the form of Fe-citrate complex displaying the importance of citrate in Fe transport in xylem (Green and Rogers 2004; Grotz and Guerinot 2006). The outflow of citrate in the roots of Arabidopsis is brought about by expression of *FRD3* gene where it attaches with Fe for transportation. Wild type plants had more concentration of Fe and citrate in xylem fluid showing role of Fe-citrate complex in transportation (Durrett et al. 2007). Important transmembrane proteins involved in citrate transport in xylem are coded by *OsFRL1* gene (*FRD* like) present in rice. These proteins assist in Fe transport from roots to shoots.

The Fe distribution from xylem to other plant tissues is dependent on their Fe need. The mechanism of Fe uptake and its transport from xylem to leaf tissues is not completely understood yet. This may happen because *FRO* genes reduce Fe^{3+} citrate complex to Fe^{2+} before Fe transport or it is absorbed by some unidentified transporters from some other sources (Jeong et al. 2008). Role of *FRO* family members has been studied in other plant organs as well which indicates the presence of Fe transport based on reduction in other plant parts other than root plasma membrane (Grillet et al. 2014).

Organic molecules called phytosiderophores (a member of the MA family) are released by plants such as rice, wheat, barley, and maize into the rhizosphere. These phytosiderophores have a strong affinity for Fe, as demonstrated by studies conducted by Kobayashi et al. (2005), Suzuki et al. (2006), Borrill et al. (2014), and Bashir et al. (2017). The process of Fe uptake and translocation is similar in both rice and wheat (Borrill et al. 2014; Connorton et al. 2017a). TOM1 (transporter of MA family) facilitates the release of MA in rhizosphere. TOM1 is associated with major facilitator family (MFS) and was first found in rice and barley. It is one of the major groups of proteins involved in protein transport across cellular membrane in reaction to chemiosmotic ion gradient (Nozoye et al. 2011). Fe-MA complex are then created, and transporters Yellow Stripe 1 (YS1) and Yellow Stripe 1-like (YSL1) take up these complexes into roots (Curie et al. 2001; Inoue et al. 2009). Then reduction of absorbed Fe(III)-DMA complex to Fe(II)-NA is brought about by ascorbate in the cytosol of root cells. These reduced complexes are then transported to the xylem and bounded with citrate or DMA for further transportation (Curie et al. 2001; Inoue et al. 2009).

Phenolic compounds also play important role in Fe procurement and its transport in plants (Clemens and Weber 2016). Caffeic acid and protocatechuic acid (PCA) are released by rice which acts as chelators for Fe^{3+} and reduces it to Fe^{2+} to make it soluble (Yoshino and Murakami 1998). Plasma membrane transporters PEZ1 and PEZ2 help in transport of PCA. These transporters are expressed in xylem and get activated when Fe is insufficient (Ogo et al. 2014). When genes for one of these transporters are knocked down by mutants, the concentration of Fe and protocatechuic acid is reduced in xylem (Bashir et al. 2011). Some other phenolic compounds including coumarins have a role in transport and supply of Fe (Schmidt et al. 2014).

5.2.4 Iron Translocation in Phloem

Iron should be transported to plant parts with active growth including root tips, shoot tips, and seeds through phloem because of ineffective xylem transpiration pull. It is also necessary to translocate Fe from old leaves to new leaves through phloem. In order to maintain solubility of Fe, the ions are attached with chelators as the pH of fluid is >7.0 in phloem (Alvarez-Fernandez et al. 2014). The attachment of Fe^{3+} ions with chelators helps in prevention of Fe precipitation and it is dependent on pH of phloem sap. In xylem, citrate- Fe binding occurs at pH 5.5 whereas in phloem Fe binds with NA at pH 7.5 (Morrissey and Guerinot 2009). The reduced form of Fe (Fe^{2+}) can be translocated through phloem to all parts of plant after binding with NA (Takahashi et al. 2003). In Arabidopsis, four genes i.e., *NAS1*, *NAS2*, *NAS3* and *NAS4* have been identified for synthesis of nicotinamine. The expression of *NAS2* and *NAS4* was enhanced under Fe deficiency conditions indicating contribution of these genes in translocation of Fe towards shoots (Kobayashi et al. 2009). In maize, Yellow Stripe 1 (*YS1*) helps in Fe translocation in internal plant tissues (Roberts et al. 2004; Ueno et al. 2009). In plants other than grasses, YSL proteins are more involved in the transfer of Fe between plant organs as compared to primary Fe uptake from roots. *YSL* genes are expressed in vascular tissues; therefore, these genes may be involved in uptake and translocation of mineral nutrients (Didonato et al. 2004; Chu et al. 2010). YSL transporter family also play role in piling up NA-complexes in phloem for Fe translocation (Zheng et al. 2012).

In rice, 18 members are included in YSL family. The transfer of Fe to distant plant tissues including young shoots and developing seeds is carried out by *OsYSL2* (Koike et al. 2004). It can transfer Fe (II)-NA complex but Fe (III)-Mas complex are not carried by this gene. *OsYSL15* helps in translocation of Fe (III)-DMA which assists in Fe absorption in roots and its transfer to other pant tissues. *OsYSL18* play important role in fertilization as it is expressed particularly in reproductive parts. On the other hand, this gene is also found in phloem present in plant laminar joints which indicates its role in Fe transport (Kobayashi and Nishizawa 2012). *OsIRT1* (ferrous transporter) and TOM1 have significant role in Fe uptake as well as in its translocation to all plant tissues as it is also expressed in vascular bundles in rice (Nozoye et al. 2011). Other efflux transporters including ENA1 (NA1) and ENA2 help in transport of nicotinamine (NA) (Nozoye et al. 2011).

The genes *OsYSL15* and *OsYSL2* for uptake of Fe(III)-DMA and Fe(II)-NA from rhizosphere respectively, were overexpressed for increased Fe uptake and translocation in rice (Lee et al. 2009a; Ishimaru et al. 2010). The results revealed an increase in Fe content up to 7.5 mg g^{-1} in polished rice. Lines with overexpression of *OsYSL15* and *OsYSL9* displayed a slight increase in Fe content (Lee et al. 2009a; Senoura et al. 2017). Another method to enhance Fe content is the over expression of NAS genes with the addition of 35-S promoter in *OsNAS2* and *OsNAS3* by T-DNA activation tagging (Lee et al. 2009b, 2012). The overexpression of *OsNAS1* gene in endosperm increased Fe content up to 19 mg g^{-1} but the Fe level was reduced in polished rice up to 5 mg g^{-1} (Zheng et al. 2010). In case of *OsNAS2* gene overexpression, polished rice displayed better Fe content (19 mg g^{-1}) as compared to Nipponbare wildtype (4.5 mg g^{-1}) (Johnson et al. 2011). In milled rice, *OsNAS2* and *OsNAS3* DNA-activation tag plants had enhanced Fe content up to 12 mg g^{-1} and 10 mg g^{-1} respectively in comparison to wild type (4 mg g^{-1}) (Lee et al. 2009b, 2012). Increased Fe content up to 55 mg g^{-1} was observed in the endosperm of japonica rice when *OsNAS1* and *HvNAAT* genes were overexpressed (Diaz-Benito et al. 2018).

5.2.5 Iron Loading and Accumulation in Seeds

Iron is transported from the roots and senescent leaves of cereal plants to emerging seeds via phloem, since xylem flow is driven by transpiration and is not available for seeds (Fig. 5.2; Morrissey and Guerinot 2009). Kruger et al. (2002) investigated a novel Fe transporter protein (ITP) in the phloem of castor bean plants and found that nicotianamine (NA) is involved in mobilizing Fe in and out of phloem, while Fe transport within the phloem occurs via a 17-kDa Fe-binding protein. In rice plants, Ishimaru et al. (2006) detected *OsIRT1-GUS* gene expression in the phloem cells of both roots and shoots, which was induced by Fe deficiency, particularly in companion cells. It is believed that *OsIRT1* carries Fe^{2+} into the phloem cells, where it is

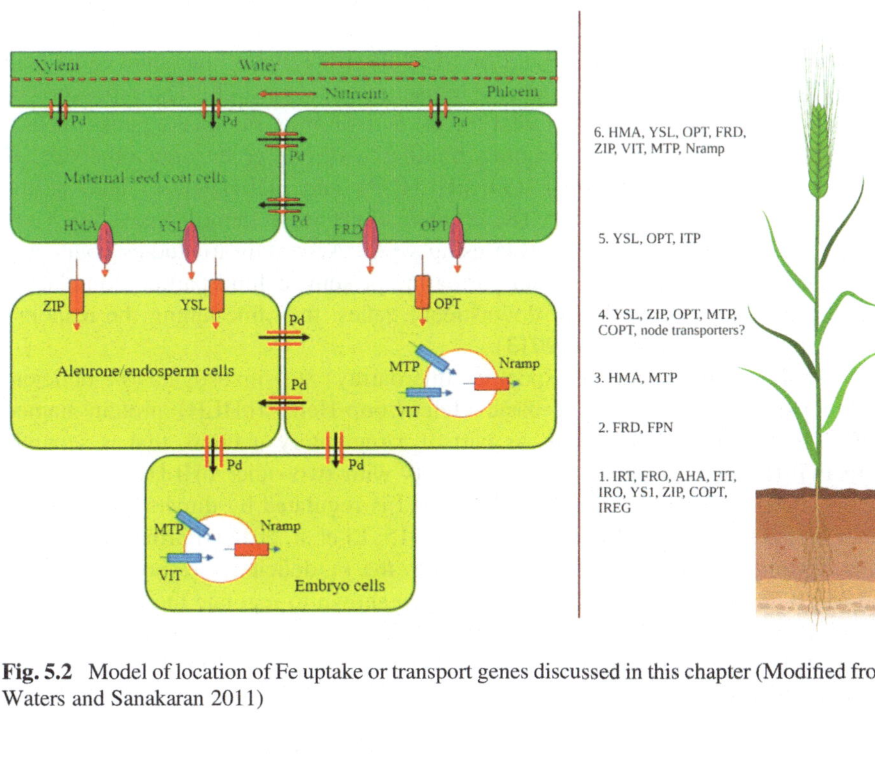

Fig. 5.2 Model of location of Fe uptake or transport genes discussed in this chapter (Modified from Waters and Sanakaran 2011)

chelated by NA (Ishimaru et al. 2006). While rice remobilizes only 4% of the Fe in its shoots to the seeds, whereas wheat transports 77% of the Fe (Morrissey and Guerinot 2009). In grains, phloem cells transport Fe into the maternal seed coat, while transport of Fe into the grain apoplast involves various types of influx and efflux transporters such as YSLs, ZIP, and Nramp (Tauris et al. 2009).

5.3 Iron Homeostasis

To ensure the appropriate distribution of Fe without reaching toxic levels, plants require an efficient regulatory system that senses their Fe status and adjusts homeostasis accordingly. Under Fe-limiting conditions, plants alter their root morphology by increasing the number of lateral roots and the density of root hairs, which provides a greater surface area for contact between the epidermis and the rhizosphere, allowing for the exploration of fresh soil (Morgan and Connolly 2013; Li et al. 2016). While modifications in response to Fe availability have been primarily investigated in the realm of plant development (Mora-Macı'as et al. 2017), the specific molecular pathways linking Fe availability to alterations in root morphology or permeability are not yet fully understood.

Significant advancements have been made in the last two decades in recognizing numerous transcriptional regulators that control Fe homeostasis. This has resulted in a complex network of basic Helix-Loop-Helix (bHLH) transcription factors that manage the Fe deficiency response. The first transcription factor to be identified was FER in tomato (Ling et al. 2002), followed by its Arabidopsis functional equivalent called FER-like iron deficiency-induced transcription factor (FIT). The expression of FIT is limited to the roots and is vital for the activation of Fe uptake genes such as *FRO2* and *IRT1*. However, it requires binding with one of the four other bHLH proteins, namely bHLH38, bHLH39, bHLH100, and bHLH101, to bind to the promoters of FRO2 and IRT1. These interactions were demonstrated by Yuan et al. (2008) and Wang et al. (2013) using yeast. According to studies conducted on mutants of *Arabidopsis*, the four partnering proteins exhibit partial redundancy, but they may regulate different downstream genes, thus fine-tuning the response (Sivitz et al. 2012; Wang et al. 2013).

A study using cell-type specific microarrays to investigate Fe-deficient Arabidopsis roots identified a basic Helix-Loop-Helix (bHLH) protein named POPEYE (PYE). PYE functions as part of a regulatory network that is separate from FIT (Long et al. 2010). PYE interacts with two other bHLH transcription factors, ILR3 (bHLH105) and bHLH115, and is regulated by dimers of ILR3 and either bHLH104 or bHLH34 (Zhang et al. 2015; Li et al. 2016). Furthermore, other transcription factors have also been linked to the Fe deficiency response. MYB10 and MYB72 from the MYB family have been shown to regulate the expression of NAS4 (Palmer et al. 2013) and the production of coumarins (Zamioudis et al. 2014). Additionally, WRKY46, a member of the WRKY family, was found to be involved in the regulation of NAS2 expression and the VIT-Like 1 gene (Yan et al. 2016).

While much progress has been made in understanding the transcriptional networks that govern Fe uptake in dicotyledonous plants such as Arabidopsis, less is known about these networks in monocotyledonous plants. To fill this research gap, scientists have concentrated on recognizing transcription factors that bind to iron-responsive motifs in the promoter of the Fe Deficiency Specific clone 2 (IDS2) (Kobayashi et al. 2003). This approach led to the discovery of two transcription factors, IDEF1 and IDEF2, in rice (Kobayashi et al. 2007), which are involved in controlling the expression of phytosiderophore biosynthesis and YSL2 (Kobayashi et al. 2010). Interestingly, studies on Arabidopsis have revealed several conserved transcriptional regulatory networks in rice, particularly members of the bHLH transcription factor family. For instance, IRO2, a homologue of bHLH39 in rice, has been shown to positively regulate the biosynthesis of phytosiderophores and YSL15 (Ogo et al. 2007). Additionally, IRO3, the rice orthologue of PYE, is known to negatively control the transcript levels of IRO2 and NAS (Zheng et al. 2010). When there is a Fe deficiency response, Fe uptake is increased, but it is important to prevent Fe overload in cases where Fe becomes suddenly available, such as after rainfall. Several mechanisms have been observed that stop Fe uptake post-transcriptionally. For example, IRT1 is continuously recycled from the plasma membrane through ubiquitination and internalization, with the identification of an E3 ligase responsible for its ubiquitination (Shin et al. 2013). In addition to degrading the uptake machinery quickly, stopping the transcriptional response is also necessary. While it is known that FIT can be degraded in response to ethylene and nitric oxide, the specific E3 ligase responsible for this degradation has not yet been identified (Sivitz et al. 2012). Overall, it is probable that all these mechanisms are connected to the Fe level within the cell, assuming the presence of iron-binding regulators comparable to Fur in bacteria, Aft1 in yeast, or IRPs and FBXL5 in mammals.

Studies conducted on plants have suggested that a group of hemerythrin E3 ligases, which include BRUTUS (BTS) in Arabidopsis and HRZ in rice, may function as negative regulators of the Fe deficiency response (Kobayashi et al. 2013; Selote et al. 2015). These ligases possess three hemerythrin domains in their N-terminus that contain conserved His-xxx-Glu motifs that are believed to bind to a di-iron center. In addition, their C-terminal domain shares 45% homology with plant and mammalian E3 ligases, which are known to target transcription factors for ubiquitination and subsequent degradation.

5.4 Iron Biofortification

The applications of our knowledge on Fe uptake, transport, and homeostasis are being further explored to enhance the nutritional value of crops. The crops that are most commonly consumed in the developing world, namely rice, wheat, and maize have been the primary focus of biofortification efforts. Plants are known to regulate Fe levels tightly to prevent unnecessary accumulation where it is not needed, which could hinder the transfer of Fe to edible parts such as seeds. To overcome this challenge, any successful biofortification approach must avoid harm to the plant while bypassing these mechanisms. Progress is being made through two primary strategies: traditional breeding and modern technology, including transgenic techniques. For centuries, traditional breeding methods have been utilized to create numerous beneficial crop varieties. However, breeders have predominantly focused on boosting crop yield, causing a decrease in Fe levels as a result of increased starch (Garg et al. 2018). Currently, crop research is aimed at reintroducing older traits into modern varieties to enhance their nutritional value. An example of this is the restoration of the NAM-B1 transcription factor in modern wheat varieties, which is absent in newer varieties but present in older ones and has been found to promote senescence resulting in increased Fe, Zn, and protein content in grains (Sultana et al. 2021). This discovery has influenced breeding programs as seen in Randhawa et al. (2013). Another approach involves exploiting natural variations to breed crops with higher Fe levels, such as in the case of pearl millet (Kodkany et al. 2013).

Iron biofortification can be achieved through transgenic and cis-genic technologies, which allow for the direct targeting of specific genes. Compared to traditional breeding methods, these approaches have the advantage of altering the expression of endogenous genes (cisgenics) or introducing genetic material from other species (transgenics) (Schaart et al. 2016). For instance, successful biofortification of rice has been accomplished by constitutively expressing endogenous *NAS* genes (Johnson et al. 2011), and Arabidopsis VIT1 has been expressed in storage roots to biofortify cassava (Narayanan et al. 2015). To biofortify wheat grain, cis-genic strategies targeting both ferritin and vacuoles have been utilized (Borg et al. 2012; Connorton et al. 2017), wherein increasing VIT expression in the wheat endosperm redirects Fe to this region of the grain. One concern regarding the use of transgenics in biofortification is the potential increase in toxic metals such as cadmium due to the low substrate specificity of the IRT1 transporter (Slamet-Loedin et al. 2015). However, the overexpression of barley Yellow Stripe 1 (*HvYS1*) has not resulted in this problem and has specifically led to increases in Fe (Banakar et al. 2016).

Current techniques for Fe biofortification have limitations in that only a limited number of crop varieties can be transformed, and introducing the developed traits into elite varieties can be time-consuming. However, modern genetic technologies like TILLING and CRISPR/Cas are not transgenic and may prove useful in developing Fe biofortified crops (Liu et al. 2021; Kumar et al. 2022). For instance, TILLING populations have been established in crops such as wheat, which allow researchers to disable the function of specific genes (Krasileva et al. 2017). This approach not only serves as a useful tool for studying gene function, but also opens up possibilities for inhibiting the function of negative regulators of Fe accumulation in specific tissues (Matres et al. 2021). Future biofortification strategies should consider all these factors to maximize bioavailability by combining high Fe in edible portions with low phytate and utilizing post-harvest processes that promote mineral bioavailability. While current research efforts tend to focus on one factor at a time, there is a growing interest in combinatorial studies.

Biofortifying cereals with Fe and other minerals is a challenging task due to the low bioavailability of these nutrients. The presence of anti-nutrient compounds such as phytate, lectins, saponins, oxalic acid and polyphenols in cereals inhibits mineral absorption in the gut (Hurrell and Egli 2010; Ram et al. 2020). Phytate, a phosphate storage compound found in the aleurone and seed coat of cereal grains, is particularly problematic (Ram et al. 2020). To reduce phytate levels in crops, researchers have explored various strategies such as breeding programs using relevant quantitative trait loci (QTL) (Singh et al. 2020) and expression of phytase genes (Wirth et al. 2009; Sheoran et al. 2022). Recently, researchers discovered that rice mutants lacking the SPDT phosphate transporter showed a significant decrease in grain phytate levels and a slight increase in Fe and other minerals (Yamaji et al. 2017). It is worth noting that post-harvest processing of plant material can also impact mineral bioavailability. For instance, Fe in white flour is generally more bioavailable than in whole meal flour (Eagling et al. 2014). Sourdough bread, produced with naturally occurring phytases, and micro milling of flour are some other methods that can enhance mineral bioavailability (Latunde-Dada et al. 2014). To optimize the bioavailability, upcoming biofortification approaches must take into account all of these variables by integrating high Fe levels in edible plant parts with low phytate content and applying post-harvest procedures that increase mineral bioavailability. Despite the present tendency for research to examine each variable independently, there is a growing emphasis on combinatorial research.

Agronomic biofortification can be used to increase the micronutrient content of cereal grains. This approach involves applying fertilizers to the soil, plant foliage or seed treatments (Zulfiqar et al. 2020). Soil-based agronomic biofortification is a cost-effective and straightforward method that provides a short-term solution. It can be particularly valuable when genetic biofortification is limited by the availability of micronutrients in the soil of the target region (Cakmak and Kutman 2017). Studies have demonstrated that the foliar application of Fe is an effective agronomic approach for enhancing the bioavailability and content of Fe in rice grains (Cakmak and Kutman 2017). Zulfiqar et al. (2020) showed that foliar application of $FeSO_4$ led to the highest increase in grain Fe concentration, followed by soil, seed priming, and seed coating, respectively. The corresponding increases in grain Fe concentration compared to the control were 61%, 37%, 15%, and 11%, respectively. Another study found that foliar application resulted in the highest grain Fe concentration, followed by osmopriming, surface broadcasting, and seed coating compared to no Fe application (Zulfiqar et al. 2021c). Foliar Fe application reduced grain phytate concentration and phytate to Fe molar ratio while improving bioavailable Fe (Zulfiqar et al. 2021c). Iron pulsing is another cost-effective and promising technique for improving grain yield and Fe concentration without causing environmental harm (Dey et al. 2021). In a study, rice seeds were treated with $FeSO_4$ and $FeCl_3$ for 72 h resulting in improved rice productivity and Fe status of the plants (Dey et al. 2019).

5.5 Conclusion

The bioavailability of micronutrients in plants has a direct impact on human health, making it essential for agriculture scientists to develop plant-based remedies to alleviate hidden hunger caused by micronutrient deficiencies. Researchers can identify potential gene targets for future alterations by conducting preliminary studies on the complex mechanisms involved in Fe transport. It is crucial to implement efficient enrichment practices to understand the molecular mechanisms of Fe accumulation in edible parts of plants without disrupting their vital life processes. Precision targeting using base editors, prime editors, and CRISPR/Cas can aid this objective. Moreover, there is a need for effective non-destructive sensors capable of analyzing nutritional stress. Agronomic biofortification can promising approach to address iron deficiency in soils. Hence, the biofortification of food through the utilization of technologies such as genetic engineering, omics, and nanotechnology holds great potential to alleviate Fe malnutrition.

Conflict of Interest The authors declare that there is no conflict of interest.

Acknowledgments The authors acknowledge the support from Princess Nourah bint Abdulrahman University Researchers Supporting Project number (PNURSP-HC2022/4), Princess Nourah bint Abdulrahman University, Riyadh, Saudi Arabia.

References

Alvarez-Fernandez A, Díaz-Benito P, Abadía A, López-Millán AF, Abadía J (2014) Metal species involved in long distance metal transport in plants. Front Plant Sci 5:105

Banakar R, Alvarez Fernandez A, Abadía J, Capell T, Christou P (2016) The expression of heterologous Fe (III) phytosiderophore transporter HvYS1 in rice increases Fe uptake, translocation and seed loading and excludes heavy metals by selective Fe transport. Plant Biotechnol J 15:423–432

Bashir K, Inoue H, Nagasaka S, Takahashi M, Nakanishi H, Mori S, Nishizawa NK (2006) Cloning and characterization of deoxymugineic acid synthase genes from graminaceous plants. J Biol Chem 281:32395–32402

Bashir K, Ishimaru Y, Shimo H, Kakei Y, Senoura T, Takahashi R, Sato Y, Sato Y, Uozumi N, Nakanishi H, Nishizawa NK (2011) Rice phenolics efflux transporter 2 (PEZ2) plays an important role in solubilizing apoplasmic iron. Soil Sci Plant Nutr 57:803–812

Bashir K, Nozoye T, Nagasaka S, Rasheed S, Miyauchi N, Seki M, Nakanishi H, Nishizawa NK (2017) Paralogs and mutants show that one DMA synthase functions in iron homeostasis in rice. J Exp Bot 68:1785–1795

Bonneau J, Baumann U, Beasley J, Li Y, Johnson AA (2016) Identification and molecular characterization of the nicotianamine synthase gene family in bread wheat. Plant Biotech J 14:2228–2239

Borg S, Brinch-Pedersen H, Tauris B, Madsen LH, Darbani B, Noeparvar S, Holm PB (2012) Wheat ferritins: improving the iron content of the wheat grain. J Cereal Sci 56:204–213

Borrill P, Connorton JM, Balk J, Miller AJ, Sanders D, Uauy C (2014) Biofortification of wheat grain with iron and zinc: integrating novel genomic resources and knowledge from model crops. Front Plant Sci 5:53

Cakmak I, Kutman UB (2017) Agronomic biofortification of cereals with zinc: a review. Eur J Soil Sci 69:172–180

Chu HH, Chiecko J, Punshon T, Lanzirotti A, Lahner B, Salt DE, Walker EL (2010) Successful reproduction requires the function of Arabidopsis YELLOW STRIPE-LIKE1 and YELLOW STRIPE-LIKE3 metal-nicotianamine transporters in both vegetative and reproductive structures. Plant Physiol 154:197–210

Clemens S, Weber M (2016) The essential role of coumarin secretion for Fe acquisition from alkaline soil. Plant Signal Behav 11:1114197

Connorton JM, Balk J, Rodríguez-Celma J (2017) Iron homeostasis in plants–a brief overview. Metallomics 9:813–823

Conte S, Walker EL (2011) Transporter contributing to iron trafficking in plants. Mol Plant 4:464–476

Curie C, Cassin G, Couch D, Divol F, Higuchi K, Le Jean M, Mission J, Schikora A, Czernic P, Mari S (2009) Metal movement within the plant: contribution of nicotianamine and yellow stripe 1-like transporters. Ann Bot 103:1–11

Curie C, Panavience Z, Loulergue C, Dellaporta SL, Briat JF, Walker EL (2001) Maize yellow stripe1 encodes a membrane protein directly involved in Fe(III) uptake. Nature 409:346–349. https://doi.org/10.1038/35053080

Dey S, Kundu R, Gopal G, Mukherjee A, Nag A, Paul S (2019) Enhancement of nitrogen assimilation and photosynthetic efficiency by novel iron pulsing technique in *Oryza sativa* L. var Pankaj. Plant Physiol Biochem 144:207–221

Dey S, Paul S, Nag A, Banerjee R, Gopal G, Mukherjee A, Kundu R (2021) Iron-pulsing, a novel seed invigoration technique to enhance crop yield in rice: a journey from lab to field aiming towards sustainable agriculture. Sci Total Environ 769:144671

Diaz-Benito P, Banakar R, Rodríguez-Menéndez S, Capell T, Pereiro R, Christou P, Abadía J, Fernández B, Alvarez-Fernández A (2018) Iron and zinc in the embryo and endosperm of rice (*Oryza sativa* L.) seeds in contrasting 2′-deoxymugineic acid/nicotianamine scenarios. Front. Plant Sci 9:1190

DiDonato RJ Jr, Roberts LA, Sanderson T, Eisley RB, Walker EL (2004) Arabidopsis yellow stripe-Like2 (YSL2): a metal-regulated gene encoding a plasma membrane transporter of nicotianamine–metal complexes. Plant J 39:403–414

Durrett TP, GassmannW REE (2007) The FRD3-mediated efflux of citrate into the root vasculature is necessary for efficient iron translocation. Plant Physiol 144:197–205

Eagling T, Wawer AA, Shewry PR, Zhao FJ, Fairweather-Tait SJ (2014) Iron bioavailability in two commercial cultivars of wheat: comparison between wholegrain and white flour and the effects of nicotianamine and 20-deoxymugineic acid on iron uptake into Caco-2 cells. J Agric Food Chem 62:10320–10325

Garg M, Sharma N, Sharma S, Kapoor P, Kumar A, Chunduri V, Arora P (2018) Biofortified crops generated by breeding, agronomy, and transgenic approaches are improving lives of millions of people around the world. Front Nutr 5:12

Green LS, Rogers EE (2004) FRD3 controls iron localization in Arabidopsis. Plant Physiol 136: 2523–2531

Grillet L, Mari S, Schmidt W (2014) Iron in seeds- loading pathways and subcellular localization. Front Plant Sci 4:535

Grotz N, Guerinot ML (2006) Molecular aspects of cu, Fe and Zn homeostasis in plants. Biochim Biophys Acta 1763:595–608

Hurrell R, Egli I (2010) Iron bioavailability and dietary reference values. Am J Clin Nutr 91:1461–1467

Inoue H, Kobayashi T, Nozoye T, Takahashi M, Kakei Y, Suzuki K, Nakazono M, Nakanishi H, Mori S, Nishizawa NK (2009) Rice *OsYSL15* is an iron-regulated iron (III)-deoxymugineic acid transporter expressed in the roots and is essential for iron uptake in early growth of the seedlings. J Biol Chem 284:3470–3479

Ishimaru Y, Kakei Y, Shimo H, Bashir K, Sato Y, Sato Y, Uozumi N, Nakanishi H, Nishizawa NK (2011) A rice phenolic efflux transporter is essential for solubilizing precipitated apoplasmic iron in the plant stele. J Biol Chem 286:24649–24655

Ishimaru Y, Masuda H, Bashir K, Inoue H, Tsukamoto T, Takahashi M, Nakanishi H, Aoki N, Hirose T, Ohsugi R, Nishizawa NK (2010) Rice metal-nicotianamine transporter, *OsYSL2*, is required for the long-distance transport of iron and manganese. Plant J 62:379–390

Ishimaru Y, Suzuki M, Tsukamoto T, Suzuki K, Nakazono M, Kobayashi T, Wada Y, Watanabe S, Matsuhashi S, Takahashi M, Nakanishi H (2006) Rice plants take up iron as an Fe^{3+}-phytosiderophore and as Fe^{2+}. Plant J 45:335–346

Ivanov R, Brumbarova T, Bauer P (2012) Fitting into the harsh reality: regulation of iron-deficiency responses in dicotyledonous plants. Mol Plant 5:27–42

Jeong J, Cohu C, Kerkeb L, Pilon M, Connolly EL, Guerinot ML (2008) Chloroplast Fe(III) chelate reductase activity is essential for seedling viability under iron limiting conditions. Proc Natl Acad Sci 105:10619–10624

Johnson AA, Kyriacou B, Callahan DL, Carruthers L, Stangoulis J, Lombi E, Tester M (2011) Constitutive overexpression of *OsNAS* gene family reveals single-gene strategies for effective iron- and zinc- biofortification of rice endosperm. PLoS One 6:24476

Kakei Y, Yamaguchi I, Kobyashi T, Takahashi M, Nakanishi H, Yamakawa T, Nishizawa NK (2009) A highly sensitive, quick and simple quantification method for nicotianamine and 2′-deoxymugienic acid from minimum samples using LC/ESITOF-MS achieves functional analysis of these components in plants. Plant Cell Physiol 50:1988–1993

Kim SA, Guerinot ML (2007) Mining iron: iron uptake and transport in plants. FEBS Lett 581: 2273–2280

Kobayashi T, Nishizawa NK (2012) Iron uptake, translocation, and regulation in higher plants. Ann Rev Plant Biol 63:31–152

Kobayashi T, Itai RN, Ogo Y, Kakei Y, Nakanishi H, Takahashi M, Nishizawa NK (2009) The rice transcription factor IDEF1 is essential for the early response to iron deficiency and induces vegetative expression of late embryogenesis abundant genes. Plant J 60:948–961

Kobayashi T, Nagasaka S, Senoura T, Itai RN, Nakanishi H, Nishizawa NK (2013) Iron-binding haemerythrin RING ubiquitin ligases regulate plant iron responses and accumulation. Nat Commun 4:2792

Kobayashi T, Nozoye T, Nishizawa NK (2018) Iron transport and it is regulation in plants. Free Rad Bio Med 10:439

Kobayashi T, Nakayama Y, Itai RN, Nakanishi H, Yoshihara T, Mori S, Nishizawa NK (2003) Identification of novel cis-acting elements, IDE1 and IDE2, of the barley IDS2 gene promoter conferring iron-deficiency-inducible, rootspecific expression in heterogeneous tobacco plants. Plant J 36:780–793

Kobayashi T, Ogo Y, Itai RN, Nakanishi H, Takahashi M, Mori S, Nishizawa NK (2007) The novel transcription factor IDEF1 regulates the response to and tolerance of iron deficiency in plants. Proc Natl Acad Sci U S A 104:19150–19155

Kobayashi T, Ogo Y, Aung MS, Nozoye T, Itai RN, Nakanishi H, Yamakawa T, Nishizawa NK (2010) The spatial expression and regulation of transcription factors IDEF1 and IDEF2. Ann Bot 105:1109–1117

Kobayashi T, Suzuki M, Inoue H, Itai RN, Takahashi M, Nakanishi H, Mori S, Nishizawa NK (2005) Expression of iron-acquisition-related genes in iron-deficient rice is coordinately induced by partially conserved iron-deficiency-responsive elements. J Exp Bot 56:1305–1316

Kodkany BS, Bellad RM, Mahantshetti NS, Westcott JE, Krebs NF, Kemp JF, Hambidge KM (2013) Biofortification of pearl millet with iron and zinc in a randomized controlled trial increases absorption of these minerals above physiologic requirements in young children. J Nutr 143:1489–1493

Koike S, Inoue H, Mizuno D, Takahashi M, Nakanishi H, Mori S, Nishizawa NK (2004) *OsYSL2* is a rice metal-nicotianamine transporter that is regulated by iron and expressed in phloem. Plant J 39:415–424

Krasileva KV, Vasquez-Gross HA, Howell T, Bailey P, Paraiso F, Clissold L, Simmonds J, Ramirez-Gonzalez RH, Wang X, Borrill P, Fosker C (2017) Uncovering hidden variation in polyploid wheat. Proc Natl Acad Sci U S A 114:913–921

Kroh GE, Pilon M (2020) Regulation of iron homeostasis and use in chloroplasts. Int J Mol Sci 21: 3395

Kruger C, Berkowitz O, StephanUdo W, Hell R (2002) A metal binding member of the late embryogenesis abundant protein family transports iron in the phloem of *Ricinus communis* L. J Biol Chem 277:25062–25069

Kumar D, Yadav A, Ahmad R, Dwivedi UN, Yadav K (2022) CRISPR-based genome editing for nutrient enrichment in crops: a promising approach toward global food security. Front Genet 12: 1650

Latunde-Dada GO, Li X, Parodi A, Edwards CH, Ellis PR, Sharp PA (2014) Micromilling enhances iron bioaccessibility from wholegrain wheat. J Agric Food Chem 62:11222–11227

Lee S, Chiecko JC, Kim SA, Walker EL, Lee Y, Guerinot ML, An G (2009a) Disruption of *OsYSL15* leads to iron inefficiency in rice plants. Plant Physiol 150:786–800

Lee S, Jeon US, Lee SJ, Kim YK, Persson DP, Husted S, Schjørring JK, Kakei Y, Masuda H, Nishizawa NK, An G (2009b) Iron fortification of rice seeds through activation of the nicotianamine synthase gene. Proc Natl Acad Sci 106:22014–22019

Lee S, Kim SA, Lee J, Guerinot ML, An G (2010) Zinc deficiency inducible *OsZIP8* encodes a plasma membrane localized zinc transporter in rice. Mol Cells 29:551–558

Lee S, Kim YS, Jeon US, Kim YK, Schjoerring JK, An G (2012) Activation of rice nicotianamine synthase 2 (*OsNAS2*) enhances iron availability for biofortification. Mol Cells 33:269–275

Li G, Kronzucker HJ, Shi W (2016) The response of the root apex in plant adaptation to iron heterogeneity in soil. Front Plant Sci 7:344

Li X, Duan X, Tan D, Zhang B, Xu A, Qiu N, Chen Z (2023) Iron deficiency and iron overload in men and woman of reproductive age, and pregnant women. Rep Toxicol 118:108381

Ling HQ, Bauer P, Bereczky Z, Keller B, Ganal M (2002) The tomato fer gene encoding a bHLH protein controls iron-uptake responses in roots. Proc Natl Acad Sci 99:13938–13943

Liu Q, Yang F, Zhang J, Liu H, Rahman S, Islam S, Ma W, She M (2021) Application of CRISPR/ Cas9 in crop quality improvement. Int J Mol Sci 22:4206

Long TA, Tsukagoshi H, Busch W, Lahner B, Salt DE, Benfey PN (2010) The bHLH transcription factor POPEYE regulates response to iron deficiency in Arabidopsis roots. Plant Cell 22:2219–2236

Ma JF, Ling HQ (2009) Iron for plants and humans. Plant Soil 325:1–3

Matres JM, Hilscher J, Datta A, Armario-Nájera V, Baysal C, He W, Huang X, Zhu C, Valizadeh-Kamran R, Trijatmiko KR, Capell T (2021) Genome editing in cereal crops: an overview. Trans Res 30:461–498

Mora-Macías J, Ojeda-Rivera JO, Gutiérrez-Alanís D, Yong-Villalobos L, Oropeza-Aburto A, Raya-González J, Jiménez-Domínguez G, Chávez-Calvillo G, Rellán-Álvarez R, Herrera-Estrella L (2017) Malate-dependent Fe accumulation is a critical checkpoint in the root developmental response to low phosphate. Proc Natl Acad Sci 114:3563–3572

Morgan JÁ, Connolly EÁ (2013) Plant-soil interactions: nutrient uptake. Nat Edu Know 4:2

Morrissey J, Guerinot ML (2009) Iron uptake and transport in plants: the good, the bad, and the ionome. Chem Rev 109:4553–4567

Narayanan N, Beyene G, Chauhan RD, Gaitán-Solis E, Grusak MA, Taylor N, Anderson P (2015) Overexpression of Arabidopsis VIT1 increases accumulation of iron in cassava roots and stems. Plant Sci 240:170–181

Nozoye T, Nagasaka S, Kobayashi T, Takahashi M, Sato Y, Sato Y, Uozumi N, Nakanishi H, Nishizawa NK (2011) Phytosiderophore efflux transporters are crucial for iron acquisition in Graminaceous plants. J Biol Chem 286:5446–5454

Ogo Y, Kakei Y, Itai RN, Kobayashi T, Nakanishi H, Takahashi H, Nakazono M, Nishizawa NK (2014) Spatial transcriptomes of iron-deficient and cadmium-stressed rice. New Phytol 201: 781–794

Ogo Y, Nakanishi Itai R, Nakanishi H, Kobayashi T, Takahashi M, Mori S, Nishizawa NK (2007) The rice bHLH protein *OsIRO2* is an essential regulator of the genes involved in Fe uptake under Fe-deficient conditions. Plant J 51:366–377

Palmer CM, Hindt MN, Schmidt H, Clemens S, Guerinot ML (2013) MYB10 and MYB72 are required for growth under iron-limiting conditions. PLoS Genet 9:1003953

Ram S, Narwal S, Gupta OP, Pandey V, Singh GP (2020) Anti-nutritional factors and bioavailability: approaches, challenges, and opportunities. In: Gupta OP, Pandey V, Narwal S, Sharma P, Ram S, Singh GP (eds) Wheat and barley grain biofortification. Elsevier Inc, Amsterdam, Netherlands, pp 101–128

Randhawa HS, Asif M, Pozniak C, Clarke JM, Graf RJ, Fox SL, Humphreys DG, Knox RE, DePauw RM, Singh AK, Cuthbert RD (2013) Application of molecular markers to wheat breeding in Canada. Plant Breed 132:458–471

Roberts LA, Pierson AJ, Panaviene Z, Walker EL (2004) Yellow stripe1. Expanded roles for the maize iron-phytosiderophore transporter. Plant Physiol 35:112–120

Santi S, Schmidt W (2008) Laser microdissection-assisted analysis of the functional fate of iron deficiency-induced root hairs in cucumber. J Exp Bot 59:697–704

Santi S, Schmidt W (2009) Dissecting iron deficiency-induced proton extrusion in Arabidopsis roots. New Phytol 183:1072–1084

Schaart JG, van de Wiel CC, Lotz LA, Smulders MJ (2016) Opportunities for products of new plant breeding techniques. Trends Plant Sci 21:438–449

Schmidt H, Günther C, Weber M, Spörlein C, Loscher S, Böttcher C, Schobert R, Clemens S (2014) Metabolome analysis of *Arabidopsis thaliana* roots identifies a key metabolic pathway for iron acquisition. PLoS One 9:102444

Selote D, Samira R, Matthiadis A, Gillikin JW, Long TA (2015) Iron-binding E3 ligase mediates iron response in plants by targeting basic helix-loop-helix transcription factors. Plant Physiol 167:273–286

Senoura T, Sakashita E, Kobayashi T, Takahashi M, Aung MS, Masuda H, Nakanishi H, Nishizawa NK (2017) The iron-chelate transporter *OsYSL9* plays a role in iron distribution in developing rice grains. Plant Mol Biol 95:375–387

Sheoran S, Kumar S, Ramtekey V, Kar P, Meena RS, Jangir CK (2022) Current status and potential of biofortification to enhance crop nutritional quality: an overview. Sustain 14:3301

Shin LJ, Lo JC, Chen GH, Callis J, Fu H, Yeh KC (2013) IRT1 degradation factor1, a ring E3 ubiquitin ligase, regulates the degradation of iron-regulated transporter1 in Arabidopsis. Plant Cell 25:3039–3051

Slamet-Loedin IH, Johnson-Beebout SE, Impa S, Tsakirpaloglou N (2015) Enriching rice with Zn and Fe while minimizing Cd risk. Front Plant Sci 6:121

Singh RK, Prasad A, Muthamilarasan M, Parida SK, Prasad M (2020) Breeding and biotechnological interventions for trait improvement: status and prospects. Planta 252:1–18

Sivitz AB, Hermand V, Curie C, Vert G (2012) Arabidopsis bHLH100 and bHLH101 control iron homeostasis via a FIT-independent pathway. PLoS One 7:44843

Sperotto RA, Ricachenevsky FK, Waldow VA, Fett JP (2012a) Iron biofortification in rice: It's a long way to the top. Plant Sci 190:24–39

Sperotto RA, Vasconcelos MW, Grusak MA, Fett JP (2012b) Effects of different Fe supplies on mineral partitioning and remobilization during the reproductive development of rice (*Oryza sativa* L.). Rice 5:27

Sultana N, Islam S, Juhasz A, Ma W (2021) Wheat leaf senescence and its regulatory gene network. Crop J 9:703–717

Suzuki M, Takahashi M, Tsukamoto T, Watanabe S, Matsuhashi S, Yazaki J, Kishimoto N, Kikuchi S, Nakanishi H, Mori S, Nishizawa NK (2006) Biosynthesis and secretion of mugineic acid family phytosiderophores in zinc-deficient barely. Plant J 48:85–97

Takahashi M, Terada Y, Nakai I, Nakanishi H, Yoshimura E, Mori S, Nishizawa NK (2003) Role of nicotianamine in the intracellular delivery of metals and plant reproductive development. Plant Cell 15:1263–1280

Tauris B, Borg S, Gregersen PL, Holm PB (2009) A roadmap for zinc trafficking in the developing barley grain based on laser capture microdissection and gene expression profiling. J Exp Bot 60: 1333–1347

Ueno D, Yamaji N, Ma JF (2009) Further characterization of ferric phytosiderophore transporters *ZmYS1* and *HvYS1* in maize and barley. J Exp Bot 60:3513–3520

Wang N, Cui Y, Liu Y, Fan H, Du J, Huang Z, Yuan Y, Wu H, Ling HQ (2013) Requirement and functional redundancy of Ib subgroup bHLH proteins for iron deficiency responses and uptake in *Arabidopsis thaliana*. Mol Plant 6:503–513

Wang YX, Specht A, Horst WJ (2011) Stable isotope labeling and zinc distribution in grains studied by laser ablation ICP-MS in an ear culture system reveals zinc transport barriers during grain filling in wheat. New Phytol 189:428–437

Waters BM, Sankaran RP (2011) Moving micronutrients from the soil to the seeds: genes and physiological processes from a biofortification perspective. Plant Sci 180:562–574

Wirth J, Poletti S, Aeschlimann B, Yakandawala N, Drosse B, Osorio S, Tohge T, Fernie AR, Günther D, Gruissem W, Sautter C (2009) Rice endosperm iron biofortification by targeted and synergistic action of nicotianamine synthase and ferritin. Plant Biotechnol J 7:631–644

Yamaji N, Ma JF (2009) A transporter at the node responsible for the intervascular transfer of silicon in rice. Plant Cell 21:2878–2883

Yamaji N, Takemoto Y, Miyaji T, Mitani-Ueno N, Yoshida KT, Ma JF (2017) Reducing phosphorus accumulation in rice grains with an impaired transporter in the node. Nature 541:92–95

Yan JY, Li CX, Sun L, Ren JY, Li GX, Ding J, Zheng SJ (2016) A WRKY transcription factor regulates Fe translocation under Fe deficiency. Plant Physiol 171:2017–2027

Yokosho K, Yamaji N, Ueno D, Mitani N, Ma JF (2009) *OsFRDL1* is a citrate transporter required for efficient translocation of iron in rice. Plant Physiol 149:297–305

Yoshino M, Murakami K (1998) Interaction of iron with polyphenolic compounds: application to antioxidant characterization. Anal Biochem 257:40–44

Yuan Y, Wu H, Wang N, Li J, Zhao W, Du J, Wang D, Ling HQ (2008) FIT interacts with *AtbHLH38* and *AtbHLH39* in regulating iron uptake gene expression for iron homeostasis in Arabidopsis. Cell Res 18:385–397

Zamioudis C, Hanson J, Pieterse CM (2014) β-Glucosidase BGLU 42 is a MYB 72-dependent key regulator of rhizobacteria-induced systemic resistance and modulates iron deficiency responses in Arabidopsis roots. New Phytol 204:368–379

Zhang J, Liu B, Li M, Feng D, Jin H, Wang P, Liu J, Xiong F, Wang J, Wang HB (2015) The bHLH transcription factor bHLH104 interacts with IAA-LEUCINE RESISTANT3 and modulates iron homeostasis in Arabidopsis. Plant Cell 27:787–805

Zheng L, Yamaji N, Yokosho K, Ma JF (2012) YSL16 is a phloem-localized transporter of the copper-nicotianamine complex that is responsible for copper distribution in rice. Plant Cell 24: 3767–3782

Zheng L, Ying Y, Wang L, Wang F, Whelan J, Shou H (2010) Identification of a novel iron regulated basic helix-loop-helix protein involved in Fe homeostasis in *Oryza sativa* L. BMC Plant Biol 10:1–9

Zulfiqar U, Hussain S, Ishfaq M, Ali N, Ahmad M, Ihsan F, Sheteiwy MS, Rauf A, Hano C, El-Esawi MA (2021a) Manganese supply improves bread wheat productivity, economic returns and grain biofortification under conventional and no tillage systems. Agriculture 11:142

Zulfiqar U, Hussain S, Ishfaq M, Ali N, Yasin MU, Ali MA (2021b) Foliar manganese supply enhances crop productivity, net benefits, and grain manganese accumulation in direct-seeded and puddled transplanted rice. J Plant Growth Regul 40:1539–1556

Zulfiqar U, Hussain S, Maqsood M, Zamir SI, Ishfaq M, Ali N, Ahmad M, Maqsood MF (2021c) Enhancing the accumulation and bioavailability of iron in rice grains via agronomic interventions. Crop Pasture Sci 73:32–43

Zulfiqar U, Maqsood M, Hussain S (2020) Biofortification of Rice with iron and zinc: Progress and prospects. In: Roychoudhury A (ed) Rice research for quality improvement: genomics and genetic engineering. Springer, Singapore, pp 605–627

Chapter 6
Selenium Biofortification in Agronomic Crops

Umair Rasheed, Abdul Sattar, Ahmad Sher, Muhammad Ijaz, Sami Ul-Allah, Jawad Ashraf, Adnan Noor Shah, and Muhammad Nawaz

Abstract Selenium (Se) is the most significant micronutrient, which is necessary for living organisms including plants, animals as well as in humans and is present in organic and inorganic forms in biomass, plants, water, and atmospheric air. The Se is beneficial to plants and mostly involved in antioxidant activity and acts as growth promoter. Amount of Se present in soils typically show its occurrence in food stuff and, as a result indicates its availability to humans. Se deficiency is a worldwide issue. Selenium plays a role in the mechanisms of response of antioxidant in organisms, detoxifies heavy metals, controls the immune as well as reproductive system, and ensures the thyroid gland's appropriate function. In humans, plants are the primary source of Se in their diets. Biofortification is an important method for increasing Se in edible plant parts. Agronomic biofortification makes it possible to increase the amount of Se in edible crop products through the application of Se-enriched fertilizers to the soil or to the leaves, Wheat, barley, maize, and rice are the cereals that humans consume the most. So, making them ideal targets for agronomic biofortification is a useful strategy. The goal of this review is to summarize the most effective form and method of applying Se through the agronomic biofortification method, supported by a huge analysis of the reports and the literatures. In the literature, foliar application performed better than soil application. In the most common cereals such as barley, maize, rice, and wheat consumed by humans, selenate appears to be the more effective form of Se for biofortification than selenite.

U. Rasheed · J. Ashraf
College of Agriculture, University of Layyah, Layyah, Pakistan

A. Sattar (✉) · A. Sher · M. Ijaz
College of Agriculture, University of Layyah, Layyah, Pakistan

Department of Agronomy, Bahauddin Zakariya University, Multan, Punjab, Pakistan

S. Ul-Allah
College of Agriculture, University of Layyah, Layyah, Pakistan

Department of Plant Breeding and Genetics, Bahauddin Zakariya University, Multan, Pakistan

A. N. Shah · M. Nawaz
Department of Agricultural Engineering, Khwaja Fareed University of Engineering and Information Technology, Rahim Yar Khan, Punjab, Pakistan

M. Hasanuzzaman et al. (eds.), *Mineral Biofortification in Crop Plants for Ensuring Food Security*, https://doi.org/10.1007/978-981-99-4090-5_6

6.1 Introduction

Selenium (Se) is an essential micronutrient for humans, plants and as well as for animals. It is required in traces for normal functioning (Germ et al. 2007). There are 25 seleno-proteins are known to regulate the biological processes in human body which are associated with Se (El-Ramady et al. 2020). The Se deficiency in the dietary product leads to a negative impact on human health which ultimately cause more than 40 different diseases in humans, such as Kashin-Beck disease, liver problems, clouding of the eye lens, and various cardiovascular and cancer disease. Intake of Se-deficient foodstuff also causes metabolic stress and reduces immune functioning (Broadley et al. 2006). Low Se in soils causes the reduction of density in plants and eventually reduces the Se level in human body (Galic et al. 2021). Unfortunately, almost ten billion people are suffering from a deficiency of selenium worldwide (Lyons et al. 2005). Although it is required in traces, the higher amount of Se may also lead to toxicity due to the abnormal folding of protein by the replacement of Se in place of the sulfur (Germ et al. 2007). The Se is present in organic and inorganic forms in nature in plants, water, and atmospheric air, (Galic et al. 2021). There are various factors that affect the availability of the Se in the soil including the Clay particles, pH, type of soil and organic matter. Dominancy of the SeO_4^{2-} over the SeO_3^{2-} increases the availability of Se in the soil because pH availability depends on the negative charge (Fernandes et al. 2014). In soils, Se is present mostly in organic form which is directly unavailable to plants (Chilimba et al. 2012). Naturally, the uptake of an inorganic form of se is elemental Se, selenate, selenide, and selenite (Hossain et al. 2021). The importance of selenium in vascular plants is still in consideration (El-Ramady et al. 2020). In recent research, it has been seen that Se can maintain the water tension in plants. Hence nowadays, it is used to reduce the drought stress in plants. A significant amount of Se is present at the surface of earth but only traces or small amount of Se is available to plants. Excess Se causes toxicity in plants resulting in harmful effects. A low quantity/amount of Se has more beneficial effects as it prevents the plant from oxidative stress by enhancing the ability to tolerate the UV (Ultraviolet). It delays the senescence in plants and enhances the growth and development of seedlings. (Germ et al. 2007).

Some agronomic strategies can be helpful for reducing Se deficiency. Se biofortification can be achieved by using agronomic microbial integration and nano selenium particles for maintaining se level in severe conditions. Biofortified food is enriched with high nutrition, healthy amount of Se and contains some valuable phytochemicals. (Schiavon et al. 2020) Agronomic biofortification has proved the most effective strategy to produce the Se-rich edible food products through the foliar and fertilizers application at plants and in soil. Wheat, maize, rice, and barley are the main staple foods which are used in different part of the world. These crops are the main objective for agronomic biofortification (Galic et al. 2021); therefore, the cultivation of certain agronomic crops along with the optimized application of Se can produce Se-rich food products which ultimately provide benefits to human health (Germ et al. 2007). Genetic biofortification of crops can also yield the Se enrich foodstuff by selecting and breeding crops which can accumulate Se. Genetic biofortification is considered as the long-term effective method for producing Se-enriched crops (Broadley et al. 2006). This chapter mainly focuses on the crucial know-how of selenium and the enhancement of selenium content in foodstuff through the most efficient method of application and suitable forms in agronomic crops, and it is seen that the Se biofortification can reduce the many diseases and disorders in human beings, plants, and animals. This chapter also discloses the importance of Se in staple food products.

6.2 Selenium Biofortification

Biofortification is the development of staple crops full of vital micronutrients to get more nutritious food. It is done with the help of agronomic practices, biotechnology, and plant breeding strategies (Combs Jr 2001). Studies have shown that selenium boosts the growth of plants and could serve as a heavy metal antagonist because it is a vital micronutrient that has some physical and anti-oxidative characteristics (Shafiq et al. 2019). Most of the staple foods lack a significant number of micronutrients causing malnutrition in many parts of the world. About 50% of the deaths occur due to malnutrition. The vital micronutrients include Se, iron (Fe), iodine (I), zinc (Zn) help to overcome malnutrition. The anti-oxidative and physicochemical properties that Se has sparked the interest of biologists over the last few years. Almost 60% population of all over the world is deficient from one of these micronutrients (Lyons 2018). One-Half of the global world, exclusively children are suffering due to malnutrition of these micronutrient (Nestel et al. 2006). In late 1950s, Se was considered as an essential nutrient as it was found to play a key role in liver functioning (Combs Jr 2001). In 1957 the significant research work on Se was started when importance of selenium revealed by Schwartz and Foltz (1957) in rats. But the importance of selenium was verified after work of certain years. Rotruck reveals the function of selenium in the plasma glutathione peroxidase p(GSH-Px) in 1973. The structure of GSH-Px was described by Ladenstein. Selenium is a crucial trace element in both animal and human health which is covalently integrated in amino acids. The most notably the amino acids selenocysteine (SeCys) and the amino acid selenomethionine (SeMet) which serves as a cofactor in antioxidant enzymes like glutathione peroxidase (Zhou et al. 2020). In-depth understanding of Se biological chemistry within humans, its clinical integration and an epidemiological level is crucial to unravel the link to Se condition and effectiveness that biological systems perform. A high amount of selenium accumulation in plant tissues can cause toxic effects of the element which is primarily attributed to the non-specific integration of selenocysteine and selenomethionin protein, instead of methionine and cysteine, respectively. It was observed that the bond between Se atom of selenocysteine protein is more extensive, weaker, and more prone to reactivity than disulfide binding and could exert a significant influence on small changes to the third-order structure of proteins and trigger disruptions to enzyme catalytic activities (Hawrylak-Nowak 2008). The amount of Se in a food can vary and is dependent on the geographical origin of the agricultural product and the soil in which the crops are being grown (Yang et al. 2003) The majority of studies focus on Se bioavailability in specific foods or isolated Se compounds. But, due to the numerous interactions that could occur, it is essential to investigate the impact of Se on the overall diet. Thus, the current study was designed to investigate the impact of different levels of Se consumption through natural sources part of the daily diet that consists of food items that are consumed regularly as well as on different Se health indicators as well as Se balance (Van-der-Torre et al. 1991). Biofortification of staple crop varieties to increase the micronutrient content in the edible components is an eating system approach to tackle dietary deficiencies which has the potential to help those most in need (Lyons 2018). In order to overcome this hurdle, numerous treatments to treat Se-deficiency illnesses were discovered when the importance of Se′s role in nutrition was proven. Selenium has been used successfully as a treatment to prevent muscular degeneration due to nutrition since the early 1960's.

6.3 Importance of Selenium in Human

Selenium was first introduced as a toxic element, but with the passing of time, it has been proved to be a beneficial element for humans. Researchers believe that ideal Se status in foodstuffs will be more beneficial however, excessive intake of selenium could be harmful for human health. (Rayman 2012). The intake of selenium by the crops and fodder depends upon the Se contents of the soil, which later becomes available to/in the food chain. There are almost 25 selenoproteins consisting of selenocysteine that play a key role in the body by providing Se which is also nutritionally important for normal functioning in the body. The formation of certain types of selenoproteins is enhanced due to lower selenium supply than the other proteins such as GPx4 and glutathione peroxidase. Many selenoproteins act as enzymes in the body; their importance cannot be neglected as mutation in only one nucleotide in selenoproteins leads to high risk of disease and mortality (Rayman 2012). According to a survey, around 35 selenoproteins and their function have been identified by researchers. (Rayman 2000).

6.3.1 Health Effects of Selenium

6.3.1.1 Viral Infection

Virulence activity, viral infection and disease advancement are enhanced by the deficiency of selenium (Rayman 2012). Selenium can be used to treat hepatitis B and C while it also helps to protect the patient from further damage that could lead to liver cancer (Rayman 2000). The advancement of HIV is associated with the progress of CD4+ helper T-cell. HIV and CD4+ helper T-cells are directly proportional to each other. In HIV, reduction in plasma Se is parallel to the loss of CD4+ T-cells which ultimately reduces the survival chances of the patient (Rayman 2000). In US adults, a test on HIV infection was conducted, which disclosed that serum with high selenium content could decrease the viral infection (Rayman 2012).

6.3.1.2 Fertility and Reproduction

Selenium plays key functioning in reproduction systems especially in female reproductive system. Most of the important redox responses occurring in male reproduction are done by glutathione peroxidase (a type of selenoprotein). The major functions of these isoform enzymes protect the cells from oxidative stress through the reduction of lipids and hydroperoxides (Mojadadi et al. 2021). Scott and his co-workers found that the motility of the sperm increases if a man takes selenium 100 mcg each day for ninety days. It has seen more than 11% of men who were consuming active selenium supplements become capable of attaining paternity. However, extra intake of selenium does not significant effect on motility but polishes the men after a specific period (Rayman 2000). The level of selenoproteins decreases due to a deficiency of selenium, which may harm fetal growth and placental functions and ultimately cause miscarriage (Mojadadi et al. 2021). Low levels of selenium in women increase inflammation during birth (Rayman 2012).

6.3.1.3 Cardiovascular Disorder

There is much evidence of the beneficial effects of selenium on the circulatory system, such as the selenoproteins protecting the body from the accumulation of platelets and reducing the oxidative changes of fatty acids, thus decreasing the inflammatory effects (Rayman 2012). It has been seen that GPx4 decreases hydroperoxide and cholesterol, which ultimately reduces the aggregation of lipoprotein in the walls of arteries. With the deficiency of Se, the amount of hydroperoxides increases which retards the function of an enzyme known as prostacyclin synthetase. The function of this enzyme includes the synthesis of prostacyclin with the help of endothelium and controls vasoconstriction by enhancing the formation of thromboxane. Research work by Suadicani revealed that men having a selenium serum level of less than 79 mcg L^{-1} were likely to have a great risk of cardiac disease (Rayman 2000). Many published studies do not clearly show the link between heart disease and low selenium status with increased Se intake, cholesterol proportion to HDL cholesterol is reduced and proves helpful against cardiovascular disease risk. (Rayman 2012).

6.3.1.4 Cancer

There are controversies about the anti-cancer activity of selenium. The recent study on selenium describes that Se has various positive effects against different types of cancer including thyroid cancer, gastric-cardia cancer, lungs cancer, esophagus cancer, liver cancer and bladder cancer. Since 1970, a number of studies demonstrate that the amount of Se intake is inversely proportional to the cancerous disease. Schrauzer research study showed that 27 different countries having an inverse relationship of Se to the cancer mortality (Rayman 2000). While some of the studies showed that there is no relation of lungs and prostate cancer risk with the selenium concentration status, excluding the natives from America and Africa. According to the recent studies, Se status has a substantial defensive approach toward the advance risk of prostate cancer rather than low level risk (Rayman 2012). The potential study of 80's and early 90's show that cancer risk increases with the decrease in Se level. A man with low Se status has twice to six times greater risk to have cancer. And it has also seen in later studies that high Se concentration has positive effects (Rayman 2000).

6.3.1.5 Immune Function

Selenium is a substantial part of the major immune tissues including liver, lymph nodes and spleen (Rayman 2000). Selenoproteins are necessary for the function of activated T cells. These T-cells are significantly having quick response toward the oxidative stress. Due to the deficiency of selenoprotein in T cells, they cannot bloom to the T-cells receptor stimulants as formation of active oxygen reduced or stopped (Rayman 2012). Intake of selenium supplements improve the activity of natural killer cells (white cells) (Rayman 2000). Intake of Se supplements by elders of Belgian 100 μg per day for 6 months in the form of Se yeast increase the response to antigen. However, supplementation of Se 400 μg day^{-1} by residents of Arizona increases the T-cells production up to 27% (Rayman 2012). The activity of selenophosphate synthetase enhanced to form selenocysteine by the activated T-cells. These selenocysteines are vital structural blocks of selenoproteins. Selenoproteins are important in activated T-cell which maintain the response of immune system (Rayman 2000).

6.4 Selenium in Soils

The amount of Se in plants can be determined by knowing the phyto-availability of selenium in the solution of soil, and is controlled by the climate, geology and soil chemical composition. Soil Se is not uniform in its distribution and accessibility levels vary from less than 100 mg kg^{-1}. The variances in the phyto-availability interpret the major deviations of Se amount in various edible plants (Kumssa et al. 2017). The plant's ability to absorb is affected by the physicochemical characteristics of soil. The amount of Se present in soil differs based on texture, type, and organic matter contents in the soil, as well as the amount of rainfall. However, the majority of the soils have Se between 1.0 and 1.5 mg kg^{-1}. In general, Se in soils is between 0.1–0.6 mg kg^{-1} (Lyons et al. 2003). In the main soil, Se could be derived from local processes, like the weathering process of rocks that are parent and distant processes involving atmospheric deposition. Selenium is generated by human activities like the burning of fossil fuels and naturally occurring resources including explosions of the volcanic or selenium volatilization caused by the occurrence of living organisms (Bañuelos et al. 2017). A significant amount of selenium can be added by the inputs of anthropogenies in the soil. These are outcomes of fossil fuels burning and the usage of fertilizers lime in the field of agriculture (Winkel et al. 2015). From the old soil and plants samples, the accumulation of Se due to industrialization has demonstrated in good manner (Bowley et al. 2017). Selenium can also be deposited in the agriculture fields due to applications of Se fertilizers or through irrigation using Se-rich waters.

6.4.1 Selenium Bioavailability Is Affected by a Variety of Factors

The bioavailability of selenium is determined by a variety of elements like soil physicochemical properties, plants conditions, climate, and agricultural methods.

6.4.1.1 Mobility of Selenium in Soil

The accessibility of Se for plants can also be determined through EC and pH of the soil, weathering of the soils as well as geography and the weather pattern. Mobile form of selenium means the form that is transported with the help of water and can be easily accessible to plants. The Se is basically present in the environment in 4 different oxidation states including the Se-IV (selenite) and Se-VI (selenate) these forms are mobile in nature; however, the selenides (Se^{2-}) and elemental form of selenium is mobile. There are two most commonly used forms of Se are Selenite and Selenate that are found in soil solutions in the semi-arid and dry zones; the absorption as well as evaluation of mobile form Se has been the focus of attention in the research literature (Dinh et al. 2019).

6.4.1.2 Chemical Behavior of se in Soil

Selenium is mainly stored in the soil profile by the rainfall, absorption, and many chemical reactions occurring in the soil. Moreover, a few processes such as desorption, dissolution and leaching as well as volatilization can remove the Se from the soil. The intensity of any reaction can be found with the help of chemical and physical properties of soil, plants conditions, and weather pattern. (Dinh et al. 2019). Numerous scientists are trying to identify the most appropriate ways that are used to determine the availability of Se more precisely and quickly from the soil (Wang et al. 2012).

6.5 Selenium Sources

There are a variety of Se sources available (i) sodium selenate (ii) selcote and (iv) Se-granules that are lime-coated. All sources, during the period of the application were effective in increasing levels of feed crops Se which is the minimum amount required for the prevention of Se deficiencies in livestock. Lime-coated Se Granules produced less Se across all crops than the other Se source. The amount of Se in soybean grain was higher than in forages and cereals which were treated with Se. Selcote caused significant increases in Se. (Gupta and MacLeod 1994). The Se basically presents in environment in 4 different oxidation states including the Se-IV (selenite) and Se-VI (selenate) these forms are mobile in nature; however, the selenides (Se^{2-}) and elemental form of Se is mobile. Pants roots can absorb selenate, organo-selenium, and selenite compounds, including SeCys and SeMet in soil. Selenium can be found in soil solutions, and they cannot absorb selenides or colloidal elements like Se (White 2016). The intensity of any reaction can be found with the help of chemical and physical properties of soil, plants conditions, and weather pattern that regulate the process of paedogenesis (Jones et al. 2017). Selenium deficiencies are more likely to develop on acid soils in areas with high rainfall. The use of Se fertilizers in crops as well as soils within China People's Republic of China is widely known. In certain countries, like Finland as well as New Zealand, animal feed crops and other crops get Se in addition to other fertilizers for agriculture (Schiavon et al. 2020). In Sweden the micronutrient mix containing Se is available and in Denmark, the British micronutrient blend is used however it is quite expensive (Gissel-Nielsen Personal Communication). The way of how Se is administered determines the effectiveness of Se for biofortification. Selenate and selenite are the bioavailable substances; however, the uptake rate of selenate could be higher 33 times more as compared to selenite (Ros et al. 2016). It is since selenite is more readily absorbed by the inner-sphere complexation on the surfaces of soil minerals oxides/hydroxides that limit its mobility and therefore the rate of plant uptake. Furthermore, selenite is limited in its mobility inside the plants therefore it is to be affixed in the roots of plants in contrast to selenate and the Se-IV is extremely mobile in the plants's xylem cells. The predominant position of the diverse species found in soils is dependent on local conditions including the geo-colloidal phase of the soil, pH, and the redox potential. The Se-IV would be predominant in soil when the pH of the soil is high with well aeration similar to the arable soils and the Se-IV content would be higher due to the well irrigated and acidic nature of soil. Se-based fertilizers are generally used in low amounts of 10–20 g Se ha^{-1} in biofortification research. To facilitate the use of the tiny volume of Se in the field, it is typically mixed with other fertilizer-based matrices offering a mix of nutrients, like Selcote Ultra as well as Top Stock or mostly macronutrients such as calcium nitrate and urea. These fertilizer matrices can be referred to "carriers" of Se. In 1993, Gupta et al. 16 examined the use of nitrogen (N) fertilizers such as ammonium nitrate (NH_4NO_3) along with urea that was doped either Se-IV or se-VI to boost levels of Se level of animals. Their main focus was on the advantages of Se-VI over Se-IV in boosting the plant's Se levels however, they also highlighted the advantages of Se. Additionally, various studies showed that Se-enriched urea Granules proved to be very effective in increasing Se concentration in rice, thus highlighting that N is a potential transporter for Se. However, rice has distinct growing conditions from cereals like wheat, which means that results from this study could or might not be adapted to other crops. We know of there has been no research that examined the effect of this type of carrier on wheat or evaluated the efficacy of different macronutrients that act as Se carriers. Some studies have evaluated the efficacy of applying Se using different methods, either to the soil, or to leaf (foliar). Results revealed that, though both methods work in raising Se levels of plants however, foliar fertilization can be up to eight times more effective than soil application of Se (Ramkissoon et al. 2020) The greater effectiveness of foliar applied fertilizers could be due to quick uptake and assimilation due the application later in the growth level, less influence of the ratio of root-to-shoot on transfer to edible parts of the crop in addition to the reduction of losses due to fixation in soils. In the average, only 12 percent of the soil-applied Se fertilizers are utilized by plants. The majority of Se applied is retained and immobilized within the soil with low residual value in subsequent crop (Mathers 2017). This means that multiple fertilizers of Se fertilizers are necessary during each growing season in the event that the effectiveness of Se fertilizers can be improved (Ramkissoon et al. 2020).

6.6 Application of Selenium to Crops

A variety of factors impact the success of Se biofortification. The two most important factors are the methods to apply selenium-based fertilizers. It could be applied to the foliar or by soil amendment. Selenium biofortification by foliar application of inorganic fertilizers plays a very important role. But only in certain conditions soil amendment of Se-based fertilizers proved more effective than foliar application, this only due the nature of soil (texture, redox potential, pH) chemical properties as well as the inherent Se amount and biological activities of soil. The efficacy of this method to provide Se supplementation is observed in various crop varieties like rice, wheat potato and soybean as well as carrot. However, if the Se concentration is greater than 100 ug mL^{-1} of water, it triggers phytotoxicity in certain crops (Hossain et al. 2021). When applying foliar spray (SeF) it is essential that the Se solution should be sprayed with a well-calibrated machine; spraying is not recommended on days with wind or rain. The application should be carried out at the end of the vegetative stage when there is a sufficient surface area that allows for maximal absorption of Se. When applied to soil, Se is efficient between the beginnings of seedling growth until the time when plants mature to absorb Se through roots.

After a lot of research on the effect of nature on the different methods of application of selenium, it has been demonstrated that the foliar application has reduced accumulation of selenium concentration in the soil. The level of Se in soil can increase by the application of Se-based fertilizers in soil as compared to foliar application (Mao et al. 2014). The foliar application is more effective. Selenium fertilizer is typically applied to soils with an acidity of 5.5 or higher. Selenium moves through the phloem cells in whole plant, and it is accumulated in the form of SeMet in the edible parts of plant. All cereal crops contain the organic form of selenium and are especially found in polished and white flour of rice. It is also easily taken up into the small intestines of animals and humans (Lyons et al. 2003). However, the process of biofortification of soils used for agriculture is an unsustainable process. When Se is applied directly in the soil, some amount of Se can be recovered by the plants up-to the 12%. When it was applied though foliar method this percentage is greater (Broadley et al. 2010). Organic selenium application is a more secure, efficient, economical, and simple way to acquire Selenium biofortification as compared to application of selenium-based fertilizer. The right amount and right time of application of Se plays a very important role in boosting the production of the crops, while decreasing waste. The right time application of Se in cereals is very crucial. The milking and booting stages of crops are the effective time to apply the selenium fertilizers because at these stages the plants are fully lush green in the field. The uptake and storage of micronutrient (Se) in plants can be increased in dry environments (Hossain et al. 2021). The green fertilizers and organic source of Se are the alternative resources to make soil and plants Se enriched. The green manures that are well decomposed helps to increase the amount of Se absorbed by different species of plants, while the decomposition process of the organic matter assists in the mobilization of other important soil nutrients (Freeman and Bañuelos 2011). For instance, When Se-rich *Stanleya pinnata* was sprayed on carrots with optimal conditions of soil water, 90 percent of organic Se was converted to inorganic selenate as well as selenite (Freeman and Bañuelos 2011). In addition, the use of SeMet as well as, SeCys made from organic materials, has increased Se intake by plants that were treated with Se that contained organic fertilizers (Bañuelos et al. 2015). The use of costly and ineffective inorganic fertilizers is not necessary to fertilize Se-rich crops. Organic matter found in soil plays an important function in absorbing Se and make sure the Se accessibility to the plants. The occurrence of such organic compounds assists in the enhancing of absorption of Se by plants (Xia et al. 2020). The addition of organic matter such as manure and crop residues to soils used for agriculture can also improve the bioavailability of Se in soils and can help to balance Se concentrations. For instance, these organic compounds reduced the level of toxicity due to occurrence of large amount of Se (Park et al. 2011). The Solubility of Se compounds also plays a very important role in the application of the Se. Selenite and selenite are both soluble, but the selenate is more water soluble as compared to selenite. In humid regions selenate is the more common type of Se (De Feudis et al. 2019; El-Ramady et al. 2015). Se levels in plants are greater when it is mixed with inorganic compounds rather than organic matter forms of Se (Bañuelos et al. 2015). Additionally, by the process of decomposition and activity of microorganisms in soil the organic compounds of Se released the Se into soil (Hossain et al. 2021).

6.7 Selenium in Plants

In plants, Se is most found as SeMet, methyl-SeCys or G-glutamyl-SeCys (Hart et al. 2011). In addition, the supplying maize and wheat with Se in amounts that exceed 100 g Se ha^{-1} had not affected yields of the crops with regard to grain yield. The study also found that the use of Se can increase yield (Broadley et al. 2010). In addition, applications of Se are known to enhance the plant's resistance to oxidative stress (Hasanuzzaman et al. 2010). Selenium biofortification of peach and pear trees increased Se concentration in the fruit and reduced the rate of softening fruit and thus extending the shelf-life of fruits. Biofortification with large amounts of Se has negatively affected the growth rate of plants within lettuce and slowed the germination rate of mustard. The same was true for mustard (Pezzarossa et al. 2012). Certain crops, like *Brassica* were not affected by the application of Se. Selenium fertilizers can be employed to reduce the effects of poisoning caused by some metalloids and toxic metals stress in fields of crops. Dosing the rice's root seedlings with a Se solution prior to transplanting can reduce the negative effects of arsenic which hinders the growth of seedlings (Hart et al. 2011). In addition, the height of rice plants as well as, the number of tillers, the amount of chlorophyll, panicle length and weight of kernels significantly increased after rice seedlings were coated with Se prior to sowing. Se-coating rice seeds reduce arsenic phytotoxicity in seedlings and improve crop yield (Moulick et al. 2018). Plants store Se within their tissues, based on Se phytoavailable in soils and there is a wide variation in the shoot Se concentrations between species, genera, as well as ecosystem types in species (White 2016). Plants are differentiated into various ecological groups based on their ability to store Se in their shoots when they grow in their natural habitat (White 2016). Se-hyperaccumulator species that thrive on soils with seleniferous elements typically have and tolerate extreme Se levels which could be harmful to grazers. However, non-hyperaccumulator species are sensitive to Se and are less sensitive to Se and have smaller than 100 micrograms Se g^{-1} dry weight (non-accumulators) as well as up to 1000 g^{-1} dry weight (Schiavon and Pilon-Smits 2017). Hyperaccumulators store Se mainly in the form of methylselenocysteine (MeSeCys) and selenocystathionine, while SeMet is the main Se organic compound identified in nonhyperaccumulators (Schiavon and Pilon-Smits 2017). Plants absorb both inorganic (selenate, selenite, and elemental Se) as well as organic (e.g., Se-amino acid) Se species and selenides, but not the selenides or colloidal Se (Chauhan et al. 2019). Their capacity to absorb Se is believed to be higher in organic species than inorganic ones. Se-amino acids are more likely to be introduced into plant cells through broad-specific amino acid transportation systems (Lima et al. 2018). However, selenate is the predominant form of Se absorbed by plants, the transport of Se through cell membranes is a process that requires energy and is controlled by the sulphate-transport system (Schiavon and Pilon-Smits 2017; Lima et al. 2018). Competitions in soil between selenate and sulphate along with the sulphate transporters in plants with different affinity of both anions, impact the rate of selenate absorption from plants (White 2016). Selenite compounds are carried across plant cell membranes through phosphorus (P) as well as silicon (Si) transporters that differ between the selenite anion (SeO_3^{2}) hydrogenselenite anion ($HSeO_3^-$) and selenous acid (H_2SeO_3). In detail, H_2SeO_3 is transported through Si transporters and aquaporins and $HSeO_3$ and the corresponding part of SeO_3^{2-} mostly make use of high-affinity and low-P transporters and high-affinity P transporters (Zhang et al. 2014).

6.7.1 Selenium Biofortification in Rice

The Se levels of grains of rice are higher than wheat and maize possibly due to the fact that various varieties of rice can be classified into cultivars with high and low Se content levels (Zhang et al. 2019). As a food staple, it is a good energy source. In almost thirty countries the majority of consumption of food occurring in the form of rice and about 3 billion people consuming the around 80 percent of the calories from it. This reason makes it a good and effective crop for biofortification of Se using agronomic techniques. The soil application of Se fertilizers enhances the low levels for Se content in edible food stuff of rice. Additionally, prolonged use of Se fertilizers could be harmful to the neighboring ecosystem. Therefore, using Se fertilizers must be done with care to prevent toxic effects. The application methods of Se are the most crucial factor that can significantly increase Se levels in the plants. Many studies have proven that foliar applications are an effective method for fertilization, despite the fact that soil amendments are the most common. Foliar spraying results in higher efficiency of absorption of Se as compared with soil application as evidenced by the absence of residues of soil. Foliar methods use the least amounts of Se compounds and have proven the most efficient and safe and suitable method to increase Se levels in rice crops (Pezzarossa et al. 2012).

6.7.2 Selenium Biofortification in Wheat

It is the fact that both inadequate and over-excessive Se intake can have adverse impacts on health of humans a safe and appropriate quantity of Se consumption must be to be maintained. According to the WHO Se intake ranges from 50 to 400 mg day^{-1}. Inorganic Se is highly toxic for the body of a human. Therefore, the addition of organic Selenium compounds are the most effective option to increase Se in foods. Se-rich wheat is the ideal food item to boost Se within the body due to its moderately steady intake, its capacity to increase Se content field Se enhancement can significantly increase the amount of Se content in grains of wheat, and after that, you can obtain Se-rich wheat food items. Many Research studies have proven that the foliar application of selenate as well as selenite has increased the effective amount of Se in wheat grains (Poblaciones et al. 2014). Application of foliar Se has shown good results and is the most frequently used method for biofortification of wheat.

6.7.3 Selenium Biofortification in Maize

Regarding biofortification in agronomic terms, Se levels of all the cereal grains may increase easily when Se is used as a selenate formulation (Broadley et al. 2006). Biofortification in maize using Se is an effective method of increasing consumption of Se (Chilimba et al. 2012). Cereals are the main source of Se as maize is consumed within the all-Western diets along with the most prominent sources of Se including wheat, rice barley, and maize (Narwal et al.2020). Maize with high Se content has been identified as a major source for consumption of additional Se for populations with Se deficiency. Numerous studies have found positive results of Se biofortification for agronomic purposes on Se levels in the maize crop (Premarathna et al. 2012). It is known that Se increases the production of the food crops like maize, wheat, and barley, as well as rice. Thus, in the cereal crops the forms Se and their application methods as well as the application time play an important part in the success of biofortification. The Se absorption and distribution, accumulation and the metabolization process in mature maize are also dependent on the type of Se that is supplied.

6.7.4 Selenium Biofortification in Barley

Barley crop is the most important small grain cereal crop among the following maize, wheat, and rice (Premarathna et al. 2012). Barley has greater capacity in Se absorption (Gibson et al. 2006), Selenate treatment in barley has led to an increase in the concentration of Se in its grains as compared to selenite. Two-row barley could be a suitable crop for inclusion as part of Se biofortification plans (Rodrigo et al. 2014). The foliar application of the sodium selenate subsequently the process of anthesis or at the germination phase of the malting process leads to the deposition of Se in food products (Gibson et al. 2006). Barley had higher efficiency in Se biofortification, compared to maize and rice; however, barley is not cultivated much more. According to a study on Se biofortification has revealed significant differences in Se accumulation in main cereal crops due to the application methods, durations and types of application (Tables 6.1, 6.2, 6.3, and 6.4).

Table 6.1 Effect of different forms of selenium at different stages on Rice

Species	Forms of selenium	Time of application	Application means	Amount of Se applied (g Se ha^{-1})	Se content in grain (μg kg^{-1} DW)	References
Rice	Selenite	At heading	Soil	30	59	Premarathna et al. (2012)
Rice	Selenate	At heading	Soil	30	79	Premarathna et al. (2012)
Rice	Selenite	At heading	Foliar	30	273	Premarathna et al. (2012)
Rice	Selenate	At heading	Foliar	30	150	Premarathna et al. (2012)

Table 6.2 Effect of different forms and concentrations of selenium applied at tillering stage of wheat on Se contents in wheat grains

Specie	Forms of selenium	Time of application	Application means	Amount of Se applied (g Se ha^{-1})	Se content in grain (μg kg^{-1} DW)	References
Wheat	Selenite	Tillering stage	Foliar	10	153.6	Poblaciones et al. (2014)
Wheat	Selenite	Tillering stage	Foliar	20	254.8	Poblaciones et al. (2014)
Wheat	Selenite	Tillering stage	Foliar	40	430.4	Poblaciones et al. (2014)
Wheat	Selenate	Tillering stage	Foliar	10	266.8	Poblaciones et al. (2014)
Wheat	Selenate	Tillering stage	Foliar	20	820	Poblaciones et al. (2014)
Wheat	Selenate	Tillering stage	Foliar	40	1383.2	Poblaciones et al. (2014)

Sources: Galic et al. (2021)

6.8 Future Perspectives

The importance of Se has been well-known for animals and humans (Kieliszek and Błażejak 2016). It is a trace element which is vital to human health and its deficiency can hinder metabolic processes (Jarzýnska and Falandysz 2011). Most crops that are biofortified with Se are rich in beneficial phytochemicals such as antioxidants and minerals (D'Amato et al. 2020). The response of plants to Se biofortification varies crop to crop that was mainly observed in crops like wheat, barley, rice, and maize. Selenium biofortification is designed to improve the quality of crops using Se-species and antioxidants with positive effects on the health and nutrition of humans and as well in animals. The significance of foods enriched with Se is crucial

Table 6.3 Effect of different forms, concentrations and methods of application of selenium applied at various growth stages of maize on Se contents maize grains

Specie	Forms of selenium	Time of application	Application means	Amount of Se applied (g Se ha^{-1})	Se content in grain (μg·kg^{-1} DW)	References
Maize	Selenite	Before sowing	Soil	150	51	Ngigi et al. (2019)
Maize	Selenite	Tasseling and one week after silking	Foliar	11	96	Ngigi et al. (2019)
Maize	Selenate	Before seeding	Soil	10	68.33	Ngigi et al. (2019)
Maize	Selenate	During the stem elongation stage	Foliar	10	205.33	Ngigi et al. (2019)

Sources: Galic et al. (2021)

Table 6.4 Effect of different forms, time and methods of application of barley on Se contents barley grains

Specie	Forms of selenium	Time of application	Application means	Amount of Se applied (g Se ha^{-1})	Se content in grain (μg kg^{-1} DW)	References
Barley	Selenite	Before seeding	Soil	20	57	Rodrigo et al. (2014)
Barley	Selenate	Before seeding	Soil	20	391	Rodrigo et al. (2014)
Barley	Selenate	End of tillering EC-39	Foliar	20	1113.9	Rodrigo et al. (2014)
Barley	Selenite	End of tillering EC-39	Foliar	20	345.5	Rodrigo et al. (2014)

Sources: Galic et al. (2021)

in the current situation of a pandemic of viral diseases and constitutes in fighting against the viruses. Attention is required for a wide-ranging study in this area of research. Selenium biofortification can be used to improve yields in optimal conditions, while minimizing the adverse effects of these conditions on the physiology of plant as well as increasing the antioxidant capabilities of plants and the number of effective phytochemicals. There are a number of studies that could be vital to the creation of the model. Hence there is a huge gap to improve the quality of research papers that focus regarding agronomic Se biofortification. One interesting task is to find out if certain plant species contain significant levels of Se starting with segetal weeds, that are usually only found in these areas, and occasionally possess medicinal characteristics, and could be in danger of being lost due to the intensified farming practices (Perrino and Calabrese 2018). In this regard, research is needed to improve

Se biofortification methods to enhance the nutritional value of food crops in challenging conditions are of immense significance worldwide. Furthermore, the Se bioavailability is controlled by interactions between soil and plants. However, many Se controlling mechanisms and the factors are unclear in widespread distribution of Se in soil (Winkel et al. 2015). The study of Joy's et al. (2015) demonstrated that environmental variables including the types of soils are a significant factor to consider. They estimated an average of 66% of the crop areas will be lost 8.7 percent Se due to moderate change in weather pattern. Due to increasing the effects of extreme change in climate, there should be introduced the climate change models that are essential for predicting the possibility deficiency of Se in various regions. The best way to deal with this managing the situation of Se deficiency includes selection of species of plants, the Se-based biofortification strategies would be determined based on the state and requirements of the specific region. According to our current understanding, we tried to determine the bioavailability of Se in the soil-plant system. Additionally, we suggested some strategies to control the bioavailability of Se which will allow the secure implementation of phytoremediation and biofortification plans in Se-deficient or Se-contaminated zones (Hossain et al. 2021).

References

Bañuelos GS, Lin ZQ, Broadley M (2017) Selenium biofortification. In: Pilon-Smits EAH, Winkel LHE, Lin ZQ (eds) Selenium in plants: molecular, physiological, ecological and evolutionary aspects. Springer International Publishing Switzerland, Cham, pp 231–255

Bañuelos GS, Arroyo I, Pickering IJ, Yang SI, Freeman JL (2015) Selenium biofortification of broccoli and carrots grown in soil amended with se-enriched hyperaccumulator *Stanleya pinnata*. Food Chem 166:603–608

Bowley HE, Mathers AW, Young SD, Macdonald AJ, Ander EL, Watts MJ, Bailey EH (2017) Historical trends in iodine and selenium in soil and herbage at the Park grass experiment, Rothamsted research UK. Soil Use Manag 33:252–262

Broadley MR, Alcock J, Alford J, Cartwright P, Foot I, Fairweather-Tait SJ, Zhao FJ (2010) Selenium biofortification of high-yielding winter wheat (*Triticum aestivum* L.) by liquid or granular se fertilization. Plant Soil 332:5–18

Broadley MR, White PJ, Bryson RJ, Meacham MC, Bowen HC, Johnson SE, Tucker M (2006) Biofortification of UK food crops with selenium. Proc Nutr Soc 65:169–181

Chauhan R, Awasthi S, Srivastava S, Dwivedi S, Pilon-Smits EAH, Dhankher OP, Tripathi RD (2019) Understanding selenium metabolism in plants and its role as a beneficial element. Crit Rev Environ Sci Technol 49:1937–1958

Chilimba ADC, Young SD, Black CR, Meacham MC, Lammel J, Broadley MR (2012) Agronomic biofortification of maize with selenium (se) in Malawi. Field Crops Res 125:118–128

Combs GF Jr (2001) Selenium in global food systems. Brit J Nutr 85:517–547

D'Amato R, Regni L, Falcinelli B, Mattioli S, Benincasa P, Dal Bosco A, Businelli D (2020) Current knowledge on selenium biofortification to improve the nutraceutical profile of food: a comprehensive review. J Agric Food Chem 68:4075–4097

De Feudis M, D'Amato R, Businelli D, Guiducci M (2019) Fate of selenium in soil: a case study in a maize (*Zea mays* L.) field under two irrigation regimes and fertilized with sodium selenite. Sci Total Environ 659:131–139

Dinh QT, Wang M, Tran TAT, Zhou F, Wang D, Zhai H, Liang D (2019) Bioavailability of selenium in soil-plant system and a regulatory approach. Crit Rev Environ Sci Technol 49:443–517

El-Ramady HR, Domokos-Szabolcsy É, Shalaby TA, Prokisch J, Fári M (2015) Selenium in agriculture: water, air, soil, plants, food, animals and nanoselenium. Environ Chem Sustain World:153–232

El-Ramady H, Faizy SED, Abdalla N, Taha H, Domokos-Szabolcsy É, Fari M, Brevik EC (2020) Selenium and nano-selenium biofortification for human health: opportunities and challenges. Soil Syst 4:57–61

Fernandes KFM, Berton RS, Coscione AR (2014) Selenium biofortification of rice and radish: effect of soil texture and efficiency of two extractants. Plant Soil Environ 60:105–110

Freeman JL, Bañuelos GS (2011) Selection of salt and boron tolerant selenium hyperaccumulator *Stanleya pinnata* genotypes and characterization of se phytoremediation from agricultural drainage sediments. Environ Sci Technol 45:9703–9710

Galic L, Vinkovic T, Ravnjak B, Loncaric Z (2021) Agronomic biofortification of significant cereal crops with selenium-a review. Agronomy 1:1015

Germ M, Stibilj V, Kreft I (2007) Metabolic importance of selenium for plants. Eur J plant Sci Biotechnol 1:91–97

Gibson C, Park YH, Myoung KH, Suh MK, Mcarthur T, Lyons G, Stewart D (2006) The bio-fortification of barley with selenium catherine. In: Proceedings of the Institute of Brewery and Distillating. Hobart, Tasmania, pp 19–24

Gupta UC, MacLeod JA (1994) Effect of various sources of selenium fertilization on the selenium concentration of feed crops. Can J Soil Sci 74:285–290

Hart DJ, Fairweather-Tait SJ, Broadleyc MR, Dickinson SJ, Foot I, Knott P, Hurst R (2011) Selenium concentration and speciation in biofortified flour and bread: retention of selenium during grain biofortification, processing and production of se-enriched food. Food Chem 126: 1771–1778

Hasanuzzaman M, Hossain MA, Fujita M (2010) Selenium in higher plants: physiological role, antioxidant metabolism and abiotic stress tolerance. J Plant Sci 5:354–375

Hawrylak-Nowak B (2008) Effect of selenium on selected macronutrients in maize plants. J Elem 13:513–519

Hossain A, Skalicky M, Brestic M, Maitra S, Sarkar S, Ahmad Z, Vemuri H, Garai S, Mondal M, Bhatt R, Kumar P (2021) Selenium biofortification: roles, mechanisms, responses, and prospects. Molecules 26:881

Jarzýnska G, Falandysz J (2011) Selenium and 17 other largely essential and toxic metals in muscle and organ meats of Red Deer (*Cervus elaphus*)-consequences to human health. Environ Int 37: 882–888

Jones GD, Droz B, Greve P, Gottschalk P, Poffet D, McGrath SP, Seneviratne SI, Smith P, Winkel LH (2017) Selenium deficiency risk predicted to increase under future climate change. Proc Natl Acad Sci U S A 114:2848–2853

Joy EJ, Broadley MR, Young SD, Black CR, Chilimba AD, Ander EL, Watts MJ (2015) Soil type influences crop mineral composition in Malawi. Sci Total Environ 505:587–595

Kieliszek M, Błażejak S (2016) Current knowledge on the importance of selenium in food for living organisms: a review. Molecules 21:609

Kumssa DB, Joy EJ, Young SD, Odee DW, Ander EL, Broadley MR (2017) Variation in the mineral element concentration of *Moringa oleifera* lam. And *M. stenopetala* (Bak. f.) Cuf.: role in human nutrition. PloS One 12:e0175503

Lima LW, Pilon-Smits EAH, Schiavon M (2018) Mechanisms of selenium hyperaccumulation in plants: a survey of molecular, biochemical, and ecological cues. Biochimica et Biophysica Acta 1862:2343–2353. https://doi.org/10.1016/j.bbagen.2018.03.028

Lyons G (2018) Biofortification of cereals with foliar selenium and iodine could reduce hypothyroidism. Front Plant Sci 9:730

Lyons G, Judson GJ, Ortiz-Monasterio I, Genc Y, Stangoulis JCR, Graham RD (2005) Selenium in Australia: selenium status and biofortification of wheat for better health. J Trace Elem Med Biol 19:75–82

Lyons G, Stangoulis J, Graham R (2003) High-selenium wheat: biofortification for better health. Nutr Res Rev 16:45–60

Mao H, Wang J, Wang Z, Zan Y, Lyons G, Zou C (2014) Using agronomic biofortification to boost zinc, selenium, and iodine concentrations of food crops grown on the loess plateau in China. J Soil Sci Plant Nutr 14:459–470

Mathers A (2017) Determining the fate of selenium in wheat biofortifcation: an isotopically labelled field trial study. Plant Soil 420:61–77

Mojadadi A, Au A, Salah W, Witting P, Ahmad G (2021) Role for selenium in metabolic homeostasis and human reproduction. Nutrients 13:3256

Moulick D, Santra SC, Ghosh D (2018) Effect of selenium induced seed priming on arsenic accumulation in rice plant and subsequent transmission in human food chain. Ecotoxicol Environ Saf 152:67–77

Nestel P, Bouis HE, Meenakshi JV, Pfeiffer W (2006) Biofortification of staple food crops. J Nutr 136:1064–1067

Ngigi PB, Lachat C, Masinde PW, Du Laing G (2019) Agronomic biofortification of maize and beans in Kenya through selenium fertilization. Environ Geochem Health 41:2577–2591

Park JH, Lamb D, Paneerselvam P, Choppala G, Bolan N, Chung JW (2011) Role of organic amendments on enhanced bioremediation of heavy metalloid contaminated soils. J Hazard Mater 185:549–574

Perrino EV, Calabrese G (2018) Endangered segetal species in southern Italy: distribution, conservation status, trends, actions and ethnobotanical notes. Genet Resour Crop Evol 65:2107–2134

Pezzarossa B, Remorini D, Gentile ML, Massai R (2012) Effects of foliar and fruit addition of sodium selenate on selenium accumulation and fruit quality. J Sci Food Agric 92:781–786

Poblaciones MJ, Rodrigo S, Santamaría O, Chen Y, McGrath SP (2014) Agronomic selenium biofortification in *Triticum durum* under Mediterranean conditions: from grain to cooked pasta. Food Chem 146:378–384

Premarathna L, McLaughlin MJ, Kirby JK, Hettiarachchi GM, Stacey S, Chittleborough DJ (2012) Selenateenriched urea granules are a highly effective fertilizer for selenium biofortification of paddy rice grain. J Agri Food Chem 60:6037–6044

Ramkissoon C, Degryse F, da Silva RC, Baird R, Young SD, Bailey EH, McLaughlin MJ (2020) Improving the efficacy of selenium fertilizers for wheat biofortification. Sci Rep 10:16963

Rayman MP (2000) The importance of selenium to human health. Lancet 356:233–241

Rayman MP (2012) Selenium and human health. Lancet 379:1256–1268

Rodrigo S, Santamaria O, Chen Y, McGrath SP, Poblaciones MJ (2014) Selenium speciation in malt, wort, and beer made from selenium-biofortified two-rowed barley grain. J Agri Food Chem 62:5948–5953

Ros G, Rotterdam A, Bussink D, Bindraban P (2016) Selenium fertilization strategies for bio-fortifcation of food: an agro-ecosystem approach. Plant Soil 404:99–112

Schiavon M, Pilon-Smits EAH (2017) Selenium biofortification and phytoremediation phytotechnologies: a review. J Environ Qual 46:10–19

Schiavon M, Nardi S, Dalla Vecchia F, Ertani A (2020) Selenium biofortification in the 21st century: status and challenges for healthy human nutrition. Plant Soil 453:245–270

Schwartz K, Foltz CM (1957) Selenium as an integral part of factor 3 against dietary necrotic liver degeneration. J Am Chem Soc 79:3292–3293

Shafiq S, Adeel M, Raza H, Iqbal R, Ahmad Z, Naeem M, Azmi UR (2019) Effects of foliar application of selenium in maize (*Zea mays* L.) under cadmium toxicity. Biol Forum-An Int J 11:27–37

Van-der-Torre HW, Van Dokkum W, Schaafsma G, Wedel M, Ockhuizen T (1991) Effect of various levels of selenium in wheat and meat on blood se status indices and on se balance in Dutch men. Brit J Nutr 65:69–80

Wang S, Liang D, Wang D, Wei W, Fu D, Lin Z (2012) Selenium fractionation and speciation in agriculture soils and accumulation in corn (*Zea mays* L.) under field conditions in Shaanxi Province. China Sci Total Environ 427–428:159–164

White PJ (2016) Selenium accumulation by plants. Ann Bot 117:217–235

Winkel LHE, Vriens B, Jones GD, Schneider LS, Pilon-Smits E, Bañuelos GS (2015) Selenium cycling across soil-plant-atmosphere interfaces: a critical review. Nutrients 7:4199–4239

Xia Q, Yang Z, Shui Y, Liu X, Chen J, Khan S, Gao Z (2020) Methods of selenium application differentially modulate plant growth, selenium accumulation and speciation, protein, anthocyanins and concentrations of mineral elements in purple-grained wheat. Front Plant Sci 11:1114

Yang F, Chen L, Hu Q, Pan G (2003) Effect of the application of selenium on selenium content of soybean and its products. Biol Trace Elem Res 93:249–256

Zhang L, Hu B, Li W, Che R, Deng K, Li H, Chu C (2014) OsPT2, a phosphate transporter, is involved in the active uptake of selenite in rice. New Phytol 201:1183–1191

Zhang M, Xing G, Tang S, Pang Y, Yi Q, Huang Q, Fu H (2019) Improving soil selenium availability as a strategy to promote selenium uptake by high-se rice cultivar. Environ Exp Bot 163:45–54

Zhou X, Yang J, Kronzucker HJ, Shi W (2020) Selenium biofortification and interaction with other elements in plants: a review. Front Plant Sci 11:586421

Chapter 7
Selenium Bio-Fortification in Cereal Crops: An Overview

Ghadeer M. Albadrani, Sadia Khalid, Attiqa Rahman, Shahid Ibni Zamir, Safdar Ali, and Saddam Hussain

Abstract Selenium is a crucial element that is present in inorganic as well as in organic nature in plants and the atmosphere. In addition to being essential for human beings and animals, selenium is helpful for plant growth and is primarily involved in antioxidant defense, and tissue regeneration. Selenium deficit in diet has become a worldwide issue, and selenium concentration in soil often correlates its availability to humans and availability in food. It helps the body's defense mechanisms against free radicals, detoxifies heavy metals, controls the immune and reproductive systems, and makes sure that the thyroid gland is functioning properly. The best tactic to increase Se in plant edible components is bio-fortification. Agronomic bio-fortification offers a practical way to raise the Se content of edible crop products through foliar or soil application of Se. Wheat, rice, maize, and barley are the cereals that people consume most frequently, agronomic bio-fortification is the best way to enrich their nutrient contents. This chapter concentrates on highlighting the Se uptake mechanisms in plants; its dual nature effect and biofortification strategies for food crops.

Keywords Selenium · Bio-fortification · Cereals · Minerals · Malnutrition

---

G. M. Albadrani
Department of Biology, Princess Nourah Bint Abdulrahman University, Riyadh, Saudi Arabia

S. Khalid · A. Rahman
Department of Botany, Faculty of Sciences, University of Agriculture Faisalabad, Faisalabad, Punjab, Pakistan

S. I. Zamir · S. Ali
Department of Agronomy, Faculty of Agriculture, University of Agriculture Faisalabad, Faisalabad, Punjab, Pakistan

S. Hussain (✉)
Department of Biology, Princess Nourah Bint Abdulrahman University, Riyadh, Saudi Arabia

Department of Agronomy, Faculty of Agriculture, University of Agriculture Faisalabad, Faisalabad, Punjab, Pakistan

M. Hasanuzzaman et al. (eds.), *Mineral Biofortification in Crop Plants for Ensuring Food Security*, https://doi.org/10.1007/978-981-99-4090-5_7

7.1 Introduction

In the past decennium, the primary objectives of breeder's projects were to boost agricultural outputs. Therefore, these initiatives primarily concentrated on choosing disease-resistant cultivars while taking the plant productivity and harvest index into account, among other crucial aspects. But increases in productivity and nutritional quality of staple foods are essential to fulfill the demands of the global population. According to estimates, one in three people on the planet suffers from the micronutrient deficit, usually called "hidden hunger," which affected 2 billion people globally (Prom-u-thai et al. 2020). Many people in emergent nations experience chronic micronutrient malnutrition and take primarily cereals on a regular basis, failing to meet their needs for vitamins and trace elements (Mayer et al. 2008).

Zinc (Zn), selenium (Se) and iron (Fe) are the elements primarily linked with nutritional deficiencies globally. A sufficient intake of the nutritionally important element Se, a micronutrient, is essential for ensuring human health (Germ and Stibilj 2007). In plants, it helps to maintain moisture content, minimizes oxidative damage, effectively reduce senescence, and stimulates plant growth (Kuznetsov and Kuznetsov 2003).

Selenium is mainly taken by plants from soil in the form of selenate, which is then transported to the chloroplasts where it follows the S absorption process (Finley 2005). Selenium is crucial for younger plants, although its importance for higher plants is currently being researched El-Ramady et al. 2020). The Se may have biological advantages in plants due to its potential to activate plant antioxidant mechanisms and enzymes, which have greater ability to eliminate ROS, particularly during extreme environments (Pilon-Smits 2019).

Human health is negatively impacted by the nutritional deficiency of Se; more than 40 different disorders, including blindness, several cancers, heart diseases, and liver disorder, have now been linked to Se insufficiency (Feng et al. 2013). Selenium deficit is a problem in many regions throughout the globe, especially where the soil Se content is inadequate which results in plants with low Se content (Schiavon et al. 2016). Bio-fortification is the method of enhancing the concentrations of key nutrients that are available in plant's parts used as food, by agronomic means or genetic improvement (Ríos et al. 2008).

More than a billion people around the world suffer from Se deficiency, which can be reduced by cultivated products that have been treated with selenium by its bio-fortification (Gupta and Gupta 2017). Dietary diversity, fortified foods, supplements, and crop bio-fortification which includes agronomic bio-fortification or genetic bio-fortification as distinct methods can all help with Se deficiencies in food (Broadley et al. 2010). Cereals are significant crops favorable for bio-fortification with increased special progressive effects in plant production because they are cultivated and consumed by humans in huge amounts than other crops (Bocchini et al. 2018; Abenavoli et al. 2021).

Rice, wheat, maize, and barley are the main cereals in the traditional diet and can be the main sources of Se, whereas Se-enriched wheat has been extensively researched to be a source of additional Se for populations lacking in Se (Finley 2005). Several studies have reported the benefits of Se agronomic bio-fortification in the Se content of cereals (Xia et al. 2020). Applying fertilizers with essential minerals that are deficient in food to enhance the availability of those elements in plants via the soil or by foliar spray is the basis of agricultural bio-fortification (Ngigi et al. 2019). The kind of treatment used, and the type of Se are the two most crucial factors in increasing the Se concentration in cereals (Deng et al. 2017). When applied foliarly rather than through soil, Se is absorbed by plants more quickly (Pezzarossa et al. 2012). The foliar method, which uses the least amount of Se salts, is the most efficient, secure, and economical way to raise the Se content of field products (Kápolna et al. 2009; Winkel et al. 2015; Márquez et al. 2020).

Selenate and selenite are the most often utilized forms of Se. Nonetheless, the utilization of nano-Se (nSe) is growing. Although there aren't yet any instances of nSe being used more widely, its production and use as a nutrient and bio-fortifier have shown to be intriguing techniques. According to studies, selenite and selenate are often the more efficient type of Se when applied directly to soil and as spray to leaves (Lyons 2018). Selenate is much more effective because Se(VI) is absorbed more quickly, moved from the root to the stem and leaves, and converted into its organic form more efficiently, whereas selenite is readily adsorbable to soil, making it harder for roots to absorb (Izydorczyk et al. 2020). Phosphate and selenite share properties, which support phosphate's ease of adsorption to soil surfaces (Cartes et al. 2005).

7.2 Selenium Uptake in Plants

The majority of nutrients are absorbed by plants through their roots, and the rhizosphere's regional characteristics might affect the plant's ability to absorb Se (Zhou et al. 2020). Many elements in the soil and the plants themselves affect how much Se plants can absorb. The kind and concentration of Se in the soil are the most crucial variables in affecting uptake. The characteristics of the soil, such as pH, soil texture, soil minerals, and the amount of antagonistic anions, are further essential determinants in regulating the acquisition of Se in plants. The potential of organic materials, clay minerals, and $Fe(OH)_2$ to chelate and attach to Se could all significantly limit the plant's ability to absorb Se (Rayman 2008). The Se can adhere on positive charge Al-octahedral sheet sites in clay particles like kaolinite, and this adsorption is very pH-dependent. It interacts with soil elements either through electrostatic force or by the formation of complexes on the mineral exterior of the soil (Natasha Shahid et al. 2018). So, the soil's plant-available Se is made up of mobile components which are readily absorbed by plants (Zhang et al. 2019). Several studies have reported the beneficial effects of Se for plant growth, particularly under both abiotic and biotic stresses (Sarwar et al. 2020). Selenium and sulphur (S) exhibit competition for the same carriers and excessive S typically inhibit Se absorption (Malagoli et al. 2015). Depending on Se absorption in cells, plants could be categorized as hyper-accumulators or non-accumulators (Gupta and Gupta 2017). Selenium hyper-accumulator species, like Princesplume, generally take Se over S, but Se non-accumulators inhibit Se uptake relative to sulphate (Deng et al. 2017). Hyperaccumulators grow better in Se-rich environments and accumulate larger concentrations of Se in their cells. Mustard greens are an example of a secondary accumulator, while *Nicotiana* and *Solanum* sp., absorb Se less than 100 mg in 1 kg DW (Wiesner-Reinhold et al. 2017).

Low Se accumulators are more common than high ones in crops (Newman et al. 2019). In comparison with selenite that is strongly attached with positive soil binding receptors, selenate form is substantially mobile, therefore frequently available to plants in the soils (Eich-Greatorex et al. 2007). From the root to the shoot, inorganic forms of Se that are absorbed by plants are carried by xylem, with transport procedure depending on the nature of external source of Se. The xylem may readily absorb and transfer Se(IV), and the phloem then distributes Se(IV) to the reproductive organs (Zhou et al. 2020).

The mechanisms by which selenate and selenite are absorbed and transported by plants differ significantly. Selenate, is higher soluble in contrast with selenite, could enter plant roots in direct contact, but selenite is likely carried through phosphate transporters (Funes-Collado et al. 2013). Through means of active transport, Se is deposited in plant cells in opposition to the gradient of electrochemical potential (White 2016). Selenite uptake, on the other hand, occurs via passive distribution. It is facilitated by active transportation because a metabolic inhibitor dramatically decreased the uptake of selenite (Gupta and Gupta 2017). Metallic selenide and elemental Se cannot be absorbed by plants directly because they are insoluble in H_2O, whereas Seleno-amines have substantially greater availability to plants (Natasha Shahid et al. 2018). Se-containing amino acids in the S assimilation route are selenocysteine (SeCys) and SeMet which are equivalents of the S-containing amino acids (Schiavon et al. 2016).

When compared to selenate or selenite, SeCys and SeMet are taken at rates up to twenty times high (Zhou et al. 2020). One of the Se species that is most efficiently collected in many organs is SeMet. Depending on the species, plant development stage and physiological state, different portions of the plant have different distributions of Se. Selenium concentrations are typically high during the seedling growth stage and drop earlier or after flowering as Se is transferred from leaves to the reproductive part (White 2016). The Se concentrations are highest in young leaves. Young leaves in Se accumulators, accumulate Se throughout the early vegetative stage of growth.

7.3 Dual-Nature Effects of Selenium on Plants

Due to different concentrations in tissues, Se has two different effects on plant physiology. It can promote plant development and mitigate a variety of environmental stresses (Feng et al. 2013). The Se helps plants to grow and function at low concentrations, supporting the maintenance of cellular components and activities and enhancing plant performance (Hasanuzzaman et al. 2020). Se released by plants has an adverse impact on Se-sensitive environmental factors, which may possibly shield plants from the attack of viruses and phytophages (Pilon-Smits 2019).

While Se absorption is associated with boosting the intake of Mg and Fe, it has been discovered that Se helps to promote potassium deposition in plants and mediates the increase in chlorophyll content. Numerous studies suggested that Se is involved in controlling the ROS as well as antioxidants, limiting the HMs absorption and transport, maintaining the structure of membranes and photosynthetic apparatus in plants (Feng et al. 2013).

Plants' stomatal conductance, transpiration rate, and net photosynthesis rate were all raised when Se added in growth media or nutritional solution (Hasanuzzaman et al. 2020). Plants respond physiologically and biochemically to Se in a variety of ways (El-Ramady et al. 2015). Rhizobium, fungus with mycorrhizae, and some bacteria are helpful soil organisms that work in symbiotic relationship with plant roots to protect plants in a variety of ways, such as by promoting nutrient availability and producing growth regulators for plants (Jha and Warkentin 2020).

The utilization of Se tolerant bacteria appears as a potential substitute for Se improvement in cereals cultivated in soils having low Se concentrations (Ye et al. 2020). Agricultural crops' ability to absorb Se is also influenced by plant type. Selenium cycling in living things is thought to be based on bio-geochemical behavior of Se in soil and plant systems (Natasha Shahid et al. 2018).

7.4 Selenium Bioavailability and Human Health

Special attention should be paid to food hygiene and nutritive values to improve the health of the population of the world (Izydorczyk et al. 2020). Almost 50% of deaths are due to diseases tied to diet, with malnutrition being the primary cause of mortality for people worldwide (Lyons 2018). Limited accessibility of micronutrients in the average intake is the main cause of micronutrient deficiencies in humans (Yang et al. 2007). Selenium could be harmful at higher levels while being one of the most important minerals for all life forms. Selenium redox chemistry also greatly affects its toxic effects, solubility, and availability. Se insufficiency is supposed to affect over a billion humans worldwide, and several croplands have inadequate Se concentrations. For most regions, se intakes through drinkable water and other non - food resources are likewise very low. Global estimates for daily dietary Se intake for individuals range from 3 to 7000 microgram (Hart et al. 2011).

The amount of tocopherol in the diet, which appears to lower the amount of required Se, as well as the kind of Se ingested, determines the lowest Se requirements. For adult women, the recommended daily intake is 40 micrograms, while for men, it is 50 micrograms (according to international criteria) (Haug et al. 2007). It has been hypothesized that mammalian systems may contain about 100 selenoproteins (Lyons et al. 2004). GPx and TrxRs reductase are two examples of selenoproteins. These enzymes serve a number of purposes, including defense against oxidative stress, control of redox potential in cell, and metabolism of the thyroid (Winkel et al. 2015). There is evidence that Se deficiency may have an adverse impact on health of people in different ways, like by suppressing the immune system making the body more susceptible to viral diseases, improper thyroid function, causing of asthma & other disorders (Hawkesford and Zhao 2007). Lately, a novel coronavirus outbreak (COVID-19) that is affecting both human health and the global economy has spread across the globe. Supplemental natural therapies, such as Se supplementation, should be taken into consideration to lower the viral diseases in hosts and strengthen immunity. Like Zn supplementation, supplementation of Se to COVID-19-affected individuals may be an alternative as natural viral cure (Schiavon et al. 2020).

Hence, scientists are particularly interested in studying Se diversification in diets that are fortified with plant-based ingredients D'Amato et al. 2020). Foods like rice and wheat, which contain Se, can have very different concentrations in different areas and countries. So, it's crucial to keep an eye on and optimize Se concentrations in different crops to prevent Se insufficiency and toxicity (Zhu et al. 2009). Selenium could be supplemented in the diet of humans by increasing Se intake by various crops throughout growth by biofortification (Newman et al. 2019).

7.5 Selenium Biofortification Strategies

The effect of bio-fortification to enhance plants with Se is dependent on a number of components, including the crop species, the Se species to be used, and the method of Se fertilization. No well-known metabolic function of Se has been found in higher plants, and specific process of Se-Cys into proteins has not been described yet. The conversation of S containing amino acids with Se containing -amino acids in proteins can result in hazardous proteins (Van Hoewyk 2013). On other side, Se could be advantageous for plants at low concentrations. Its benefits were first noted in Se hyper-accumulators (Pilon-Smits and LeDuc 2009). Nevertheless, more research having a number of non-accumulators demonstrated that Se boosts growth, phytochemicals and antioxidants (D'Amato et al. 2020). Se sources are rare, nonrenewable, and vulnerable to depletion due to excessive use, so they must be carefully conserved (Lyons 2018). The foliar spray of Se in rice is transformed into Se containing amino acids in endosperm that is bioavailable for humans (Wang et al. 2013). Since 80 percent of the total Se in rice grain is available in forms namely, as Se-confined proteins, foliar treatment has proven to be the best method for Se-enriching cereal grains.

7.5.1 Conventional Approaches

Fortification involves adding Se compounds to food while it is processed. It is an effective substitute to agriculture biofortification by Se fertilizers (Haug et al. 2007). The biofortification tactics are influenced by a number of factors, including the method for Se application, Se quantity, Se in soil and harvesting systems, weather conditions and crop varieties, and application with micro-nutrients as well (D'Amato et al. 2020).

The majority of investigation now on Se biofortification has focused on using Se either alone or infrequently in combination with micronutrients, typically I or Zn (Cakmak et al. 2020; Golob et al. 2020). Although the lack of various trace elements is a major concern worldwide, crop biofortification with several micronutrients has only been the subject of a few research investigations. A recent study with a different type of soil and climate showed that *T. aestivum* L. combination of Se, iron, iodine, and Zn was effective for grains when foliar spray was applied (Zou et al. 2019).

In order to rise the nutritional importance of diets that people ultimately consume, food crops that are "biofortified" with necessary micronutrients and other substances that promote health (Jha and Warkentin 2020). It is a novel approach to combat micronutrient deficiencies that are manageable, inexpensive, and effective in long term (Ros et al. 2016). Farmers must use biofortified crops in quantities sufficient to enhance the nutritional value of the food (Miller and Welch 2013).

Earlier studies revealed that crop biofortification is beneficial for iodine and Se, while gene-based biofortification might be appropriate for iron fortification, carotene, and vitamin A. Genetic as well as agronomic bio-fortification techniques for Zn have both been effectively proven, either separately or in combination. Due to the chemical similarity between S and Se, Se can be transported by the S transport pathway and is hence suitable for agronomic (fertilizer) biofortification (White 2016). Agriculture biofortification is frequently used in soils with less Se concentration and primarily involves in addition of Se to the soil and spraying leaves with Se, which is normally accomplished using fertilizers with selenate (Wu et al. 2015). To meet biofortification goals, Se fertilizers are often used in low quantities (10 to 20 grams Se ha^{-1}). They serve as "carriers" of Se and are typically blended with other nutrients to facilitate their administration (Ramkissoon et al. 2019).

According to estimates, 12% of soil applications of Se are typically absorbed in plants since the majority of Se is fixed in the soil and hence not accessible (Broadley et al. 2010). The effectiveness of foliar Se, in contrast, is eight times greater than that of soil application (Ros et al. 2016). The faster the uptake of Se and assimilation, the lack of Se translocation from root to shoot (Ramkissoon et al. 2019).

Due to their ability to lower oxidized as well as methylated kinds of Se beneficial rhizosphere bacteria may increase soil Se phyto-availability and the ability of plants to utilize applied nutrients (White and Broadley 2009). Se levels in wheat grain were increased by inoculating them with certain microorganisms, either alone or in combination with arbuscular mycorrhizal fungus (Yasin et al. 2015). Corresponding to this, onion bulbs inoculated with fungus had increased selenocystine and selenate levels that were 36% and 21% higher than those of control (Golubkina et al. 2019).

To a lesser extent, foliar fertilization with sodium selenite or selenate can be used to investigate agronomic biofortification with Se. The genotypes with the highest selenite content in flour of rice varied certainly. Additionally, the increase of Se in rice crops directly affects the content of proteins, carbohydrates, and fatty acids, stimulating positive effects such as lessening the consumption of saturated fats (Lidon et al. 2018).

Se-hyperaccumulator plants, particularly in organic nature that contain high amounts of Se, may be used as green compost (Wan et al. 2018). One disadvantage of this method is that it may speed the discharge of Se into the environment in some circumstances, especially when Se is concentrated by means of evapotranspiration, which could lead to an ecotoxicological hazard.

7.5.2 Nano-Biofortification of Selenium

Selenium in the form of Se nanoparticles as a substitute to traditional Se fertilizers for supplementing plants has only recently been recommended by cutting-edge technologies (Juárez-Maldonado et al. 2019). Se-NPs are formed from precursor Se salts, mostly selenite and selenate, and can take on a variety of shapes and sizes when they are combined with reducing agents such as phenolics, alcohol and amines that are produced by microorganisms, fungi, and plants (Nayantara 2018). Using Se-NPs the opportunity for synchronized Se management in relation to elimination and absorption by plants, while reducing Se in agro-ecosystems (Babajani et al. 2019). According to studies on tobacco, garlic, and *Vigna radiata* plants ISeNPs appear to be very less hazardous as compared with selenite and selenate (Bărbieru et al. 2019).

Three different types of peanuts (*Arachis hypogaea*) were exposed to solutions with applications of control, 20 and 40 mg in 1 L Se NPs, which were produced using chemical methods, had a size of 10–30 nano-meter, and showed no toxicity. The weight of seeds increased by 204%, the fresh weight of plant increased by 129%, and quantity of pods/plants increased by 200% at 20 mg L^{-1}. Such results might be due to better photosynthetic pigments, maintaining the ionic content, cell equilibrium, high photochemical efficiency, and decreased ROS levels. It has demonstrated that SeNPs increase plant growth more than inorganic forms of Se, which raises the possibility that SeNPs contribute to the buildup of signaling hormones like C_2H_4, salicylic acid, gibberellic acid, cytokinins that are essential in tissue differentiation and growth of crop (Sotoodehnia-Korani et al. 2020). Due to their ability to progressively release Se, SeNPs are primarily employed in medicinal applications to rise the bioavailability of medications and target specific organ (Constantinescu-Aruxandei et al. 2018).

7.5.3 Genetic Biofortification of Selenium

The process, known as "genetic biofortification," tries to choose plant cultivars with a high or moderate ability to take as well as translocate Se in food or with a preference for organic Se (SeMet and/or MetSeCys) uptake (Wu et al. 2015). Crops breeding having a high potential to store Se is a practicable process since there is substantial genetic diversity in grain Se level of numerous cereals and legume crops, chicory, lettuce tomato, fruits, and potato (White 2016).

There have been numerous chromosomal loci (QTLs) linked to excessive Se in grains and in the leaves of various crops (Ates et al. 2016). Marker-assisted breeding can be utilized to allocate Se-QTLs to high-yielding cultivars with low Se by choosing plant varieties with high Se in the edible parts (Wu et al. 2015). Yet, due to limitations on the use of transgenics still in place in many countries, genetic engineering is still not as widely used and recognized as agricultural bio-fortification and conventional breeding methods. There are a few transgenics that are good candidates for biofortification, with improved capability to obtain and release Se, preferably in organic nature. The majority of these express sulfate carriers, enzymes that are involved in Se acclimatization, including ATP-sulphuryls, proteins that are associated with the mechanisms that stop Se from being misincorporated with proteins (Zhu et al. 2009).

The investigators discovered higher grain Se in 2n wheat when compared with modern wheat cultivars, but insignificant results were identified among wheat cultivars, indicating Se biofortification through specialized breeding programs. In soils with low Se content, the genotypic changes could not be noticeable, whereas they may be noticeable in the soil having higher Se. Due to stronger environmental impacts, two field tests on 14 kinds of winter wheat conducted in the USA revealed considerable differences in grain Se (Garvin et al. 2006). El Mehdawi et al. (2018) discovered selenate/sulphate transporters and one isoform of ATP-sulphurylase (APS2) as new targets for genetic modification in Se hyperaccumulator Princeplume (Jiang et al. 2018). It has high affinity sulfate or selenate transporter in roots, which is highly expressed and not exhibited to normal suppression caused by high-level sulphate in non-hyper-accumulators (Wang et al. 2013). Moreover, the expression of the low-affinity sulphate transporter which is important for sulphate or selenate root to shoot transferring is elevated in this hyperaccumulator (Jiang et al. 2018).

7.6 Impacts of Selenium Bio-Fortification on the Nutritional Value of Crops

Selenium is an important micronutrient that has a role in the dietary value of crops in various ways. Crop nutrition can be improved by Se; as it has been proven to raise the amounts of many nutrients in crops, including protein, vitamins, and minerals. For instance, a Chinese study reported that wheat fertilized with Se had higher amounts of protein, Fe, and Zn than unfertilized wheat (Li et al. 2017). Selenium-fortified potatoes also had higher levels of other minerals like vitamin C and carotenoids, according to a study carried out in Finland (Kaseva et al. 2000).

Even though Se is necessary for plant growth, excessive levels of the micronutrient can be detrimental to crops. Reduced growth and output in crops can result from high soil Se levels, along with other damaging effects such leaf damage and low chlorophyll levels (Neuhierl et al. 2019). To avoid toxicity, it is crucial to properly monitor soil Se levels. Various levels of Se in crops can affect the food of both animals and humans. Selenium is a crucial vitamin for both people and animals, and eating foods that are deficient in Se can cause deficiency. Consuming foods that contain an excessive amount of Se, however, can also be harmful. For instance, a study carried out in the United States discovered that feeding high-Se alfalfa to cattle led to higher Se levels in their tissues, which might cause selenosis in people who eat the meat (Ullrey et al. 1983).

7.7 Conclusions

The most common food in the global human dietary intake is cereals. Around one billion people are Se deficient, which causes many diseases in the human body. The importance of having a healthy immune system has recently received more attention. The Se concentration in cereals can be enhanced by different biofortification techniques. Among them, traditional breeding and agronomic biofortification with Se are perhaps the most popular and widely accepted strategies for eradicating Se deficiency globally. The primary goal of Se biofortification is to enhance crops with Se and various antioxidants that have a good impact on the nutrition and health of people and animals. In crux, biofortification of Se particularly through foliar fertilization should be done to reduce the Se malnutrition.

Conflict of Interest The authors declare that there is no conflict of interest.

Acknowledgments The authors acknowledge the support from Princess Nourah bint Abdulrahman University Researchers Supporting Project number (PNURSP-HC2022/4), Princess Nourah bint Abdulrahman University, Riyadh, Saudi Arabia.

References

Abenavoli L, Milanovic M, Procopio AC, Spampinato G, Maruca G, Perrino EV, Mannino GC, Fagoonee S, Luzza F, Musarella CM (2021) Ancient wheats: beneficial effects on insulin resistance. Minerva Med 112(5):641–650

Ates D, Sever T, Aldemir S, Yagmur B, Temel HY, Kaya HB, Alsaleh A, Kahraman A, Ozkan H, Vandenberg A, Tanyolac B (2016) Identification QTLs controlling genes for se uptake in lentil seeds. PLoS One 11(3):e0149210

Babajani A, Iranbakhsh A, Oraghi Ardebili Z, Eslami B (2019) Differential growth, nutrition, physiology, and gene expression in *Melissa officinalis* mediated by zinc oxide and elemental selenium nanoparticles. Environ Sci Pollut Res Int 26(24):24430–24444

Bărbieru OG, Dimitriu L, Călin M, Raut I, Constantinescu Aruxandei D, Oancea F (2019) Plant biostimulants based on selenium nanoparticles biosynthesized by *Trichoderma* strains. Proceedings 29(1):95

Bocchini M, D'Amato R, Ciancaleoni S, Fontanella MC, Palmerini CA, Beone GM, Onofri A, Negri V, Marconi G, Albertini E et al (2018) Soil selenium (Se) biofortification changes the physiological, biochemical and epigenetic responses to water stress in *Zea mays* L. by inducing a higher drought tolerance. Front Plant Sci 9:1–14

Broadley MR, Alcock J, Alford J, Cartwright P, Foot I, Fairweather-Tait SJ, Hart DJ, Hurst R, Knott P, McGrath SP, Meacham MC (2010) Selenium biofortification of high yielding winter wheat (*Triticum aestivum* L.) by liquid or granular Se fertilisation. Plant Soil 332:5–18

Cakmak I, Marzorati M, Van den Abbeele P, Hora K, Tjalling Holwerda H, Yazici MA, Savasli E, Neri J, Du Laing G (2020) Fate and bioaccessibility of iodine in food prepared from agronomically biofortified wheat and rice and impact of cofertilization with zinc and selenium. J Agric Food Chem 68(6):1525–1535

Cartes P, Gianfreda L, Mora ML (2005) Uptake of selenium and its antioxidant activity in ryegrass when applied as selenate and selenite forms. Plant Soil 276:359–367

Constantinescu-Aruxandei D, Frîncu RM, Capră L, Oancea F (2018) Selenium analysis and speciation in dietary supplements based on next-generation selenium ingredients. Nutrients 10:1466

Curie C, Panaviene Z, Loulergue C, Dellaporta SL, Briat JF, Walker EL (2001) Maize yellow stripe1 encodes a membrane protein directly involved in Fe (III) uptake. Nature 409(6818): 346–349. https://doi.org/10.1038/35053080

D'Amato R, Regni L, Falcinelli B, Mattioli S, Benincasa P, Dal Bosco A, Pacheco P, Proietti P, Troni E, Santi C, Businelli D (2020) Current knowledge on selenium biofortification to improve the nutraceutical profile of food: a comprehensive review. J Agric Food Chem 68:4075–4097

Deng X, Liu K, Li M, Zhang W, Zhao X, Zhao Z (2017) Field crops research difference of selenium uptake and distribution in the plant and selenium form in the grains of rice with foliar spray of selenite or selenate at different stages. Field Crop Res 211:165–171

Eich-Greatorex S, Sogn TA, Øgaard AF, Aasen I (2007) Plant availability of inorganic and organic selenium fertiliser as influenced by soil organic matter content and pH. Nutr Cycl Agroecosyst 79:221–231

El Mehdawi AF, Jiang Y, Guignardi ZS, Esmat A, Pilon M, Pilon- Smits EAH, Schiavon M (2018) Influence of sulfate supply on selenium uptake dynamics and expression of sulfate/ selenate transporters in selenium hyperaccumulator and non-hyper-accumulator Brassicaceae. New Phytol 217:194–205

El-Ramady H, Faizy SED, Abdalla N, Taha H, Domokos-Szabolcsy É, Fari M, Elsakhawy T, Omara AED, Shalaby T, Bayoumi Y (2020) Selenium and nano-selenium biofortification for human health: opportunities and challenges. Soil Syst 4:57–58

El-Ramady H, Abdalla N, Alshaal T, El-Henawy A, Faizy SEDA, Shams MS, Shalaby T, Bayoumi Y, Elhawat N, Shehata S et al (2015) Selenium and its role in higher plants. In: Lichtfouse E, Schwarzbauer J, Robert D (eds) Pollutants in buildings, water and living organisms, Environmental Chemistry for a Sustainable World. Springer, Cham, pp 235–296

Feng R, Wei C, Tu S (2013) The roles of selenium in protecting plants against abiotic stresses. Environ Exp Bot 87:58–68

Finley JW (2005) Selenium accumulation in plant foods. Nutr Rev 63:196–202

Funes-Collado V, Morell-Garcia A, Rubio R, López-Sánchez JF (2013) Selenium uptake by edible plants from enriched peat. Sci Hortic 164:428–433

Garvin DF, Welch RM, Finley JW (2006) Historical shifts in the seed mineral micronutrient concentration of US hard red winter wheat germplasm. J Sci Food Agric 86:2213–2220

Germ M, Stibilj V (2007) Selenium and plants. Acta Agric Slov 89:1

Golob A, Novak T, Maršić NK, Šircelj H, Stibilj V, Jerše A, Kroflič A, Germ M (2020) Biofortification with selenium and iodine changes morphological properties of *Brassica oleracea* L. var. gongylodes and increases their contents in tubers. Plant Physiol Biochem 150:234–243

Golubkina N, Zamana S, Seredin T, Poluboyarinov P, Sokolov S, Baranova H, Krivenkov L, Pietrantonio L, Caruso G (2019) Effect of selenium biofortification and beneficial microorganism inoculation on yield, quality and antioxidant properties of shallot bulbs. Plan Theory 8(4): 102

Gupta M, Gupta S (2017) An overview of selenium uptake, metabolism, and toxicity in plants. Front Plant Sci 7:1–14

Hart DJ, Fairweather-Tait SJ, Broadley MR, Dickinson SJ, Foot I, Knott P, McGrath SP, Mowat H, Norman K, Scott PR et al (2011) Selenium concentration and speciation in biofortified flour and bread: retention of selenium during grain biofortification, processing and production of se-enriched food. Food Chem 126:1771–1778

Hasanuzzaman M, Bhuyan MHMB, Raza A, Hawrylak-Nowak B, Al M-GR, Mahmud J, Nahar K, Fujita M (2020) Selenium in plants: boon or bane? Environ Exp Bot 178:104170

Haug A, Graham R, Christophersen O, Lyons G (2007) How to use the world's scarce selenium resources efficiently to increase the selenium concentration in food. Microbiol Ecol Health Dis 19(4):209–228

Hawkesford MJ, Zhao FJ (2007) Strategies for increasing the selenium content of wheat. J Cereal Sci 46:282–292

Izydorczyk G, Ligas B, Mikula K, Witek-Krowiak A, Moustakas K, Chojnacka K (2020) Biofortification of edible plants with selenium and iodine—a systematic literature review. Sci Total Environ 754:141983

Jha AB, Warkentin TD (2020) Biofortification of pulse crops: status and future perspectives. Plan Theory 9:73

Jiang Y, Schiavon M, Lima LW, Tripti JRR, El Mehdawi AF, Royer S, Zeng Z, Hu Y, Pilon-Smits EAH, Pilon M (2018) Comparison of ATP sulfurylase 2 from seleniumhyperaccumulator *Stanleya pinnata* and non-accumulator *Stanleya elata* reveals differential intracellular localization and enzyme activity levels. Biochim Biophys Acta Gen Subj. S0304-4165(18)30076-X

Juárez-Maldonado A, Ortega-Ortíz H, Morales-Díaz AB, González-Morales S, Morelos-Moreno Á, Cabrera-De la Fuente M, Sandoval-Rangel A, Cadenas-Pliego G, Benavides-Mendoza A (2019) Nanoparticles and nanomaterials as plant biostimulants. Int J Mol Sci 20(1):162

Kápolna E, Hillestrøm PR, Laursen KH, Husted S, Larsen EH (2009) Effect of foliar application of selenium on its uptake and speciation in carrot. Food Chem 115:1357–1363

Kaseva ME et al (2000) Fortification of potatoes with selenium using agronomic biofortification. J Agric Food Chem 48:5300–5304

Kuznetsov V, Kuznetsov V (2003) Selenium regulates the water status of plants exposed to drought. Dokl Biol Sci 390:266–268

Li HF et al (2017) Effect of selenium fertilization on yield, quality, and selenium content of winter wheat (*Triticum aestivum* L.) in China. Field Crops Res 201:121–127

Lidon FC, Oliveira K, Ribeiro MM, Pelica J, Pataco I, Ramalho JC, Leitão AE, Almeida AS, Campos PS, Ribeiro-Barros AI, Pais IP (2018) Selenium biofortification of rice grains and implications on macronutrients quality. J Cereal Sci 81:22–29

Lyons G (2018) Biofortification of cereals with foliar selenium and iodine could reduce hypothyroidism. Front Plant Sci 9:1–8

Lyons GH, Stangoulis JCR, Graham RD (2004) Exploiting micronutrient interaction to optimize biofortification programs: the case for inclusion of selenium and iodine in the HarvestPlus program. Nutr Rev 62:247–252

Malagoli M, Schiavon M, Dall' Acqua S, Pilon-Smits EAH (2015) Effects of selenium biofortification on crop nutritional quality. Front Plant Sci 6:1–5

Márquez VG, Moreno ÁM, Mendoza AB, Macías JM (2020) Ionic selenium and nanoselenium as biofortifiers and stimulators of plant metabolism. Agronomy 10:1399

Mayer JE, Pfeifer WH, Beyer P (2008) Biofortified crops to alleviate micronutrient malnutrition. Curr Opin Plant Biol 11:166–179

Miller DD, Welch RM (2013) Food system strategies for preventing micronutrient malnutrition. Food Policy 42:115–128

Natasha Shahid M, Niazi NK, Khalid S, Murtaza B, Bibi I, Rashid MI (2018) A critical review of selenium biogeochemical behavior in soil-plant system with an inference to human health. Environ Pollut 234:915–934

Nayantara KP (2018) Biosynthesis of nanoparticles using ecofriendly factories and their role in plant pathogenicity: a review. Biotech Res Innov 2:63–73

Neuhierl B et al (2019) Impact of selenium on growth and quality of crops. Plan Theory 8:217

Newman R, Waterland N, Moon Y, Tou JC (2019) Selenium biofortification of agricultural crops and effects on plant nutrients and bioactive compounds important for human health and disease prevention—a review. Plant Foods Hum Nutr 74:449–460

Ngigi PB, Lachat C, Masinde PW, Du Laing G (2019) Agronomic biofortification of maize and beans in Kenya through selenium fertilization. Environ Geochem Health 41:2577–2591

Pezzarossa B, Remorini D, Gentile ML, Massai R (2012) Effects of foliar and fruit addition of sodium selenate on selenium accumulation and fruit quality. J Sci Food Agric 92:781–786

Pilon-Smits EA, LeDuc DL (2009) Phytoremediation of selenium using transgenic plants. Curr Opin Biotechnol 20(2):207–212

Pilon-Smits EAH (2019) On the ecology of selenium accumulation in plants. Plan Theory 8:197

Prom-u-thai C, Rashid A, Ram H (2020) Simultaneous biofortifcation of rice with zinc, iodine, iron and selenium through foliar treatment of a micronutrient cocktail in five countries. Front Plant Sci 11:589835

Ramkissoon C, Degryse F, da Silva RC, Baird R, Young SD, Bailey EH, McLaughlin MJ (2019) Improving the efficacy of selenium fertilizers for wheat biofortification. Sci Rep 9:19520

Rayman MP (2008) Food-chain selenium and human health: emphasis on intake. Braz J Nutr 100: 254–268

Ríos JJ, Rosales MA, Blasco B, Cervilla LM, Romero L, Ruiz JM (2008) Biofortification of se and induction of the antioxidant capacity in lettuce plants. Sci Hortic 116:248–255

Ros G, van Rotterdam A, Bussink D, Bindraban P (2016) Selenium fertilization strategies for bio-fortification of food: an agro-ecosystem approach. Plant Soil 404:99–112

Sarwar N, Akhtar M, Kamran MA, Imran M, Riaz MA, Kamran K, Hussain S (2020) Selenium biofortification in food crops: key mechanisms and future perspectives. J Food Compos Anal 93:103615

Schiavon M, Berto C, Malagoli M, Trentin A, Sambo P, Dall'Acqua S, Pilon-Smits EAH (2016) Selenium biofortification in radish enhances nutritional quality via accumulation of methyl-selenocysteine and promotion of transcripts and metabolites related to glucosinolates, phenolics amino acids. Front Plant Sci 7:1371

Schiavon M, Nardi S, dalla Vecchia F, Ertani A (2020) Selenium biofortification in the 21st century: status and challenges for healthy human nutrition. Plant Soil 453:245–270

Sotoodehnia-Korani S, Iranbakhsh A, Ebadi M, Majd A, Ardebili ZO (2020) Selenium nanoparticles induced variations in growth, morphology, anatomy, biochemistry, gene expression, and epigenetic DNA methylation in *Capsicum annuum*; an in vitro study. Environ Pollut 265:114727

Ullrey DE et al (1983) Elevated selenium in tissues of beef cattle fed high-selenium alfalfa. J Animal Sci 57:926–993

Van Hoewyk D (2013) A tale of two toxicities: malformed selenoproteins and oxidative stress both contribute to selenium stress in plants. Ann Bot 112:965–972

Wan J, Zhang M, Adhikari B (2018) Advances in selenium- enriched foods: from the farm to the fork. Trends Food Sci Tech 76:1–5

Wang YD, Wang X, Wong YS (2013) Generation of selenium-enriched rice with enhanced grain yield, selenium content and bioavailability through fertilization with selenite. Food Chem 141: 2385–2393

White PJ (2016) Selenium accumulation by plants. Ann Bot 117:217–235

White PJ, Broadley MR (2009) Biofortification of crops with seven mineral elements often lacking in human diets–iron, zinc, copper, calcium, magnesium, selenium and iodine. New Phytol 182(1):49–84

Wiesner-Reinhold M, Schreiner M, Baldermann S, Schwarz D, Hanschen FS, Kipp AP, Rowan DD, Bentley-Hewitt KL, McKenzie MJ (2017) Mechanisms of selenium enrichment and measurement in brassicaceous vegetables, and their application to human health. Front Plant Sci 8:1365

Winkel LH, Vriens B, Jones GD, Schneider LS, Pilon-Smits E, Bañuelos GS (2015) Selenium cycling across soil-plant-atmosphere interfaces: a critical review. Nutrients 7:4199–4239

Wu Z, Bañuelos GS, Lin ZQ, Liu Y, Yuan L, Yin X, Li M (2015) Biofortification and phytoremediation of selenium in China. Front Plant Sci 6:136

Xia Q, Yang Z, Shui Y, Liu X, Chen J, Khan S, Wang J, Gao Z (2020) Methods of selenium application differentially modulate plant growth, selenium accumulation and speciation, protein, anthocyanins and concentrations of mineral elements in purple-grained wheat. Front Plant Sci 11:1–12

Yang XE, Chen WR, Feng Y (2007) Improving human micronutrient nutrition through biofortification in the soil-plant system: China as a case study. Environ Geochem Health 29: 413–428

Yasin M, El-Mehdawi AF, Anwar A, Pilon-Smits EAH, Faisal M (2015) Microbial-enhanced selenium and iron biofortification of wheat (*Triticum aestivum* L.): applications in phytoremediation and biofortification. Int J Phytoremediation 17:341–347

Ye Y, Qu J, Pu Y, Rao S, Xu F, Wu C (2020) Selenium biofortification of crop food by beneficial microorganisms. J Fungi 6:59

Zhang M, Xing G, Tang S, Pang Y, Yi Q, Huang Q, Huang X, Huang J, Li P, Fu H (2019) Improving soil selenium availability as a strategy to promote selenium uptake by high-se rice cultivar. Environ Exp Bot 163:45–54

Zhou X, Yang J, Kronzucker HJ, Shi W (2020) Selenium biofortification and interaction with other elements in plants: a review. Front Plant Sci 11:1–18

Zhu YG, Pilon-Smits EAH, Zhao FJ, Williams PN, Meharg AA (2009) Selenium in higher plants: understanding mechanisms for biofortification and phytoremediation. Trends Plant Sci 14:436–442

Zou C, Du Y, Rashid A, Ram H, Savasli E, Pieterse PJ, Ortiz-Monasterio I, Yazici A, Kaur C, Mahmood K, Singh S, Le Roux MR, Kuang W, Onder O, Kalayci M, Cakmak I (2019) Simultaneous biofortification of wheat with zinc, iodine, selenium, and iron through foliar treatment of a micronutrient cocktail in six countries. J Agric Food Chem 67:8096–8106

Chapter 8
Nanoparticles Based Biofortification in Food Crops: Overview, Implications, and Prospects

Kaleem ul din, Muhammad Shahbaz Naeem, Usman Zulifqar, Ghadeer M. Albadrani, Ejaz Ahmad Waraich, and Saddam Hussain

Abstract Human health is being seriously affected by nutrient deficiencies in food crops, particularly in underdeveloped and remote areas. The availability of sufficient and safe nutrition to prevent malnutrition and cure various diseases may be an aspect of improving human health. To overcome the present issue; biofortification, a method to enhance nutritional status of food crops, can address the issue of hidden hunger. Nanotechnology may contribute to improving the quality of food through biofortification and may prove to be an effective and sustainable remedy to this issue by foliar application of essentials nutrients (Zn, Cu, Fe and Se) nanoparticles and their nano-based fertilizers in the soil to improve nutrient deficiency. This review highlights the use of nanomaterials in biofortification of plant nutrients, specifically their absorption and translocation, which have positive outcomes for alleviation of hidden hunger and improving nutrient levels. Additionally, the significance of nano-biofortification is discussed in relation to the COVID-19 pandemic and the problem of nutritional security. By understanding the various issues related to the safe use of nanoparticles and their future prospects, we can improve their effectiveness in fulfilling dietary needs through nano-biofortification.

K. ul din
Department of Botany, Faculty of Sciences, University of Agriculture Faisalabad, Faisalabad, Pakistan

M. S. Naeem · E. A. Waraich
Department of Agronomy, Faculty of Agriculture, University of Agriculture Faisalabad, Faisalabad, Pakistan

U. Zulifqar
Department of Agronomy, The Islamia University of Bahawalpur, Bahawalpur, Pakistan

G. M. Albadrani
Department of Biology, Princess Nourah Bint Abdulrahman University, Riyadh, Saudi Arabia

S. Hussain (✉)
Department of Agronomy, Faculty of Agriculture, University of Agriculture Faisalabad, Faisalabad, Pakistan

Department of Biology, Princess Nourah Bint Abdulrahman University, Riyadh, Saudi Arabia

M. Hasanuzzaman et al. (eds.), *Mineral Biofortification in Crop Plants for Ensuring Food Security*, https://doi.org/10.1007/978-981-99-4090-5_8

Keywords Agronomic biofortification · Essential minerals · Malnutrition · Nanotechnology

8.1 Introduction

Nutrient deficiencies in food crops, particularly in underprivileged populations, greatly compromise human health (Kapoor et al. 2022). Inadequate dietary diversity, along with an increasing global population, further threatens the sustainability of nutrition through the use of cereal-based crops with low mineral supplements, posing a worldwide nutritional security threat (Dhaliwal et al. 2022). Inadequacies of micronutrients including zinc (Zn), selenium (Se), iodine (I), iron (Fe), and vitamin A, pose major worldwide health concerns for one-third of the population due to inaccessibility (Mullen 2019; Godecke et al. 2018). The absence of these essential micronutrients in the diet has adverse effects on human health, including stunted development, dementia, prenatal problems, and increased mortality (Bailey et al. 2015; Bailey et al. 2011; Zulfiqar et al. 2020a). Undernourishment occurs when the body does not receive enough of any necessary nutrient to repair its tissues. Currently, 2 billion people suffer from micronutrient deficiencies (Zn and Fe), are overweight or obese, and over 820 million people are undernourished (UNEP 2021; Huang et al. 2020; FAO 2019).

Iron and Zn are two essential micronutrients required for human nutrition, and their deficiency is a severe health issue globally. Zinc deficiency is prevalent in many regions, especially in South Asian and Sub-Saharan African countries (Shekari et al. 2015; Shahzad et al. 2014). There are also deficiencies of iodine and vitamin A in these regions (Stevens et al. 2015). The lack of micronutrients is typically an issue in regions where soil contains low plant accessible micronutrients in food items (Cakmak et al. 2017; Zulfiqar et al. 2021a, 2021b). Therefore, practical and affordable strategies are necessary to address the nutritional needs of these regions within the global food framework.

The provision of nutrients via fertilization or other methods to edible plants is known as biofortification. This interaction is essential for maintaining human health (Tiozon et al. 2021). Cereal crops such as wheat, rice and maize (Al-Juthery et al. 2022; Zulfiqar et al. 2020b, 2020c; Cheah et al. 2020), pulse crops (Jha and Warkentin 2020) and horticultural crops like strawberry (Budke 2020) are among the most critical biofortified food crops. Nanotechnology is one of the technological solutions that can be gradually used in agricultural production to fortify crops (Thakur et al. 2018).

Biofortification using nano-based micronutrients such as Zn (Du et al. 2019), copper (Lopez-Vargas et al. 2018), Fe (Guha et al. 2021) and selenium (El-Ramady et al. 2020b) fertilizers can improve the bioavailability of these micronutrients due to their tiny size and larger surface area, which also facilitates the ability for them to concentrate in sink areas (Liu and Lal 2015). In order to attain precision farming, which aims for higher production with fewer resources, the present agricultural sector is seeking alternatives to the use of agrochemicals through the application of nanotechnology with the biological synthesis of nanomaterials (Chen et al. 2018). To overcome the transnational agriculture issues, a variety of nanomaterials applications, such as, nano detectors, nano herbicides and nano fertilizers derived from natural polymeric nanoparticles are used to address nutrient deficiency and enhance crop productivity (Valencia et al. 2019).

This review focused on the advancement in the biofortification of crops using various nanomaterials and modes of application to overcome malnutrition. Moreover, the use of nano-biofortification as a potential remedy for the COVID-19 and climate change world crises is also discussed.

8.2 Nano-Biofortification

A novel strategy nano-biofortification is being utilized to complement human diets with an adjusted diet utilizing nutrients to prevent malnutrition. This innovative strategy has various advantages and drawbacks including nanoparticles-based fertilizers or nutrients (El-Ramady et al. 2020a, 2020b, 2020c). The significant advantages might include improved agricultural productivity and soil and water treatment using nanotechnology, while nano-contamination and toxicity are likely to be the primary drawbacks (Martinez et al. 2020; Rizwan and Ali 2021). Biogenic synthesis of nanoparticles plays a vital role in human beings as compared to chemical and physical processes because it is more efficient, cheap, and ecologically friendly (Stephen et al. 2021). Various types of metal and metal oxides nanoparticles are synthesized such as sulfur, Fe and copper from basil, green and black tea, mint, and eucalyptus leaves extracts respectively (Mareedu et al. 2021; Iliger et al. 2021; Ragab and Saad-Allah 2020). Zinc oxide and nickel oxide nanoparticles are prepared from *Nilgirian tusciliantus* leaf and seeds of black cumin respectively through the green approach (Resmi et al. 2021; Boudiaf et al. 2021). The nano-based nutrients play a significant role in human beings curing most the diseases in human beings, such as ZnO and selenium, particularly in fighting against COVID-19 (He et al. 2019; Gatadi et al. 2021).

8.3 Agronomic Biofortification Vs Nano Biofortification

In developing countries, cereal grains such as wheat, rice and maize are the main food crops that provide almost 60% of daily calorie needs. However, these whole grain products that have a high amount of nutrients including carbohydrates, fiber, lipids, proteins, minerals, phytochemicals, and vitamins, lose the majority of their beneficial components, especially micronutrients, due to conventional handling (Laskowski et al. 2019). On a large scale, agronomic biofortification is considered one of the approaches to enhance the mineral concentration in these cereal crops. Unfortunately, agronomic biofortification is unable to meet expectations owing to either low soil quality or an ineffective drainage system, resulting in the loss of most applied fertilizers (Yashveer et al. 2014). The quality of the crop grains is reduced by the uncontrolled discharge of nutrients caused by chemical fertilizer. One of the major obstacles that adversely affect the environment through agronomic biofortification, is when chemical fertilizers are used in higher quantities such as heavy metals and nitrate into the water and soil. Considering this, the long-term effects on soil fertility, soil structure, and disturbances in the balance of soil nutrients using chemical fertilizers (Elemike et al. 2019).

Nanotechnology could provide potential solutions for some of the few downsides of agronomic biofortification. Nanomaterials possess several important characteristics such as slow and controlled release at target sites, high surface-to-volume ratio, and high sorption capacity, making them more viable for the synthesis of nano-based fertilizers. This consistent long-term supply of plants using nanofertilizers enables improved crop development compared to traditional fertilizers (Feregrino-Perez et al. 2018). Furthermore, the use of nanofertilizers in low amounts can potentially prevent the soil from being overburdened with chemical fertilizer byproducts, thereby overcoming environmental toxicity (León-Silva et al. 2018). In contrast to synthetic fertilizer, biosensors may be used to create and deliver nanofertilizers based on the crops' nutritional needs and the state of the soils' nutrients. (Kah et al. 2018). The application of nano-based fertilizers at the right times and in the right doses can maximize the use of natural resources and enable precise farming. Farmers often use conventional fertilizers to boost yields multiple times during a growing season, but the slow release of nutrients from nanofertilizers reduces the number of doses required, ultimately reducing the cost of the application (Fellet et al. 2021). When used in extensive field projects, nanofertilizers can be particularly cost-effective since they are effective enough to overcome these limitations. Consequently, nanofertilizers are now consistently favored for biofortification because of the benefits over synthetic fertilizers.

8.3.1 Limitations of Traditional Agronomic Methods

Inorganic fertilizers are a widely used method to biofortify crops in both soil and soilless systems. However, this approach has several limitations. One of the primary concerns is the negative impact of overuse, which leads to environmental problems such as water pollution, algal blooms, and reduced biodiversity in natural systems (Fernández-Luqueño and López-Valdez 2018; Zhang et al. 2019; Schier et al. 2019). Another limitation is that traditional agronomic methods can be labor-intensive and expensive, which can be challenging for small-scale farmers in developing countries who may lack the necessary resources to implement them effectively.

Another challenge is identifying the optimal timing for fertilizer application to achieve the highest nutrient content (Phattarakul et al. 2012; Rodrigo et al. 2014; Mabesa et al. 2013). For example, foliar Zn application to rice during the early milk plus dough stages increased grain Zn content more than earlier applications

(Phattarakul et al. 2012), while foliar selenium application to wheat pre-flowering and in between the booting and heading stages increased selenium grain content the most (Rodrigo et al. 2014). The success of soilless cultivation for biofortification also depends on factors such as timing, chemical form, and amount of fertilizers (Rouphael and Kyriacou 2018). When biofortifying carrots hydroponically with iodine, the same rate of fertilization as field foliar fertilizer applications resulted in cumulative toxic levels of iodine in the hydroponically biofortified carrots (Signore et al. 2018). Additionally, the biofortification of multiple micronutrients at once may be problematic in soilless cultivation due to potential antagonistic effects on crop accumulation (Germ et al. 2019).

Other factors that can influence the effectiveness of fertilizers to biofortify crops include crop phenotype, genotype, and soil conditions (Izydorczyk et al. 2021). Plant genotype can result in differences in nutrient uptake, translocation, and accumulation, which affects the effectiveness of fertilizers (Clark 1983; Rosa et al. 2019). Soil factors can also impact nutrient accumulation differences between genotypes (Mabesa et al. 2013). Moreover, Fe biofortification via Fe fertilization of calcareous soil with high pH is ineffective due to the reduced mobility of Fe and rapid conversion into unavailable forms (Ramzani et al. 2016).

Soilless biofortification has several unique limitations, including its expense due to necessary equipment and energy costs (Lages Barbosa et al. 2015), its inapplicability to all crop types, and its limited capacity due to physical constraints of the system being used (e.g., size of the greenhouse, pots, etc.). Some regions may also consider it unsustainable (Lages Barbosa et al. 2015). In conclusion, traditional agronomic methods have several limitations that can limit their effectiveness in addressing global malnutrition and health.

8.3.2 How Nano Technology Can Overcome these Limitations

Nanotechnology has emerged as a promising approach for overcoming the limitations of traditional agronomic methods for crop biofortification. The use of nanotechnology-based approaches can help improve nutrient uptake and absorption in plants, leading to better crop yields and increased nutritional value (Shang et al. 2019).

One way in which nanotechnology can overcome the limitations of traditional agronomic methods is through the use of nanoparticles. Nanoparticles are incredibly small particles, typically between 1 and 100 nanometers in size, with unique physical and chemical properties that can be manipulated to enhance nutrient uptake and absorption in plants (Mittal et al. 2020). Plant nutrient absorption can be optimized by using nanomaterials to encapsulate nutrients. This is because the slow or controlled release of nanoparticles enhances nutrient uptake, while their small size allows them to easily penetrate biological barriers and enter the plant's vascular system (Al-Mamun et al. 2021). For example, nanoparticles can be engineered to release micronutrients slowly over time, providing a more sustained source of nutrition for crops (Sun et al. 2022). They can also be designed to penetrate the plant's cell walls, allowing for more efficient uptake of nutrients (Mittal et al. 2020).

Nanoparticles can also help address challenges related to micronutrient bioavailability. Traditional agronomic methods often involve adding micronutrients to soil or fertilizer, but these nutrients may not be taken up by crops effectively. Nanoparticle-based approaches can help overcome this challenge by providing micronutrients in a more bioavailable form. For example, nanoparticles can be designed to encapsulate micronutrients, protecting them from environmental degradation and improving their absorption in plants (Elemike et al. 2019). Furthermore, nanotechnology-based approaches can help improve the efficiency of fertilizer use. Traditional agronomic methods often involve the application of excessive amounts of fertilizer, leading to nutrient leaching and environmental pollution. Nanoparticle-based approaches can help reduce fertilizer use by enhancing the efficiency of nutrient uptake and absorption in plants (Abdel-Hakim et al. 2023).

In conclusion, nanotechnology-based approaches have the potential to overcome the limitations of traditional agronomic methods for crop biofortification. Nanoparticle-based approaches can improve nutrient uptake and absorption in plants, address challenges related to micronutrient bioavailability, and reduce the environmental impact of traditional agronomic methods. With further research and development, nanotechnology-based approaches could play a critical role in enhancing food security and improving the nutritional quality of crops.

8.4 Nanocarriers-Based Nutrient Delivery Approach

To increase the availability of nutrients to roots and reduce nutrient loss, slow and controlled distribution of nutrient carrier materials affects the nutrient delivery rate in soil and in the edible portions of crops (Dutta et al. 2022). Compared to traditional fertilizers, cellulosic biopolymer-based delayed release nanocarriers can now improve the nutrient transformation proportion by breaking down more slowly in the soil and releasing nutrients gradually over time, resulting in a sustained release profile (Hess 2012). For biofortification efforts, it is crucial to give plants enough Zn through foliar or soil application under field conditions (Ludwig and Slamet-Loedin 2019; Kumar et al. 2016). Waste materials from maize, such as cellulose and lignin, can be found in nanocomposite biochar prepared from corn cob and effectively utilized as transporters of micronutrients.

Recently, a variety of nanoscale nutrition delivery approaches and their interactions with active substances have been discovered to work through emulsion, surface adsorption, entrapment, encapsulation, and release of nutrients from nanoscale delivery systems (De Matos et al. 2018). Several micronutrients may be delivered to field crops using chitosan nanoparticles, and they also possess antibacterial properties (Malerba and Cerana 2016). The acetamiprid insecticide, which is soluble in water, the fast release is delayed when it is encapsulated in a chitosan alginate nanocarrier act as a nano-encapsulation in chitosan nanoparticles to mitigate the toxicity of herbicides (Kumar et al. 2015). Chitosan nanoparticles are used as a covering for controlled release fertilizers and insecticides due to the appropriate absorption on leaves. They also act as a nanocarrier for a wide range of substances by possessing important properties such as biodegradable biopolymer, biocompatible and nontoxic (Nadendla et al. 2018; Deshpande et al. 2018).

8.4.1 Nanocellulose in Precision Farming

Utilizing methylcellulose and hydrolyzed polyacrylamide to create highly adsorbent nanocomposites, nanocellulose is currently employed in fertilizers for smart agriculture (Bortolin et al. 2016). Most delivery agents in agriculture are made from nanocellulose, which exists in different morphological types such as nanoparticles, nanofibers, and films, contributing to agricultural applications through nanocapsulation (Khaledian et al. 2019; Davidson et al. 2013). Nanoencapsulation processes involve coating chitosan, liposomes, polylactides, and lipids, and can also encapsulates protein.

8.5 Essential Nano Micronutrients in Biofortification

Nanotechnology-enabled precision agriculture has the potential to significantly increase the yield and quality of food crops through biofortification (Sharma et al. 2017; Xiong et al. 2017; Datta and Vitolins 2016; Clemens, 2014). The excessive use of fertilizers to achieve high-yielding crop varieties and the lack of micronutrient supplementation are the main causes of micronutrient-deficient soil (Sillanpaa 1990). Foliage spray or soil application can enhance the level of micronutrients and also play an important role in increasing the growth and production of crops. (Sharma et al. 2017; Buzea et al. 2007).

8.5.1 Zinc

Zinc sulphate and zinc oxide (ZnO) are two sources of Zn. Due to their lower volatilization and high surface area to volume ratio, ZnO nanoparticles are known to be effectively absorbed, stored, and metabolized in plants to treat the deficiency of this essential micronutrient (Milani et al. 2015). It has been reported that foliar spray of ZnO nanoparticles boosts the concentration of Zn up to 82% in maize (*Zea mays* L.) as compared to the zinc sulphate ($ZnSO_4$) used as a source of conventional Zn fertilizer (Umar et al. 2021). As compared to the soil applied zinc sulphate the foliar application of zinc oxide nanoparticles positively increases the concentration of Zn in the kernels (Javaid et al. 2020). Plants treated with ZnO nanoparticles by different application methods such as seed priming, foliar or in soil have higher protein contents and reduced levels of cadmium, as well as increased concentrations of Zn in the grains compared to conventional fertilizers (Hussain et al. 2018; Rizwan et al. 2019a; Sheoran et al. 2021). The use of these nanoscale ZnO based fertilizers plays an important role in the biofortification approach (Ivanov et al. 2021).

8.5.2 Copper

Copper (Cu) is an essential micronutrient that contributes to metabolic syndrome, cardiovascular disease, diabetes, and associated problems due to their oxidative properties, have been shown to increase the inflammatory response. The deficiency of this nutrient cause disease of ruminants in the animals due to copper deficient soils which ultimately reduce their concentration in the forage grasses (Menzir and Dessie 2017). Copper oxide-based nanoparticles are being utilized in the plant growth promoter, fertilizers and for soil remediation (Xiong et al. 2017). Fsoliar application of Cu nanoparticles on tomato plants at a concentration of 10–250 mg L^{-1} increased the level of β-carotenoid and also 36% of vitamin C (Hernandez-Hernandez et al. 2019). It has been reported that the soil applied nanoparticles of Cu (50–500 mg kg^{-1}) play a potential role to enhance the morphological attributes and concertation of copper in the seeds which are 1.8 times more than other tissues of the soybean plant (Yusefi-Tanha et al. 2020a). These studies reveal that the use of Cu nanoparticles in agriculture is an important strategy for Cu-deficient crop biofortification. However, it is important to note that the inaccurate use of these nanoparticles can cause serious health problems, as a higher amount of Cu is toxic and thus must be used appropriately.

8.5.3 Iron

One-third of the world's population is suffering from Fe deficiency, which is the most important micronutrient for metabolism, growth, and development in humans and plants (Chugh et al. 2022). The lack of dietary Fe affects around 14% of the global populace (Matres et al. 2021). Wheat and rice are two main crops used as food worldwide, but the removal of the outer bran layers during processing results in an insufficient amount of Fe (Ludwig and Slamet-Loedin 2019). Studies have shown that the application of Fe-chelated nanofertilizers in rice and basil plants can improve growth and yield contributing indices and nutrient concentration, particularly NPK (Parande and Mirza 2011; Fakharzadeh et al. 2020). The seed priming of finger millet with nanoparticles of iron oxide at a level of 100 ppm has been found to increase the concentration of Fe in the grains compared to iron sulfate (Chandra et al.

2021). When these nanoparticles are applied to the soil at a level of 50 and 500 mg kg^{-1} positively improved the biomass and a gradual increase in the contents of Fe, amino acids such as tyrosine and cysteine in wheat plants (Wang et al. 2019a). These studies clearly demonstrate that the application of Fe nanoparticles is a novel and environmentally friendly approach to reducing the deficiency of Fe in plants and achieving effective biofortification.

8.5.4 Selenium

Selenium (Se) is a fundamental micronutrient essential for both plant and human health. It is a biological compound with 25 identified selenoproteins serving biological purposes that are relevant to human health. The human genome contains a total of 25 selenoproteins, several of which are oxido-reductases and include SeCys as a catalytic residue (Shu et al. 2020; Baclaocos et al. 2019). These selenoproteins are essential for controlling human immunity because they are primarily involved in several redox processes (Avery and Hoffmann 2018). The main sources of Se for humans are Se dietary supplements, foods enriched with selenium, and plants that uptake Se from the soil (El-Ramady et al. 2020b). Selenium application through the soil and foliar is a biofortification strategy to overcome the deficiency of this nutrient in the crops which can be grown in the low Se soils (Lopes et al. 2017). Selenium biofortification in the cereals crops such as wheat, rice and maize are achieved through the use of Se based fertilizers, which is a well-known method. Several studies have shown the positive effect of Se fertilizers on different crops (Brevik et al. 2020), such as rice (Reis et al. 2020; Farooq et al. 2019), maize (Joy et al. 2019; Bocchini et al. 2018), wheat (Du et al. 2019; Ramkissoon et al. 2019; Zou et al. 2019). In addition, selenium nanoparticles show potential in mitigating the low concentration of this nutrient in different edible crops (Golubkina et al. 2017).

These nano-based selenium have great potential compared to other bulk selenium because of their unique properties such as higher bioavailability, antimicrobial, anticancer, higher surface area to volume ratio, and improved antioxidants (Hosnedlova et al. 2018). Selenium nanoparticles may be used to increase the biochemical synthesis of proteins, phenolics, amino acids, and glucosinolates and they are considered an inexpensive way to biofortify field crops (Carvalho et al. 2003). Biofortification of cereals, legumes, and vegetables through a variety of nanomaterials such as Zn, Fe, selenium and copper-based nanoparticles are summarized in Tables 8.1, 8.2, and 8.3 respectively. These tables include size, concentration, mode of applications and positive response of these nanoparticles on plants discussed in previous studies.

Table 8.1 Effect of different types, size and concentration of nanoparticles (NPs) are used in the biofortification of cereal crops

Nano-particles type	Plant Species	Size	Dose of application	Mode of application	Micronutrient enhancement	Other positive responses	Reference
ZnO	Wheat (*Triticum aestivum* L.)	20–30 nm	100 mg kg^{-1}	Soil	Increased Zn content in grains by 40–185% and 190%	Enhanced the grain Yield and photosynthesis reduced the oxidative stress Reduce the cd toxicity	Hussain et al. (2018)
		18 nm	3.5 and 1.7 mg kg^{-1}	Soil	Increased (29%) grain Zn concentration	Increased chlorophyll levels Accelerated panicle emergence under drought	Dimkpa et al. (2020)
		< 100 nm	20 to 1000 mgL^{-1}	Soil	The concentration of Zn in grains increased by 3.3 times	–	Du et al. (2019)
		–	0.96 kg ha^{-1}	Foliar	Zn concentrations increased by approx. 30-fold in grain endosperm	Enhanced activity of catalase and per-oxidase enzymes	Sun et al. (2020)
	Maize (*Zea mays* L.)	25 nm	50–2000 mg L^{-1}	Foliar	37% increase in grain Zn content	Improved growth, yield and enhanced the zinc content of maize grains	Subbaiah et al. (2016)
		–	8 kg Zn ha^{-1}	Soil	The maximum increase in grain Zn concentration (82%)	Improved growth and physiological parameters of plant	Umar et al. (2021)
		20–30 nm	50–100 mg L^{-1}	Foliar	–	Improved the height of maize plants, number of leaves, shoot and roots dry biomass, chlorophyll concentrations and gas exchange attributes	Rizwan et al. (2019a)
		10–30 nm	25 and 100 mg kg^{-1}	Soil	Enhanced Zn concentration 13.5–39.4%	Enhanced the grain yield, nutrient acquisition, and grain Zn fortification	

	Rice (*Oryza sativa* L.)						Yang et al. (2021)
		31 nm	20 and 60 mg L^{-1}	Foliar	Enhanced Zn concentration from 17.7 to 50% in rice grains.	Improved agronomic and physiological features	Elshayb et al. (2021)
FeO/Fe	Wheat (*Triticum aestivum* L.)	20-40 nm	500 mg L^{-1}	Hydroponic	–	Enhanced chlorophyll content and biomass	Al-Amri et al. (2020)
		50–100 nm	100 mg kg^{-1}	Soils	Increase 25% of Fe concentration in grains	Improved growth improved the photosynthesis, yield and Fe concentrations	Adrees et al. (2020)
		–	25–600 mg L^{-1}	Seed priming	Grain Fe content increased by 26.8 and 45.7%	Increased in seed germination and shoot length	Sundaria et al. (2019)
		50–100 nm	5 to 20 mg L^{-1}	Seed priming	Almost 38% increased the grain Fe concentration	Positively affected photosynthesis. Reduced electrolyte leakage and superoxide dismutase and peroxidase activities.	Rizwan et al. (2019b)
Se	*Maize (Zea mays* L.)	25 nm	2–20 mg L^{-1}	Foliar	–	Increased chlorophyll content, net photosynthetic rate, and leaf biomass	Wang et al. (2023)
	Rice (*Oryza sativa* L.)	–	25–100 μmol L^{-1}	Foliar	Enrichment of se up to 218.9–1096.6 μg kg^{-1}	Decrease cd accumulation, improved photosynthesis, and organic se contents in rice grains	Wang et al. (2021)
Cu/CuO	Wheat (*Triticum aestivum* L.)	14.85 nm	50 and 500 mg kg^{-1}	Soil	Grain cu concentration enhanced by 18.84%–30.45%	–	Wang et al. (2019b)
		17–38 nm	25–100 mg kg^{-1}	Soil	–	Increased the biomass and height of wheat plants by reduced cd influx and improved nutrient acquisition	Noman et al. (2020)

Table 8.2 Effect of different types, size and concentration of nanoparticles (NPs) are used in the biofortification of pulses crops

Nano-particles type	Plant Species	Size	Dose of application	Mode of application	Micronutrient enhancement	Other positive responses	Reference
ZnO	Common bean (*Phaseolus Vulgaris* L.)	30 nm	0.05%, 0.1%, or 0.15% w/v	Foliar	Enrichmentof Zn up to 49.00–58.50 mg kg^{-1}	Improved vegetative characteristics, yield and quality	Mahdieh et al. (2018)
		<20 nm	20, and 40 ppm	Foliar	–	Improved morphological indices and enhance yield	Salama et al. (2019)
		–	50 mg L^{-1}	Foliar	–	Improved vegetative growth, fresh pod yield, pods physical quality and nutritional value.	Marzouk et al. (2019)
	Soybean (*Glycine max* L.)	38 nm	40 to 400 mg kg^{-1}	Soil	–	Enhanced seed yield, lipid peroxidation, and various antioxidant	Yusefi-Tanha et al. (2020b)
	Green gram (*Vigna radiata* L.)	8 nm	10, 20, 50 mg L^{-1}	Foliar	3.2% higher zinc (Zn) content in seed	Improved in yield and quality of green gram	Sahoo et al. (2021)
Fe	Soybean (*Glycine max* L.)	50–100 nm	100, and 200 ppm	Foliar	–	Improved drought tolerance, yield, and seed quality of soybean	Dola et al. (2022)
Se	Groundnut (*Arachis hypogaea* L.)	10–30 nm	20–40 mg L^{-1}	Foliar	–	Enhanced the antioxidant defense systems. Improved yield components and oil Production of seeds	Hussein et al. (2019)
Cu	Soybean (*Glycine max* L.)	25 nm	50, 100, 200, and 500 mg kg^{-1}	Soil	1.8 times higher (6.55 mg/kg) than the control	Improved seed nutritional cu value	Yusefi-Tanha et al. (2020a)

Table 8.3 Effect of different types, size and concentration of nanoparticles (NPs) are used in the biofortification of vegetables

Nanoparticles type	Plant Species	Size	Dose of application	Mode of application	Positive responses	Reference
ZnO	Eggplant (*Solanum melongena* L.)	–	50, and 100 ppm	Foliar	Improved growth characteristics and increased fruit yield	Semida et al. (2021)
Se	Cucumber (*Cucumus sativus* L.)	50–200 nm	25 mg L^{-1}	Foliar	Increased nutrient uptake. Increase in marketable fruit yield and yield quality	Shalaby et al. (2021)
	Tomato (*Solanum lycopersicum* L.)	–	10 mg L^{-1}	Foliar	Increases the chlorophyll content and improves the enzymatic and non-enzymatic antioxidant systems. Improve yield	Hernandez-Hernandez et al. (2019)
	Bitter melon (*Momordica charantia* L.)	10–45 nm	1 to 50 mg L^{-1}	In vitro experiment	Enhanced the activity of leaf nitrate reductase	Rajaee Behbahani et al. (2020)
Cu/ CuO	Bell pepper (*Capsicum annuum* L.)	50 nm	100 and 500 mg L^{-1}	Foliar	Increased the content of chlorophylls, lycopene and GPX activity in the leaves.	Gonzalez-Garcia et al. (2021)
	Tomato (*Solanum lycopersicum* L.)	50 nm	10 and 50 mg L^{-1}	Foliar	Increased the content of chlorophyll and activity of the antioxidant enzymes. Improved the nutraceutical characteristics	Quiterio-Gutierrez et al. (2019)
	Green onion (*Allium fistulosum*)	10–100 nm	75, 150, 300, and 600 mg kg^{-1}	Soil	Improved their nutrient quality by increasing the allicin and essential element	Wang et al. (2020)
	Lettuce (*Lactuca sativa* L.)	10–100 nm	200 to 400 mg kg^{-1}	Soil	Modification in plant photosynthesis system with increased transpiration rate and higher stomatal conductance	Wang et al. (2019a)

8.6 Nanoparticle Synthesis and Characterization

Nanoparticle synthesis and characterization are critical steps in the development of nanotechnology, with numerous applications in various fields including medicine, environmental remediation, and energy production (Khan et al. 2019). Recent advances in the field of nanotechnology have led to the development of novel and innovative techniques for the synthesis and characterization of nanoparticles.

Green synthesis of nanoparticles has been the subject of intense research in recent years due to its cost-effectiveness and eco-friendly nature. Various natural sources such as plants, bacteria, and fungi have been used for the green synthesis of nanoparticles (Singh et al. 2018). Additionally, several studies have reported the use of biocompatible polymers and surfactants for the synthesis of nanoparticles, which offer several advantages such as biodegradability and reduced toxicity (Kucuk et al. 2023). Physical methods such as ball milling, laser ablation, and plasma synthesis have also been employed for the synthesis of nanoparticles, with each method offering unique advantages (Baig et al. 2021). Chemical methods such as sol-gel, co-precipitation, and hydrothermal synthesis have been extensively studied and are widely used due to their ease of use, reproducibility, and scalability (Baig et al. 2021).

The characterization of nanoparticles is equally important as the synthesis itself. Several techniques such as scanning electron microscopy (SEM), transmission electron microscopy (TEM), dynamic light scattering (DLS), and X-ray diffraction (XRD) are commonly used for the characterization of nanoparticles (Khan et al. 2019). Additionally, several recent studies have reported on the use of spectroscopic techniques such as Raman spectroscopy and Fourier-transform infrared spectroscopy (FTIR) for the characterization of nanoparticles (Lopez-Lorente and Mizaikoff 2022). In recent years, concerns have been raised regarding the safety of nanoparticles and their potential impact on human health and the environment. Several studies have focused on investigating the toxicological effects of nanoparticles and their possible interactions with biological systems (Missaoui et al. 2018). The development of safe and effective nanoparticle-based technologies requires a thorough understanding of their properties and potential risks. Nanoparticle synthesis and characterization are crucial for nanotechnology-based applications. Recent research has emphasized the development of new synthesis methods, their characterization, and safety concerns. Green synthesis methods, biocompatible materials, and various techniques have been investigated. Characterization techniques are essential to determine properties and potential risks.

8.7 Mode of Uptake and Translocation of Nanoparticles

The most important issue to consider when estimating the influence of nanoparticles on the metabolism of plants and their translocation from the environment to the plant body. Because the plant is one of the most important components which uptake and transfer various type of nanoparticles to the food chain (Wang et al. 2013a). Several factors, including plant species, chemical properties, size, and stability, can impact the uptake, transport, and accumulation of these nanoparticles (Rico et al. 2011).

8.7.1 Foliar Uptake and Translocation

There are two main pathways for the foliar uptake of nanoparticles, the first stomatal and the second cuticular pathway (Lv et al. 2019). Leaves of higher plants are covered with a waxy cuticle which protects against the loss of water and other solutes exchange and is also thought to be the first natural barrier for the entry of nanoparticles into the leaves tissue (Pollard et al. 2008). The nature of the cuticle changes with variations in the life stages of the leaves, so nanoparticles can enter the tissues without this initial barrier in some cases such as diseases caused by the pest and environmental conditions that disturb the cuticle nature. In the cuticular pathway, there are two main ways through which the solutes are taken up and cross this waxy layer one is the hydrophilic pathway through which polar solutes enter via aqueous pores, and the lipophilic pathway, through which nonpolar solutes enter via diffusion. Both pathways have effective sizes between 0.6 and 4 nm (Eichert et al. 2008). So, the nanoparticles that have a lower size (4.8 nm) can penetrate through the cuticular pathway, but it has also been reported that larger nanoparticles (5 nm) can penetrate this pathway, when foliar is applied. Several studies prefer the stomatal pathway for the uptake of nanoparticles because small pore size of the cuticle (Schreiber 2005; Eichert et al. 2008; Ali et al. 2012; Wang et al. 2021). The capacity of nanoparticles foliar uptake depends on the variation in the morphology of the leaves of different plant species, the number, and size of the stomata (Uzu et al. 2010). Long distance transportation of nanoparticles occurs through the vascular system when they are absorbed by the stomatal pathway and enter the apoplast of the leaf. There are two essential conductive tissues in the vascular system named as xylem and phloem through which long distance transportation occurs in higher plants, while the direction of flow from top to bottom occurs in the phloem system. Therefore, the only feasible foliar absorption channel for nanoparticles to translocate from the upper to lower part of the plants (leaf to root) is the phloem system (Lough and Lucas 2006). Although both stomatal and cuticular pathways are involved in the foliar absorbance of the Zn based nanoparticles have separate importance (Singh et al. 2018). It was also observed that the uptake of Fe_3O_4 nanoparticles from the leaves of the wheat seedling and transport to the roots through the phloem system (Cai et al. 2020; Lu et al. 2020). Foliar uptake of metal oxide nanoparticles has the range of size 24–47 nm was also observed in the watermelon, detection of different metal elements in the roots revealed that these nanoparticles are transported from shoots to stem and finally in roots through the phloem vascular system (Wang et al. 2013b). Foliar application of ZnO based nanoparticles enhanced their accumulation in the treated coffee plant leaves compared to the untreated plants (Rossi et al. 2019).

8.7.2 Root Uptake and Translocation

The uptake of nanoparticles via root is preferred according to several studies because they remain available and exist for a longer time in the soil compared to the leaves of plants. However, the soil uptake and transport of nanoparticles face more difficulties compared to the foliar uptake. During transport from root to shoot via the xylem vascular system these nanoparticles cross many barriers, some of which are root cuticles and Casparian strips (Kapoor et al. 2022). Soil application of nanoparticles in the form of nano fertilizers or directly firstly attaches them to the root surface and then penetrate through various ways, such as carrier protein, aquaporins, endocytosis and ion channels. One of the most recognized routes for the uptake of nanoparticles is endocytosis, which allows them to cross the membranes (Rico et al. 2011; Schwab et al. 2016; Wong et al. 2016). Root hairs on the surface of the plant roots secrete negatively charge tiny molecules like organic acids that attract positively charged nanoparticles, enhancing absorbance and accumulation in the roots of the plants (Zhou et al. 2011). In the regions of the roots where the waxy layer of the cuticle is underdeveloped like root hairs and root tips, nanoparticles can easily penetrate (Schwab et al. 2016). Apoplastic and symplastic pathways are also used in the translocation of nanoparticles which involves the transport via the spaces between the cells and plasmodesmatal passage for cell-to-cell movement respectively (Kapoor et al. 2022). Casparian strips are the main barrier to protect the entry of nanoparticles in the roots when they come from the epidermis to cortex and then reaches the endodermis, all occurring in the apoplastic pathway. However, under some conditions nanoparticles enter the vascular system to bypass this barrier through the regions where these strips are not formed and like root tips (Lv et al. 2015; Schymura et al. 2017) or the regions where the disconnection of the Casparian strips like the junction of lateral roots (Dietz and Herth 2011).

Several studies have indicated that the apoplastic pathway is the more likely route for nanoparticles to be transported in root tissues, as observed by the presence of nanoparticles in the spaces between the cells of roots (Li et al. 2016; Tombuloglu et al. 2019; Wang et al. 2012; Zhu et al. 2008). It has also been observed by using techniques like Transmission electron microscopy (TEM) and Confocal laser scanning microscopy (CLMS) that the nanoparticles are present in intercellular spaces of the cells of the root. For instance, rye grass contains 20 nm of zinc oxide (ZnO) nanoparticles in their roots (Lin and Xing 2008), silver (Ag) and gold (Au) nanoparticles having the size of 20–80 nm and 12 nm respectively in the roots of *Arabidopsis thaliana* (Geisler-Lee et al. 2012; Avellan et al. 2017) the roots of *Elsholtzia splendens* contain the copper oxide (CuO) nanoparticles of size 43 nm (Shi et al. 2014) and in maize plant, the junction of lateral roots and xylem contain a high amount of ZnO nanoparticles was reported. So, this junction of lateral roots plays a significant role in the apoplastic movement of the nanoparticles (Lv et al. 2015).

8.8 Nano Biofortification Is Helpful for Human Wellbeing during the COVID-19 Pandemic Case

Several studies in the fields of medicine, agriculture, and industry are on improving human wellbeing. Providing adequate and safe nutrition to prevent starvation and fight against respiratory infections like COVID-19 may be a part of strengthening the health of humans (El-Ramady et al. 2021). Additionally, the widespread malnutrition and the COVID-19 pandemic have made life more stressful for humans. According to reports, the COVID-19 pandemic may result in a significant rise in all types of malnutrition as well as an increased risk of the particularly worrying dual load of the malnutrition epidemic (Littlejohn and Finlay 2021).

Production of essential nutrients rich crops for the health of humans through nano-biofortification in which the utilization of metal or metal oxide nanoparticles is the vital and cost-effective approach (Elemike et al. 2019). The production of nanoparticles via green way is gaining interest now a day due to t its eco-friendly and cost-effective nature. (Velazquez-Gamboa et al. 2021). Biofortifying crops with essential micronutrients, such as Zn and selenium, can positively impact respiratory health and help alleviate diseases caused by viral infections like COVID-19. (El-Ramady et al. 2021; Gastelum-Estrada et al. 2021; Okwuonu et al. 2021). Selenium is one of the fundamental micronutrients and its nano scale form has unique qualities including less toxicity, making it a promising choice for treating diseases caused by viral infections, Huntington's disease, and cancers and has a direct relationship with COVID-19 (He et al. 2021).

Production of pharmaceuticals based on nanoparticles (ZnO-NPs, Ag-NPs, Co-NPs and Cu-NPs) has garnered significant attention recently. These medications based on the nanoparticles (anti-COVID-19 nanoparticles) can mitigate the detrimental effects of viral diseases e.g., coronavirus by preventing viral replication and growth, obstructing receptor cell entrance, and directly inactivating the virus (Gatadi et al. 2021). This is achieved through two distinct methods: first, by biofortifying crops with nanonutrients; second, by employing these essential nutrients in pharmaceutical applications to enhance their bioavailability in medications.

8.9 Environmental and Safety Concerns of Nanotechnology-Assisted Biofortification

Nanotechnology-assisted biofortification is a technique that utilizes nanomaterials to enhance the nutrient content of crops, making them more nutritious and resistant to disease (Saritha et al. 2022). While this approach holds significant promise for addressing global food insecurity, it also raises concerns about its potential environmental and safety impacts.

One of the main environmental concerns associated with nanotechnology-assisted biofortification is the risk of nanoparticles leaching into the soil and water systems (Rizwan and Ali 2021). Nanoparticles have been shown to have negative effects on the environment, including toxicity to aquatic organisms and changes in microbial communities in soil (Hegde et al. 2016). Additionally, nanoparticles could accumulate in the food chain, potentially leading to negative effects on human health (Maharramov et al. 2019). There is also concern about the safety of nanotechnology-assisted biofortification. The potential for human exposure to nanoparticles during production and consumption of biofortified crops raises questions about their safety. Studies have shown that nanoparticles can be absorbed through the skin, lungs, and gastrointestinal tract, potentially leading to health problems (Ferdous and Nemmar 2020). Moreover, the long-term effects of exposure to nanoparticles are not yet fully understood, and more research is needed to ensure their safety. Another safety concern is the potential for nanoparticles to cross the blood-brain barrier and affect brain function. Recent studies have suggested that exposure to nanoparticles may be linked to neurological disorders such as Alzheimer's and Parkinson's disease (Waris et al. 2022). To mitigate the environmental and safety risks associated with nanotechnology-assisted biofortification, it is essential to adopt a precautionary approach. This approach involves identifying and evaluating potential risks before introducing nanomaterials into the environment and taking measures to reduce or eliminate those risks (Groso et al. 2010). Additionally, it is essential to conduct more research on the safety of nanoparticles and their environmental impact to ensure that nanotechnology-assisted biofortification is safe and sustainable.

Nanotechnology-assisted biofortification has the potential to improve global food security and address malnutrition, but also raises environmental and safety concerns. The risks of nanoparticles need to be considered, and a precautionary approach is necessary to ensure safety and sustainability. More research is needed to understand the safety of nanoparticles and their environmental impact.

8.10 Conclusions

Micronutrient malnutrition is a widespread issue that negatively affects both plants and humans. The parent soil contains low amounts of micronutrients (Zn, Cu, Fe and Se) and uptake of these essential nutrients by plants is affected by various soil factors such as pH and soil texture. This causes dietary nutrient deficiencies in humans. The nano-biofortification strategy can be viewed as a useful tool in the fight against malnutrition. This strategy involves the application of different types of nanoparticles, which have various benefits, including enrichment of nutrients in the edible portion of plants, effectiveness, and minimal usage. Nanoparticles play a vital role in the smooth uptake of micronutrients and their accumulation in the grains of cereal crops such as wheat, rice, and maize, which helps to reduce the deficiency of essential micronutrients in humans. In addition to improving the quality attributes of crops, biofortification through nanoparticles increased agricultural productivity and reduces hidden hunger. This demonstrates that it is a sustainable and economical strategy. Although the application of nanomaterials makes the grain more nutrient-rich in plants, improving human health and reducing malnutrition, their delivery and degree of accumulation should be carefully regulated as their excessive intake might have negative health effects. Moreover, green nanomaterials synthesized from different plant parts and microorganisms may be preferable for use in the biofortification of vegetables, leguminous and cereal crops to be the most environmentally sustainable and non-toxic forms. Therefore, utilizing nanomaterials to biofortify food crops in agricultural areas can increase nutritional health and sustainable food availability in both developed and developing countries.

Conflict of Interest The authors declare that there is no conflict of interest.

Acknowledgments The authors acknowledge the support from Princess Nourah bint Abdulrahman University Researchers Supporting Project number (PNURSP-HC2022/4), Princess Nourah bint Abdulrahman University, Riyadh, Saudi Arabia.

References

Abdel-Hakim SG, Shehata AS, Moghannem SA, Qadri M, El-Ghany MF, Abdeldaym EA, Darwish OS (2023) Nanoparticulate fertilizers increase nutrient absorption efficiency and agro-physiological properties of lettuce plant. Agronomy 13:691. https://doi.org/10.3390/agronomy13030691

Adrees M, Khan ZS, Ali S, Hafeez M, Khalid S, ur Rehman MZ, Hussain A, Hussain K, Chatha SA, Rizwan M (2020) Simultaneous mitigation of cadmium and drought stress in wheat by soil application of iron nanoparticles. Chemosphere 238:124681. https://doi.org/10.1016/j.chemosphere.2019.124681

Ali S, Mehmood A, Khan N (2012) Uptake, translocation, and consequences of nanomaterials on plant growth and stress adaptation. J Nanomater 2021:1–7. https://doi.org/10.1155/2021/6677616

Al-Amri N, Tombuloglu H, Slimani Y, Akhtar S, Barghouthi M, Almessiere M, Alshammari T, Baykal A, Sabit H, Ercan I, Ozcelik S (2020) Size effect of iron (III) oxide nanomaterials on the growth, and their uptake and translocation in common wheat (*Triticum aestivum* L.). Ecotoxicol Environ Saf 194:110377. https://doi.org/10.1016/j.ecoenv.2020.110377

Al-Juthery HW, Lahmoud NR, Alhasan AS, Al-Jassani NA, Houria A (2022) Nano-fertilizers as a novel technique for maximum yield in wheat biofortification (Article Review). IOP Conf Ser Earth Environ Sci 1060:012043. https://doi.org/10.1088/1755-1315/1060/1/012043

Al-Mamun MR, Hasan MR, Ahommed MS, Bacchu MS, Ali MR, Khan MZ (2021) Nanofertilizers towards sustainable agriculture and environment. Environ Technol Innov 23:101658. https://doi.org/10.1016/j.eti.2021.101658

Avellan A, Schwab F, Masion A, Chaurand P, Borschneck D, Vidal V, Rose J, Santaella C, Levard C (2017) Nanoparticle uptake in plants: gold nanomaterial localized in roots of *Arabidopsis thaliana* by X-ray computed nanotomography and hyperspectral imaging. Environ Sci Technol Aug 51:8682–8691

Avery JC, Hoffmann PR (2018) Selenium, selenoproteins, and immunity. Nutrients 10:1203. https://doi.org/10.3390/nu10091203

Baclaocos J, Santesmasses D, Mariotti M, Bierła K, Vetick MB, Lynch S, McAllen R, Mackrill JJ, Loughran G, Guigo R, Szpunar J (2019) Processive recoding and metazoan evolution of selenoprotein P: up to 132 UGAs in molluscs. J Mol Biol 431:4381–4407

Baig N, Kammakakam I, Falath W (2021) Nanomaterials: a review of synthesis methods, properties, recent progress, and challenges. Adv Mater 2:1821–1871

Bailey RL, Gahche JJ, Lentino CV, Dwyer JT, Engel JS, Thomas PR, Betz JM, Sempos CT, Picciano MF (2011) Dietary supplement use in the United States, 2003–2006. J Nutr 141:261–266

Bailey RL, West KP Jr, Black RE (2015) The epidemiology of global micronutrient deficiencies. Ann Nutr Metab 66:22–33

Bocchini M, D'Amato R, Ciancaleoni S, Fontanella MC, Palmerini CA, Beone GM, Onofri A, Negri V, Marconi G, Albertini E, Businelli D (2018) Soil selenium (se) biofortification changes the physiological, biochemical and epigenetic responses to water stress in *Zea mays* L. by inducing a higher drought tolerance. Front. Plant Sci 9:389. https://doi.org/10.3389/fpls.2018.00389

Bortolin A, Serafim AR, Aouada FA, Mattoso LH, Ribeiro C (2016) Macro-and micronutrient simultaneous slow release from highly swellable nanocomposite hydrogels. J Agric Food Chem 64:3133–3140

Boudiaf M, Messai Y, Bentouhami E, Schmutz M, Blanck C, Ruhlmann L, Bezzi H, Tairi L, Mekki DE (2021) Green synthesis of NiO nanoparticles using *Nigella sativa* extract and their enhanced electro-catalytic activity for the 4-nitrophenol degradation. J Phys Chem Solids 153:110020. https://doi.org/10.1016/j.jpcs.2021.110020

Brevik EC, Slaughter L, Singh BR, Steffan JJ, Collier D, Barnhart P, Pereira P (2020) Soil and human health: current status and future needs. Air Soil Water Res 13:1–23

Budke C, thor Straten S, Muehling KH, Broll G, Daum D (2020) Iodine biofortification of field-grown strawberries–approaches and their limitations. Sci Hortic 269:109317. https://doi.org/10.1016/j.scienta.2020.109317

Buzea C, Pacheco II, Robbie K (2007) Nanomaterials and nanoparticles: sources and toxicity. Biointerphases 2:26–27

Cai L, Cai L, Jia H, Liu C, Wang D, Sun X (2020) Foliar exposure of Fe3O4 nanoparticles on Nicotiana benthamiana: evidence for nanoparticles uptake, plant growth promoter and defense response elicitor against plant virus. J Hazard Mater 393:122415. https://doi.org/10.1016/j.jhazmat.2020.122415

Cakmak I, McLaughlin MJ, White P (2017) Zinc for better crop production and human health. Plant Soil 411:1–4

Carvalho KM, Gallardo-Williams MT, Benson RF, Martin DF (2003) Effects of selenium supplementation on four agricultural crops. J Agric Food Chem 51:704–709

Chandra AK, Pandey D, Tiwari A, Gururani K, Agarwal A, Dhasmana A, Kumar A (2021) Metal based nanoparticles trigger the differential expression of key regulatory genes which regulate iron and zinc homeostasis mechanism in finger millet. J Cereal Sci 100:103235. https://doi.org/10.1016/j.jcs.2021.103235

Cheah ZX, O'Hare TJ, Harper SM, Kochanek J, Bell MJ (2020) Zinc biofortification of immature maize and sweetcorn (*Zea mays* L.) kernels for human health. Sci Hortic 272:109559. https://doi.org/10.1016/j.scienta.2020.109559

Chen J, Lu S, Zhang Z, Zhao X, Li X, Ning P, Liu M (2018) Environmentally friendly fertilizers: a review of materials used and their effects on the environment. Sci Total 613:829–839

Chugh G, Siddique KH, Solaiman ZM (2022) Iron fortification of food crops through nanofertilisation. Crop Pasture Sci 73:736–748

Clark RB (1983) Plant genotype differences in the uptake, translocation, accumulation, and use of mineral elements required for plant growth. Plant Soil 72:175–196

Clemens S (2014) Zn and Fe biofortification: the right chemical environment for human bioavailability. Plant Sci 225:52–57. https://doi.org/10.1016/j.plantsci.2014.05.014

Datta M, Vitolins MZ (2016) Food fortification and supplement use-are there health implications? Crit Rev Food Sci Nutr 56:2149–2159

Davidson DW, Verma MS, Gu FX (2013) Controlled root targeted delivery of fertilizer using an ionically crosslinked carboxymethyl cellulose hydrogel matrix. Springer plus 2:1–9

De Matos M, Mattos BD, Tardy BL, Rojas OJ (2018) Magalhães WL (2018) use of biogenic silica in porous alginate matrices for sustainable fertilization with tailored nutrient delivery. ACS Sustain Chem Eng 6:2716–2723

Deshpande P, Dapkekar A, Oak M, Paknikar K, Rajwade J (2018) Nanocarrier-mediated foliar zinc fertilization influences expression of metal homeostasis related genes in flag leaves and enhances gluten content in durum wheat. PLoS One 17:e0191035. https://doi.org/10.1371/journal.pone.0191035

Dhaliwal SS, Sharma V, Shukla AK, Verma V, Kaur M, Shivay YS, Nisar S, Gaber A, Brestic M, Barek V (2022) Skalicky M (2022) biofortification—a frontier novel approach to enrich micronutrients in field crops to encounter the nutritional security. Molecules 27:1340. https://doi.org/10.3390/molecules27041340

Dietz KJ, Herth S (2011) Plant nanotoxicology. Trends Plant Sci 16:582–589

Dimkpa CO, Andrews J, Sanabria J, Bindraban PS, Singh U, Elmer WH, Gardea-Torresdey JL, White JC (2020) Interactive effects of drought, organic fertilizer, and zinc oxide nanoscale and bulk particles on wheat performance and grain nutrient accumulation. Sci Total Environ 722: 137808. https://doi.org/10.1016/j.scitotenv.2020.137808

Dola DB, Mannan MA, Sarker U, Al Mamun MA, Islam T, Ercisli S, Saleem MH, Ali B, Pop OL, Marc RA (2022) Nano-iron oxide accelerates growth, yield, and quality of *Glycine max* seed in water deficits. Front Plant Sci 13:992535. https://doi.org/10.3389/fpls.2022.992535

Du W, Yang J, Peng Q, Liang X, Mao H (2019) Comparison study of zinc nanoparticles and zinc sulphate on wheat growth: from toxicity and zinc biofortification. Chemosphere 227:109–116

Dutta S, Pal S, Panwar P, Sharma RK, Bhutia PL (2022) Biopolymeric nanocarriers for nutrient delivery and crop biofortification. ACS Omega 7:25909–259020

Eichert T, Kurtz A, Steiner U, Goldbach HE (2008) Size exclusion limits and lateral heterogeneity of the stomatal foliar uptake pathway for aqueous solutes and water-suspended nanoparticles. Physiol Plant 134:151–160

Elemike EE, Uzoh IM, Onwudiwe DC, Babalola OO (2019) The role of nanotechnology in the fortification of plant nutrients and improvement of crop production. Appl Sci 9:499. https://doi.org/10.3390/app9030499

El-Ramady H, Brevik E, Amer MM, Elsakhawy T, Omara Ahmed AE, Elbasiouny H, Elbehiry F, Mosa AA, El-Ghamry A, Bayoumi Y, Shalaby TA (2020a) Soil and air pollution in the era of COVID-19: a global issue. Egypt J Soil Sci 60:437–448

El-Ramady H, Eid YZ, Brevik EC (2020c) New pollution challenges in groundwater and wastewater due to COVID-19. J Sustain Agric Sci 46:61–76

El-Ramady H, Faizy SE, Abdalla N, Taha H, Domokos-Szabolcsy É, Fari M, Elsakhawy T, Omara AE, Shalaby T, Bayoumi Y, Shehata S (2020b) Selenium and nano-selenium biofortification for human health: opportunities and challenges. Soil Syst 4:57. https://doi.org/10.3390/soilsystems4030057

El-Ramady HR, Singh A, Rajput VD, Amer MM, Omara AE, Elsakhawy T, Elbehiry F, Elbasiouny H, Abdalla N (2021) Environment, biodiversity and soil security: a new dimension in the era of COVID-19. Env Biodivers Soil Secur 5:1–4

Elshayb OM, Farroh KY, Amin HE, Atta AM (2021) Green synthesis of zinc oxide nanoparticles: fortification for rice grain yield and nutrients uptake enhancement. Molecules 26:584. https://doi.org/10.3390/molecules26030584

Fakharzadeh S, Hafizi M, Baghaei MA, Etesami M, Khayamzadeh M, Kalanaky S, Akbari ME, Nazaran MH (2020) Using nanochelating technology for biofortification and yield increase in rice. Sci Rep 10:4351. https://doi.org/10.1038/s41598-020-60189

FAO (2019) The state of food security and nutrition in the world. http://www. fao.org/3/ca5162en/ca5162en. pdf. Accessed 29 Sept 2021

Farooq MU, Tang Z, Zeng R, Liang Y, Zhang Y, Zheng T, Ei HH, Ye X, Jia X, Zhu J (2019) Accumulation, mobilization, and transformation of selenium in rice grain provided with foliar sodium selenite. J Sci Food Agric 99:2892–2900

Fellet G, Pilotto L, Marchiol L, Braidot E (2021) Tools for nano-enabled agriculture: fertilizers based on calcium phosphate, silicon, and chitosan nanostructures. Agronomy 11:1239. https://doi.org/10.3390/agronomy11061239

Ferdous Z, Nemmar A (2020) Health impact of silver nanoparticles: a review of the biodistribution and toxicity following various routes of exposure. Int J Mol Sci 21:2375. https://doi.org/10.3390/ijms21072375

Feregrino-Perez AA, Magaña-López E, Guzmán C, Esquivel K (2018) A general overview of the benefits and possible negative effects of the nanotechnology in horticulture. Sci Hortic 238:126–137

Fernández-Luqueño F, López-Valdez F (2018) Agricultural nanobiotechnology: modern agriculture for a sustainable future. Springer

Gastelum-Estrada A, Serna-Saldívar SO, Jacobo-Velázquez DA (2021) Fighting the COVID-19 pandemic through biofortification: innovative approaches to improve the immunomodulating capacity of foods. Food Sci Technol 1:480–486

Gatadi S, Madhavi YV, Nanduri S (2021) Nanoparticle drug conjugates treating microbial and viral infections: a review. J Mol Struct 1228:129750. https://doi.org/10.1016/j.molstruc.2020.129750

Geisler-Lee J, Wang Q, Yao Y, Zhang W, Geisler M, Li K, Huang Y, Chen Y, Kolmakov A, Ma X (2012) Phytotoxicity, accumulation and transport of silver nanoparticles by *Arabidopsis thaliana*. Nanotoxicology 7:323–337

Germ M, Stibilj V, Šircelj H, Jerše A, Kroflič A, Golob A, Maršić NK (2019) Biofortification of common buckwheat microgreens and seeds with different forms of selenium and iodine. J Sci Food Agric 99:4353–4362

Godecke T, Stein AJ, Qaim M (2018) The global burden of chronic and hidden hunger: trends and determinants. Glob Food Sec 17:21–29

Golubkina NA, Kosheleva OV, Krivenkov LV, Dobrutskaya HG, Nadezhkin S, Caruso G (2017) Intersexual differences in plant growth, yield, mineral composition and antioxidants of spinach (*Spinacia oleracea* L.) as affected by selenium form. Sci Hortic 225:350–358

Gonzalez-Garcia Y, Cárdenas-Alvarez C, Cadenas-Pliego G, Benavides-Mendoza A, Cabrera-de-la-Fuente M, Sandoval-Rangel A, Valdés-Reyna J, Juárez-Maldonado A (2021) Effect of three nanoparticles (se, Si and cu) on the bioactive compounds of bell pepper fruits under saline stress. Plan Theory 10:217. https://doi.org/10.3390/plants10020217

Groso A, Petri-Fink A, Magrez A, Riediker M, Meyer T (2010) Management of nanomaterials safety in research environment. Part Fibre Toxicol 7:1–8

Guha T, Mukherjee A, Kundu R (2021) Nano-scale zero valent iron (nZVI) priming enhances yield, alters mineral distribution and grain nutrient content of *Oryza sativa* L. cv. Gobindobhog: a field study. J Plant Growth Regul 41:710–733

He L, Zhao J, Wang L, Liu Q, Fan Y, Li B, Yu YL, Chen C, Li YF (2019) Using nano-selenium to combat coronavirus disease (COVID-19)? Nano Today 36:101037. https://doi.org/10.1016/j.nantod.2020.101037

He L, Zhao J, Wang L, Liu Q, Fan Y, Li B, Yu YL, Chen C, Li YF (2021) Using nano-selenium to combat Coronavirus Disease 2019 (COVID-19)? Nano Today 36:101037. https://doi.org/10.1016/j.nantod.2020.101037

Hegde K, Brar SK, Verma M, Surampalli RY (2016) Current understandings of toxicity, risks and regulations of engineered nanoparticles with respect to environmental microorganisms. Nanotechnol Environ Eng 1:1–2

Hernandez-Hernandez H, Quiterio-Gutiérrez T, Cadenas-Pliego G, Ortega-Ortiz H, Hernández-Fuentes AD, Cabrera de la Fuente M, Valdés-Reyna J, Juárez-Maldonado A (2019) Impact of selenium and copper nanoparticles on yield, antioxidant system, and fruit quality of tomato plants. Plan Theory 8:355. https://doi.org/10.3390/plants8100355

Hess SY (2012) Deficiency disorders and prevention programs. Encycl Human Nutr 2(1629):431

Hosnedlova B, Kepinska M, Skalickova S, Fernandez C, Ruttkay-Nedecky B, Peng Q, Baron M, Melcova M, Opatrilova R, Zidkova J, Bjørklund G (2018) Nano-selenium and its nanomedicine applications: a critical review. Int J Nanomedicine 13:2107–2128

Huang S, Wang P, Yamaji N, Ma JF (2020) Plant nutrition for human nutrition: hints from rice research and future perspectives. Mol Plant 13:825–835

Hussain A, Ali S, Rizwan M, ur Rehman MZ, Javed MR, Imran M, Chatha SA, Nazir R (2018) Zinc oxide nanoparticles alter the wheat physiological response and reduce the cadmium uptake by plants. Environ Pollut 242:1518–1526

Hussein HA, Darwesh OM, Mekki BB (2019) Environmentally friendly nano-selenium to improve antioxidant system and growth of groundnut cultivars under sandy soil conditions. Biocatal Agric Biotechnol 18:101080. https://doi.org/10.1016/j.bcab.2019.101080

Iliger KS, Sofi TA, Bhat NA, Ahanger FA, Sekhar JC, Elhendi AZ, Al-Huqail AA, Khan F (2021) Copper nanoparticles: green synthesis and managing fruit rot disease of chilli caused by *Colletotrichum capsici*. Saudi J Biol Sci 28:1477–1486

Ivanov K, Vasilev A, Mitkov A, Nguyen N, Tonev T (2021) Application of Zn-containing foliar fertilisers for recovery of the grain productivity potential of Zn-deficient maize plants. Ital J Agron 16:1759. https://doi.org/10.4081/ija.2021.1759

Izydorczyk G, Ligas B, Mikula K, Witek-Krowiak A, Moustakas K, Chojnacka K (2021) Biofortification of edible plants with selenium and iodine–a systematic literature review. Sci Total Environ 754:141983. https://doi.org/10.1016/j.scitotenv.2020.141983

Javaid S, Ayyaz MK, Rehman ZU, Naveed U, Raza AA, Jamil A, Sharif M, Rasheed S (2020) Biofortification through zinc oxide nanoparticles, a sustainable approach for enhancing the productivity of rice: a review. Agric Food Secur 2:1–12

Jha AB, Warkentin TD (2020) Biofortification of pulse crops: status and future perspectives. Plan Theory 9:73. https://doi.org/10.3390/plants9010073

Joy EJ, Kalimbira AA, Gashu D, Ferguson EL, Sturgess J, Dangour AD, Banda L, Chiutsi-Phiri G, Bailey EH, Langley-Evans SC, Lark RM (2019) Can selenium deficiency in Malawi be alleviated through consumption of agro-biofortified maize flour? Study protocol for a randomised, double-blind, controlled trial. Trials 20(1):1–9

Kah M, Kookana RS, Gogos A, Bucheli TD (2018) A critical evaluation of nanopesticides and nanofertilizers against their conventional analogues. Nat Nanotechnol 13:677–684

Kapoor P, Dhaka RK, Sihag P, Mehla S, Sagwal V, Singh Y, Langaya S, Balyan P, Singh KP, Xing B, White JC (2022) Nanotechnology-enabled biofortification strategies for micronutrients enrichment of food crops: current understanding and future scope. NanoImpact 26:100407. https://doi.org/10.1016/j.impact.2022.100407

Khaledian Y, Pajohi-Alamoti M, Bazargani-Gilani B (2019) Development of cellulose nanofibers coating incorporated with ginger essential oil and citric acid to extend the shelf life of ready-to-cook barbecue chicken. J Food Process Preserv 43:14114. https://doi.org/10.1111/jfpp.14114

Khan MR, Adam V, Rizvi TF, Zhang B, Ahamad F, Jośko I, Zhu Y, Yang M, Mao C (2019) Nanoparticle–plant interactions: two-way traffic. Small 15:1901794. https://doi.org/10.1002/smll.201901794

Kucuk N, Primozic M, Knez Z, Leitgeb M (2023) Sustainable biodegradable biopolymer-based nanoparticles for healthcare applications. Int J Mol Sci 24:3188. https://doi.org/10.3390/ijms24043188

Kumar J, Gupta DS, Kumar S, Gupta S, Singh NP (2016) Current knowledge on genetic biofortification in lentil. J Agric Food Chem 64:6383–6396

Kumar S, Chauhan N, Gopal M, Kumar R, Dilbaghi N (2015) Development and evaluation of alginate–chitosan nanocapsules for controlled release of acetamiprid. Int J Biol Macromol 81: 631–637

Kumar A (2021) Metal based nanoparticles trigger the differential expression of key regulatory genes which regulate iron and zinc homeostasis mechanism in finger millet. J Cereal Sci 100: 103235. https://doi.org/10.1016/j.jcs.2021.103235

Lages Barbosa G, Almeida Gadelha FD, Kublik N, Proctor A, Reichelm L, Weissinger E, Wohlleb GM, Halden RU (2015) Comparison of land, water, and energy requirements of lettuce grown using hydroponic vs. conventional agricultural methods. Int J Environ Res Public Health 12:6879–6891. https://doi.org/10.3390/ijerph120606879

Laskowski W, Górska-Warsewicz H, Rejman K, Czeczotko M, Zwolińska J (2019) How important are cereals and cereal products in the average polish diet? Nutrients 11:679. https://doi.org/10.3390/nu11030679

León-Silva S, Arrieta-Cortes R, Fernández-Luqueño F, López-Valdez F (2018) Design and production of Nanofertilizers. In: López-Valdez F, Fernández-Luqueño F (eds) Agricultural Nanobiotechnology. Springer, Cham, pp 17–31. https://doi.org/10.1007/978-3-319-96719-6_2

Li J, Hu J, Ma C, Wang Y, Wu C, Huang J, Xing B (2016) Uptake, translocation and physiological effects of magnetic iron oxide (γ-Fe2O3) nanoparticles in corn (Zea mays L.). Chemosphere 159:326–334. https://doi.org/10.1016/j.chemosphere.2016.05.083

Lin D, Xing B (2008) Root uptake and phytotoxicity of ZnO nanoparticles. Environ Sci Technol 42: 5580–5585

Littlejohn P, Finlay BB (2021) When a pandemic and an epidemic collide: COVID-19, gut microbiota, and the double burden of malnutrition. BMC Med 19:1–8

Liu R, Lal R (2015) Potentials of engineered nanoparticles as fertilizers for increasing agronomic productions. Sci Total Environ 514:131–139

Lopes G, Avila FW, Guilherme LR (2017) Selenium behavior in the soil environment and its implication for human health. Cienc Agrotecnologia 41:605–615

Lopez-Lorente AI, Mizaikoff B (2022) Recent advances on the characterization of nanoparticles using infrared spectroscopy. Trends Analyt Chem 84:97–106

Lopez-Vargas ER, Ortega-Ortíz H, Cadenas-Pliego G, de Alba RK, Cabrera de la Fuente M, Benavides-Mendoza A, Juárez-Maldonado A (2018) Foliar application of copper nanoparticles increases the fruit quality and the content of bioactive compounds in tomatoes. Appl Sci 8:1020. https://doi.org/10.3390/app8071020

Lough TJ, Lucas WJ (2006) Integrative plant biology: role of phloem long-distance macromolecular trafficking. Annu Rev Plant Biol 57:203–232. https://doi.org/10.1146/annurev.arplant.56.032604.144145

Lu K, Shen D, Liu X, Dong S, Jing X, Wu W, Tong Y, Gao S, Mao L (2020) Uptake of iron oxide nanoparticles inhibits the photosynthesis of the wheat after foliar exposure. Chemosphere 259:127445. https://doi.org/10.1016/j.chemosphere.2020.127445

Ludwig Y, Slamet-Loedin IH (2019) Genetic biofortification to enrich rice and wheat grain iron: from genes to product. Front Plant Sci 10:833. https://doi.org/10.3389/fpls.2019.00833

Lv J, Zhang S, Luo L, Zhang J, Yang K, Christie P (2015) Accumulation, speciation and uptake pathway of ZnO nanoparticles in maize. Environ Sci: Nano 2:68–77. https://doi.org/10.1039/C4EN00064A

Lv J, Christie P, Zhang S (2019) Uptake, translocation, and transformation of metal-based nanoparticles in plants: recent advances and methodological challenges. Environ Sci: Nano 6:41–59. https://doi.org/10.1039/C8EN00645H

Mabesa RA, Impa SA, Grewal D, Johnson-Beebout SE (2013) Contrasting grain-Zn response of biofortification rice (*Oryza sativa* L.) breeding lines to foliar Zn application. Field Crops Res 149:223–233

Maharramov AM, Hasanova UA, Suleymanova IA, Osmanova GE, Hajiyeva NE (2019) The engineered nanoparticles in food chain: Potential toxicity and effects. SN Appl Sci 1:1–25

Mahdieh M, Sangi MR, Bamdad F, Ghanem A (2018) Effect of seed and foliar application of nano-zinc oxide, zinc chelate, and zinc sulphate rates on yield and growth of pinto bean (*Phaseolus vulgaris*) cultivars. J Plant Nutr 41:2401–2412

Malerba M, Cerana R (2016) Chitosan effects on plant systems. Int J Mol Sci 17:996. https://doi.org/10.3390/ijms17070996

Mareedu T, Poiba V, Vangalapati M (2021) Green synthesis of iron nanoparticles by green tea and black tea leaves extract. Mater Today Proc 42:1498–1501

Martinez G, Merinero M, Perez-Aranda M, Perez-Soriano EM, Ortiz T, Villamor E, Begines B, Alcudia A (2020) Environmental impact of nanoparticles' application as an emerging technology: a review. Materials 14:166. https://doi.org/10.3390/ma14010166

Marzouk NM, Abd-Alrahman HA, AM EL-T, Mahmoud SH (2019) Impact of foliar spraying of nano micronutrient fertilizers on the growth, yield, physical quality, and nutritional value of two snap bean cultivars in sandy soils. Bull Natl Res Cent 43:1–9

Matres JM, Arcillas E, Cueto-Reaño MF, Sallan-Gonzales R, Trijatmiko KR, Slamet-Loedin I (2021) Biofortification of rice grains for increased iron content. Rice Imp 5:471

Menzir A, Dessie D (2017) Review on copper deficiency in domestic ruminants Int J Adv Res 1: 101–110

Milani N, Hettiarachchi GM, Kirby JK, Beak DG, Stacey SP, McLaughlin MJ (2015) Fate of zinc oxide nanoparticles coated onto macronutrient fertilizers in an alkaline calcareous soil. PLoS One 10:126275. https://doi.org/10.1371/journal.pone.0126275

Missaoui WN, Arnold RD, Cummings BS (2018) Toxicological status of nanoparticles: what we know and what we don't know. Chem Biol Interact 295:1–12. https://doi.org/10.1016/j.cbi.2018.07.015

Mittal D, Kaur G, Singh P, Yadav K, Ali SA (2020) Nanoparticle-based sustainable agriculture and food science: recent advances and future outlook. Front Nanosci 2:579954. https://doi.org/10.3389/fnano.2020.579954

Mullen A (2019) Expectations from nano in agriculture. Nat Nanotechnol 14(6):515–516

Nadendla SR, Rani TS, Vaikuntapu PR, Maddu RR, Podile AR (2018) HarpinPss encapsulation in chitosan nanoparticles for improved bioavailability and disease resistance in tomato. Carbohydr Polym 199:11–19

Noman M, Ahmed T, Hussain S, Niazi MB, Shahid M, Song F (2020) Biogenic copper nanoparticles synthesized by using a copper-resistant strain *shigella flexneri* SNT22 reduced the translocation of cadmium from soil to wheat plants. J Hazard Mater 398:123175. https://doi.org/10.1016/j.jhazmat.2020.123175

Okwuonu IC, Narayanan NN, Egesi CN, Taylor NJ (2021) Opportunities and challenges for biofortification of cassava to address iron and zinc deficiency in Nigeria. Glob Food Sec 28: 100478

Parande H, Mirza M (2011) Comparison of nano Fe chelate with Fe chelate effect on growth parameters and antioxidant enzymes activity of *Ocimum basilicum*. New Cell Mol Biotechnol J 1:89–98

Phattarakul N, Rerkasem B, Li LJ, Wu LH, Zou CQ, Ram H, Sohu VS, Kang BS, Surek H, Kalayci M, Yazici A (2012) Biofortification of rice grain with zinc through zinc fertilization in different countries. Plant Soil 361:131–141

Pollard M, Beisson F, Li Y, Ohlrogge JB (2008) Building lipid barriers: biosynthesis of cutin and suberin. Trends Plant Sci 13:236–246

Quiterio-Gutiérrez T, Ortega-Ortiz H, Cadenas-Pliego G, Hernández-Fuentes AD, Sandoval-Rangel A, Benavides-Mendoza A, Cabrera-de la Fuente M, Juárez-Maldonado A (2019). The application of selenium and copper nanoparticles modifies the biochemical responses of tomato plants under stress by Alternaria solani. Int J Mol Sci 20(8): 1950.

Ragab GA, Saad-Allah KM (2020) Green synthesis of sulfur nanoparticles using *Ocimum basilicum* leaves and its prospective effect on manganese-stressed *Helianthus annuus* (L.) seedlings. Ecotoxicol Environ Saf 191:110242

Rajaee Behbahani S, Iranbakhsh A, Ebadi M, Majd A, Ardebili ZO (2020) Red elemental selenium nanoparticles mediated substantial variations in growth, tissue differentiation, metabolism, gene transcription, epigenetic cytosine DNA methylation, and callogenesis in bittermelon (*Momordica charantia*); an in vitro experiment. PLoS One 15(7):e0235556

Ramkissoon C, Degryse F, da Silva RC, Baird R, Young SD, Bailey EH, McLaughlin MJ (2019) Improving the efficacy of selenium fertilizers for wheat biofortification. Sci Rep 9:1–9

Ramzani PM, Khalid M, Naveed M, Irum A, Khan WU, Kausar S (2016) Iron biofortification of cereals grown under calcareous soils: problems and solutions. Soil Sci Agricultural and environmental prospectives:232–258

Reis HP, de Queiroz Barcelos JP, Silva VM, Santos EF, Tavanti RF, Putti FF, Young SD, Broadley MR, White PJ, Dos Reis AR (2020) Agronomic biofortification with selenium impacts storage proteins in grains of upland rice. J Sci Food Agric 100:1990–1997

Resmi R, Yoonus J, Beena B (2021) A novel greener synthesis of ZnO nanoparticles from *Nilgiri antusciliantus* leaf extract and evaluation of its biomedical applications. Mater Today Pro 46: 3062–3068

Rico CM, Majumdar S, Duarte-Gardea M, Peralta-Videa JR, Gardea-Torresdey JL (2011) Interaction of nanoparticles with edible plants and their possible implications in the food chain. J Agric Food Chem 59:3485–3498

Rizwan M, Ali S, ur Rehman MZ, Adrees M, Arshad M, Qayyum MF, Ali L, Hussain A, Chatha SA, Imran M (2019a) Alleviation of cadmium accumulation in maize (*Zea mays* L.) by foliar spray of zinc oxide nanoparticles and biochar to contaminated soil. Environ Pollut 248:358–367

Rizwan M, Ali S, Ali B, Adrees M, Arshad M, Hussain A, ur Rehman MZ, Waris AA (2019b) Zinc and iron oxide nanoparticles improved the plant growth and reduced the oxidative stress and cadmium concentration in wheat. Chemosphere 214:269–277

Rizwan M, Ali S, ur Rehman MZ, Riaz M, Adrees M, Hussain A, Zahir ZA, Rinklebe J (2021) Effects of nanoparticles on trace element uptake and toxicity in plants: a review. Ecotoxicol Environ Saf 221:112437. https://doi.org/10.1016/j.ecoenv.2021.112437

Rodrigo S, Santamaria O, Poblaciones MJ (2014) Selenium application timing: Influence in wheat grain and flour selenium accumulation under mediterranean conditions J Agric Sci 6:23. https://doi.org/10.5539/jas.v6n3p23

Rosa AT, Diaz DR, Hansel FD, Sebastian JS, Adee EA (2019) Genotypic variation on root growth and nutrient uptake in corn and soybean. Agrosyst Geosci Environ 2:1–2

Rossi L, Fedenia LN, Sharifan H, Ma X, Lombardini L (2019) Effects of foliar application of zinc sulfate and zinc nanoparticles in coffee (*Coffea arabica* L.) plants. Plant Physiol Biochem 135: 160–166

Rouphael Y, Kyriacou MC (2018) Enhancing quality of fresh vegetables through salinity eustress and biofortification applications facilitated by soilless cultivation. Front Plant Sci 9:1–6

Sahoo SK, Dwivedi GK, Dey P, Praharaj S (2021) Green synthesized ZnO nanoparticles for sustainable production and nutritional biofortification of green gram. Environ Technol Innov 24:101957

Salama DM, Osman SA, Abd El-Aziz ME, Abd Elwahed MS, Shaaban EA (2019) Effect of zinc oxide nanoparticles on the growth, genomic DNA, production and the quality of common dry bean (*Phaseolus vulgaris*). Biocatal Agric Biotechnol 18:101083

Saritha GN, Anju T, Kumar A (2022) Nanotechnology-big impact: how nanotechnology is changing the future of agriculture? J Sci Food Agric 19:100457. https://doi.org/10.1016/j.jafr.2022.100457

Schier HE, Eliot KA, Herron SA, Landfried LK, Migicovsky Z, Rubin MJ, Miller AJ (2019) Comparative analysis of perennial and annual *Phaseolus* seed nutrient concentrations. Sustainability 11(10):2787

Schreiber L (2005) Polar paths of diffusion across plant cuticles: new evidence for an old hypothesis. Ann Bot 95:1069–1073. https://doi.org/10.1093/aob/mci122

Schwab F, Zhai G, Kern M, Turner A, Schnoor JL, Wiesner MR (2016) Barriers, pathways and processes for uptake, translocation and accumulation of nanomaterials in plants–critical review. Nanotoxicology 10:257–278

Schymura S, Fricke T, Hildebrand H, Franke K (2017) Elucidating the role of dissolution in CeO_2 nanoparticle plant uptake by smart radiolabeling. Angew Chem Int Ed Engl 56:7411–7414

Semida WM, Abdelkhalik A, Mohamed GF, Abd El-Mageed TA, Abd El-Mageed SA, Rady MM, Ali EF (2021) Foliar application of zinc oxide nanoparticles promotes drought stress tolerance in eggplant (*Solanum melongena* L.). Plan Theory 10(2):421

Shahzad Z, Rouached H, Rakha A (2014) Combating mineral malnutrition through iron and zinc biofortification of cereals. Compr Rev Food Sci Food Saf 13:329–346

Shalaby TA, Abd-Alkarim E, El-Aidy F, Hamed ES, Sharaf-Eldin M, Taha N, El-Ramady H, Bayoumi Y, Dos Reis AR (2021) Nano-selenium, silicon and H_2O_2 boost growth and productivity of cucumber under combined salinity and heat. Ecotoxicol Environ Saf 212:111962

Shang Y, Hasan MK, Ahammed GJ, Li M, Yin H, Zhou J (2019) Applications of nanotechnology in plant growth and crop protection: a review. Molecules 24:2558. https://doi.org/10.3390/molecules24142558

Sharma P, Aggarwal P, Kaur A (2017) Biofortification: a new approach to eradicate hidden hunger. Food Rev Int 33:1–21

Shekari F, Mohammadi H, Pourmohammad A, Avanes A, Benam MB (2015) Spring wheat yielding and the content of protein and zinc in its grain depending on zinc fertilisation. Electron J Pol Agric Univ 18:1–12

Sheoran P, Grewal S, Kumari S, Goel S (2021) Enhancement of growth and yield, leaching reduction in Triticum aestivum using biogenic synthesized zinc oxide nanofertilizer. Biocatal Agric Biotechnol 32:101938

Shi J, Peng C, Yang Y, Yang J, Zhang H, Yuan X, Chen Y, Hu T (2014) Phytotoxicity and accumulation of copper oxide nanoparticles to the cu-tolerant plant *Elsholtzia splendens*. Nanotoxicology 8:179–188

Shu N, Cheng Q, Arner ES, Davies MJ (2020) Inhibition and crosslinking of the selenoprotein thioredoxin reductase-1 by p-benzoquinone. Redox Biol 28:101335

Signore A, Renna M, D'Imperio M, Serio F, Santamaria P (2018) Preliminary evidences of biofortification with iodine of "Carota di Polignano", an Italian carrot landrace. Front Plant Sci 2018:170. https://doi.org/10.1093/aob/mci122

Sillanpaa M (1990) Micronutrient assessment at the country level: Int Stud:1990

Singh J, Dutta T, Kim KH, Rawat M, Samddar P, Kumar P (2018) Green synthesis of metals and their oxide nanoparticles: applications for environmental remediation. J Nanobiotechnol 16:1–24

Stephen BJ, Sharma MM, Jain D, Singh A (2021) Biogenic synthesized nanoparticles a boon to human health. Mater Today Proc 43:3211–3213

Stevens GA, Bennett JE, Hennocq Q, Lu Y, De-Regil LM, Rogers L, Danaei G, Li G, White RA, Flaxman SR, Oehrle SP (2015) Trends and mortality effects of vitamin a deficiency in children in 138 low-income and middle-income countries between 1991 and 2013: a pooled analysis of population-based surveys. Lancet Glob Health 3(9):e528–e536

Subbaiah LV, Prasad TN, Krishna TG, Sudhakar P, Reddy BR, Pradeep T (2016) Novel effects of nanoparticulate delivery of zinc on growth, productivity, and zinc biofortification in maize (*Zea mays* L.). J Agric Food Chem 64:3778–3788

Sun H, Du W, Peng Q, Lv Z, Mao H, Kopittke PM (2020) Development of ZnO nanoparticles as an efficient Zn fertilizer: using synchrotron-based techniques and laser ablation to examine elemental distribution in wheat grain. J Agric Food Chem 68:5068–5075

Sun Y, Zhu G, Zhao W, Jiang Y, Wang Q, Wang Q, Rui Y, Zhang P, Gao L (2022) Engineered nanomaterials for improving the nutritional quality of agricultural products: a review. Nano 12: 4219. https://doi.org/10.3390/nano12234219

Sundaria N, Singh M, Upreti P, Chauhan RP, Jaiswal JP, Kumar A (2019) Seed priming with iron oxide nanoparticles triggers iron acquisition and biofortification in wheat (*Triticum aestivum* L.) grains. J Plant Growth Regul 38:122–131

Thakur S, Thakur S, Kumar R (2018) Bio-nanotechnology and its role in agriculture and food industry. J Mol Genet Med 12:1747–0862

Tiozon RN, Fernie AR, Sreenivasulu N (2021) Meeting human dietary vitamin requirements in the staple rice via strategies of biofortification and post-harvest fortification. Trends Food Sci Technol 109:65–82

Tombuloglu H, Slimani Y, Tombuloglu G, Almessiere M, Baykal A, Ercan I, Sozeri H (2019) Tracking of NiFe2O4 nanoparticles in barley (Hordeum vulgare L.) and their impact on plant growth, biomass, pigmentation, catalase activity, and mineral uptake. Environ Nanotechnol Monit Manage 11:100223. https://doi.org/10.1016/j.enmm.2019.100223

Umar W, Hameed MK, Aziz T, Maqsood MA, Bilal HM, Rasheed N (2021) Synthesis, characterization and application of ZnO nanoparticles for improved growth and Zn biofortification in maize. Arch Agron Soil Sci 67:1164–1176

UNEP (2021) Food systems hold key to ending world hunger. Https://www. Unep. Org/ news- and-stori es/ story/ food- syste ms- hold- keyending- world- hunger. Accessed 29 Sept 2021

Uzu G, Sobanska S, Sarret G, Munoz M, Dumat C (2010) Foliar lead uptake by lettuce exposed to atmospheric fallouts. Environ Sci Technol 44:1036–1042

Valencia GA, Zare EN, Makvandi P, Gutiérrez TJ (2019) Self-assembled carbohydrate polymers for food applications: a review. Compr Rev Food Sci Food Saf 18:2009–2024

Velazquez-Gamboa MC, Rodriguez-Hernandez L, Abud-Archila M, Gutiérrez-Miceli FA, González-Mendoza D, Valdez-Salas B, González-Terreros E, Lujan-Hidalgo MC (2021) Agronomic biofortification of *Stevia rebaudiana* with zinc oxide (ZnO) phytonanoparticles and antioxidant compounds. Sugar Tech 23:453–460

Wang C, Cheng T, Liu H, Zhou F, Zhang J, Zhang M, Liu X, Shi W, Cao T (2021) Nano-selenium controlled cadmium accumulation and improved photosynthesis in indica rice cultivated in lead and cadmium combined paddy soils. J Environ Sci 103:336–346

Wang M, Wang Y, Ge C, Wu H, Jing F, Wu S, Li H, Zhou D (2023) Foliar selenium nanoparticles application promotes the growth of maize (*Zea mays* L.) seedlings by regulating carbon, nitrogen and oxidative stress metabolism. Sci Hortic 311:111816

Wang Z, Xie X, Zhao J, Liu X, Feng W, White JC, Xing B (2012) Xylem-and phloem-based transport of CuO nanoparticles in maize (Zea mays L.). Environ Sci Technol 46:4434–4441. https://doi.org/10.1021/es204212z

Wang P, Menzies NW, Lombi E, McKenna BA, Johannessen B, Glover CJ, Kappen P, Kopittke PM (2013a) Fate of ZnO nanoparticles in soils and cowpea (*Vigna unguiculata*). Environ Sci Technol 47:13822–13830

Wang WN, Tarafdar JC, Biswas P (2013b) Nanoparticle synthesis and delivery by an aerosol route for watermelon plant foliar uptake. J Nanopart Res 15:1–3

Wang Y, Deng C, Cota-Ruiz K, Peralta-Videa JR, Sun Y, Rawat S, Tan W, Reyes A, Hernandez-Viezcas JA, Niu G, Li C (2020) Improvement of nutrient elements and allicin content in green onion (*Allium fistulosum*) plants exposed to CuO nanoparticles. Sci Total Environ 725:138387

Wang Y, Jiang F, Ma C, Rui Y, Tsang DC, Xing B (2019a) Effect of metal oxide nanoparticles on amino acids in wheat grains (*Triticum aestivum*) in a life cycle study. J Environ Manag 241: 319–327

Wang Y, Lin Y, Xu Y, Yin Y, Guo H, Du W (2019b) Divergence in response of lettuce (var. ramosa Hort.) to copper oxide nanoparticles/microparticles as potential agricultural fertilizer. Environ Pollut Bioavailab 31:80–84

Waris A, Ali A, Khan AU, Asim M, Zamel D, Fatima K, Raziq A, Khan MA, Akbar N, Baset A, Abourehab MA (2022) Applications of various types of nanomaterials for the treatment of neurological disorders. Nano 12:2140. https://doi.org/10.3390/nano12132140

Wong MH, Misra RP, Giraldo JP, Kwak SY, Son Y, Landry MP, Swan JW, Blankschtein D, Strano MS (2016) Lipid exchange envelope penetration (LEEP) of nanoparticles for plant engineering: a universal localization mechanism. Nano Lett 16:1161–1172

Xiong T, Dumat C, Dappe V, Vezin H, Schreck E, Shahid M, Pierart A, Sobanska S (2017) Copper oxide nanoparticle foliar uptake, phytotoxicity, and consequences for sustainable urban agriculture. Environ Sci Technol 51:5242–5251

Yang G, Yuan H, Ji H, Liu H, Zhang Y, Wang G, Chen L, Guo Z (2021) Effect of ZnO nanoparticles on the productivity, Zn biofortification, and nutritional quality of rice in a life cycle study. Plant Physiol Biochem 163:87–94

Yashveer S, Singh V, Kaswan V, Kaushik A, Tokas J (2014) Green biotechnology, nanotechnology and bio-fortification: perspectives on novel environment-friendly crop improvement strategies. Biotechnol Genet Eng Rev 30:113–126

Yusefi-Tanha E, Fallah S, Rostamnejadi A, Pokhrel LR (2020a) Root system architecture, copper uptake and tissue distribution in soybean (*Glycine max* (L.) Merr.) grown in copper oxide nanoparticle (CuONP)-amended soil and implications for human nutrition. Plan Theory 9(10): 1326

Yusefi-Tanha E, Fallah S, Rostamnejadi A, Pokhrel LR (2020b) Zinc oxide nanoparticles (ZnONPs) as a novel nanofertilizer: influence on seed yield and antioxidant defense system in soil grown soybean (*Glycine max* cv. Kowsar). Sci Total Environ 738:140240

Zhang H, Yasmin F, Song BH (2019) Neglected treasures in the wild—legume wild relatives in food security and human health. Curr Opin Plant Biol 49:17–26

Zhou D, Jin S, Li L, Wang Y, Weng N (2011) Quantifying the adsorption and uptake of CuO nanoparticles by wheat root based on chemical extractions. J Environ Sci 23:1852–1857

Zhu H, Han J, Xiao JQ, Jin Y (2008) Uptake, translocation, and accumulation of manufactured iron oxide nanoparticles by pumpkin plants. J Environ Monit 10:713–717. https://doi.org/10.1039/B805998E

Zou C, Du Y, Rashid A, Ram H, Savasli E, Pieterse PJ, Ortiz-Monasterio I, Yazici A, Kaur C, Mahmood K, Singh S (2019) Simultaneous biofortification of wheat with zinc, iodine, selenium, and iron through foliar treatment of a micronutrient cocktail in six countries. J Agric Food Chem 67:8096–8106

Zulfiqar U, Hussain S, Ishfaq M, Matloob A, Ali N, Ahmad M, Alyemeni MN, Ahmad P (2020c) Zinc-induced effects on productivity, zinc use efficiency, and grain biofortification of bread wheat under different tillage permutations. Agronomy 10:1566. https://doi.org/10.3390/agronomy10101566

Zulfiqar U, Hussain S, Maqsood M, Ishfaq M, Ali N (2021b) Zinc nutrition to enhance rice productivity, zinc use efficiency, and grain biofortification under different production systems. Crop Sci 61:739–749

Zulfiqar U, Hussain S, Maqsood M, Zamir SI, Ishfaq M, Ali N, Ahmad M, Maqsood MF (2021a) Enhancing the accumulation and bioavailability of iron in rice grains via agronomic interventions. Crop Pasture Sci. https://doi.org/10.1071/CP21140

Zulfiqar U, Maqsood M, Hussain S (2020a) Biofortiefication of rice with iron and zinc: progress and prospects. Rice Research for Quality Improvement: Genomics and Genetic Engineering: Nutrient Biofortification and Herbicide and Biotic Stress Resistance in Rice 2:605–627

Zulfiqar U, Maqsood M, Hussain S, Anwar-ul-Haq M (2020b) Iron nutrition improves productivity, profitability, and biofortification of bread wheat under conventional and conservation tillage systems. J Soil Sci Plant Nutr 20:1298–1310

Chapter 9
Role of Nanoparticles in Improving Biofortification

Hafiz Zulqurnain Raza, Anis Ali Shah, Sheeraz Usman, and Adnan Noor Shah

Abstract Application of nanoparticles (NPs) have been proven as a promising technique to improve plant performance under different environmental conditions, Application of NPs not only improves abiotic stress tolerance but also regulates the uptake of nutrients via regulating ion transporters. The application of NPs in biofortifying different essential nutrients has been shown as the most effective technique. In this book chapter, we have discussed the role of different NPs such as Fe NPs, Zn NPs, Cu NPs, I NPs and Se NPs in biofortifying these nutrients in plants.

Keywords Micronutrients · Biofortification · Nanoparticles · Grains · Human health

9.1 Introduction

The United Nations Food and Agriculture Organization reported that more than 700 million people would experience malnutrition globally in 2020 (FAO 2021). Another type of mineral deficiency known as food insecurity, which affects more than 2 billion people, happens due to scarcity of minerals and vitamins (Garg et al. 2018). In underdeveloped countries, a large number of individuals suffer from deficiencies of micronutrients, about more than half of the world's population is known to suffer from micronutrient deficiencies, which is one of the greatest threats to humanity on a worldwide scale (Stein 2010). Malnutrition in humans and animals is a hidden epidemic in many impoverished nations across the world. The diet plan must provide vital elements, such as carbohydrates, fat, proteins, vitamins, and minerals, in order to end malnutrition (Lean 2019). Two vital nutrients which have

H. Z. Raza · A. A. Shah (✉) · S. Usman
Department of Botany, Division of Science and Technology, University of Education, Lahore, Pakistan

A. N. Shah
Department of Agricultural Engineering, Khwaja Fareed University of Engineering and Information Technology, Rahim Yar Khan, Punjab, Pakistan

M. Hasanuzzaman et al. (eds.), *Mineral Biofortification in Crop Plants for Ensuring Food Security*, https://doi.org/10.1007/978-981-99-4090-5_9

beneficial impacts on human health are iron (Fe) and zinc (Zn), they are also considered as deficient nutrients worldwide. A severe issue related to health is Zn deficiency (Shekari et al. 2015). Micronutrient imbalances can be seen in neutral and alkaline soil as well as those soil with anoxic environments in arid zones (Tanveer and Shabala 2022). Micronutrient insufficiency has negative effects on intelligence, and academic performances. Additionally, micronutrient deficiency has far-reaching long-term effects on development of the nation (Bouis et al. 2011). Consumers need a variety of minerals and other nutrients, which must be present in their diets, in order to sustain a healthy lifestyle, these minerals play important role in many physiological processes (Gupta and Gupta 2014). Less intake of micronutrients can seriously harm the health of people (Tardy et al. 2019). According to research, people consume insufficient amounts of calcium and potassium (Castiglione et al. 2018). The primary food crops in underdeveloped countries that provide around 60% of the total average calorie needs per day are simple carbohydrates like wheat, maize, and rice (Ritchie et al. 2017). Furthermore, adding those substances and nutrients to diet could be viewed as a tactic to combat malnutrition (Vlaic et al. 2019). The production and food quality are being affected extensively by several causes, including soil pollution, soil salinity, arid zone, and soil degradation. Micronutrients are important for growth and development of humans because they support several metabolic processes. These micronutrients include such as Fe, Zn, selenium (Se), iodine (I), carotenoids, and folates (Waters and Grusak 2008).

According to definitions, a vitamin is an organic molecule that an organism requires in small amounts but cannot create on its own, hence it must be consumed through diet. As a coenzyme for various metabolic enzymes, vitamin B6 plays an important role in cells. Additionally, it has been demonstrated to be a powerful antioxidant agent. The model plant Arabidopsis's vitamin B6 level has been successfully increased by genetic engineering and biofortification (Raschke et al. 2011). Two parameters should preferably be considered when choosing suitable food crops for vitamin B6 biofortification procedures, the first one is the crop must be economically significant and broadly utilized and the other one is limitations in physiology or development should not prevent vitamin B6 from accumulating in the section of the crop plant that is ingested. The need for plant-based medicines is expanding along with the population (Ekor 2014). Bioactive compounds found in medicinal plants are utilized as replacements for synthetic medications all over the world (Anand et al. 2019). In order to lessen its impacts on agriculture's output, we need strategies or tactics to improve both amount and content of crops (Xie et al. 2020).

The process of nutrient enrichment in the consumable parts of plants by the external administration of nutrients in the form of fertilizer is referred to as "biofortification" (De Valença et al. 2017; Tanveer et al. 2020). Biofortification is especially important in the current economic situation, where price hikes for non-essential foods have further hampered dietary diversification and reduced food for the poor people (Thorne-Lyman et al. 2010). Staple foods that have been biofortified may have the ability to significantly enhance micronutrient consumption and have a positive effect on the lives of people (Jiang et al. 2019). Providing fertilizers of mineral nutrients to crops that are deficient in the diet by soil or foliar spray is known as agronomic biofortification (Ngigi et al. 2019). Widespread use of biofortification has been made to boost the mineral content of basic foods. Biofortification is a way for growing nutritionally rich crops at a reduced cost as well as it boosts the variety of the crop's nutritional content. Nutritional profile of agricultural crops can be enhanced by agronomic and genetic biofortification (Athar et al. 2020). Mostly because of unsatisfactory leadership, fortification techniques are frequently unsustainable in developing countries. Strategies for biofortification often concentrate on basic micronutrients, which are either organic substances (vitamins) or minerals. Antioxidant food fortification reflects a fusion of two quite different environments. Specifically, adding important nutrients with antioxidant capabilities to common diets i.e., vitamins C and E and fortifying meals with antioxidants that promote health, but these antioxidants are not necessary, such flavonoids. Biofortification boosts the synthesis of certain antioxidants without harming the development of plant (Farre et al. 2011). Biofortification of plants increased agricultural output to reduce nutritional deficiencies. The major functions of plant biofortification are to treat hidden hunger, nourish plants with desired nutrients, and alleviate micronutrient deficiencies (Díaz-Gómez et al. 2017).

Both traditional and recombinant breeding methods are utilized in genetic biofortification (Pérez-Massot et al. 2013). The expression of genes lowers the action of antinutrients, it also modulates the accessibility of nutrients to plants (Garg et al. 2018). Modern agricultural approaches are being used more frequently in biofortification operations due to the inadequacies of genetic techniques for biofortification (Athar et al. 2020). Leaching is a significant problem with NPK fertilizer since it is simple for them to be lost after being absorbed by plants. Plants utilize NPK and other agrochemicals inefficiently due to fixation, leakage, biological degradation (Raj et al. 2016). Use of fertilizers, which are usually in the form of salts or in ionic form, results in ready availability and rapid loss from the soil. Sources for agricultural land and water are becoming scarcer. Use of technological tools are required to boost efficiency of resource utilization with the minimum environmental harm to maintain the development of the agricultural area (Manjunatha et al. 2016). Agronomic biofortification is the most straightforward method using nowadays. Agronomic and genetic biofortification are economical methods to provide nourishing meals to those who are deprived of proper food (White and Broadley 2009). However, each of these approaches has drawbacks of its own.

Biofortification may raise the level of micronutrients (such Fe, Zn, and Se) in plants, it is a new technique to produce crops with full of nutrients and ensure food safety (Bouis and Saltzman 2017). Through the mechanism of "biofortification," some beneficial nutrients may be inexpensively and sustainably supplied to plants. This method may be used to supply food crops with certain minerals (Fe, I, Zn, Ca, Se, etc.), vitamins (A, E, B1, B6, etc.), and folate (Singh et al. 2016). A food that has been enriched with nutrients lowers the risk of cardiovascular disorders, malnutrition and reduces mortality rates (Hossain and Mohiuddin 2012).

Nanotechnology is a relatively recent strategy in the field of study. These days, a variety of disciplines employ this technology extensively (Cristiano José de Andrade et al. 2021). In the last 10 years, reports on the application of nanotechnology for agricultural purposes have increased (Lowry et al. 2019). Nanotechnology developments have enabled the use of nanostructured materials in a wide range of industrial applications, including food and agriculture (Yin et al. 2018). This technology provides the framework for creative and environmentally friendly methods of nutrient delivery that use the nano porous surfaces of plant components. The application of nano fertilizer is more effective as it improves soil richness with nutrients and environmental, agroecological degradation is reduced (Manjunatha et al. 2016).

Nano fertilizers are preferred to use in biofortification, because they increase nutrient bioavailability by three times and decrease the need for synthetic fertilizers, they also produce environmentally friendly crops and can tolerate disease and dry conditions. These nutrients are taken by the plants as they develop, helping to fill any nutritional gaps. When compared to traditional fertilizers, nano fertilizers might be seen as a type of fertilizer that is more productive. These nano fertilizers have the capability to boost plant nutrition, increase the effectiveness of utilization of nutrients, and defend plants from numerous climatic challenges (Sarlak and Taherifar 2017). A nano fertilizer termed as "Nano-Leucite Fertilizer" has just been created by several researchers. It is pollution free, might prevent leaching of nutrients in food, and could generally maximize crop and food output (Kamran et al. 2016). Various multi-element nano fertilizers have augmented the amount of minerals in plant tissues. For instance, $NiFe_2O_4$ boosted the concentration of nickel and Fe (Tombuloglu et al. 2019). Manganese and Fe concentrations were increased via $MnFe_2O_4$ nanoparticles (NPs) (Tombuloglu et al. 2018). Hybrid organic NPs have been created to use as a highly practical method to increase the nutritive value of many microelements in crops (Elemike et al. 2019). These metallic NPs, produced using several chemical processes (Aguilera et al. 2016). They can be used in multiple fields as well as in biofortification. NPs are the results of improvements in nanotechnology, it is widely employed in agricultural purposes in the form of nano fertilizer, nano sensor etc. Agriproducts like nano pesticides, nano herbicides and nano fertilizers are developing by nanotechnology they have a less influence on the environment and boost crop output (Younes et al. 2020). Metal NPs of any size and shape can be produced with ease. Metallic NPs, one of the fundamentals nanoproducts, have been utilized most in the manufacturing of nano fertilizers (Kolesnikov et al. 2019). Both the physical and natural sciences are benefiting greatly from nanotechnology (Roco 2001). NPs are Smaller in dimension and having a greater surface area. NPs are absorbed by plants by a number of mechanisms, such as stomata, water pores, ion channels, protein carriers, and endocytosis. NPs possess special structural, refractive, electrical, thermal, chemical, and magnetic characteristics (Hatami et al. 2016). NPs may represent a viable technique with favorable effects on germination of seeds, plant development, and plant metabolism (Seabra et al. 2014). Growth of a plant and tolerance to biotic and abiotic challenges can be supported by exposing to NPs (Rajput et al. 2018).

9.2 Silicon Nanoparticles

Silicon (Si) nanoparticles have vast application in agriculture due to greater surface area, absorption of nutrient and plant development also stimulated by Si nanoparticles (Rajput et al. 2021). NPs have complex effects, it's crucial to keep in mind that they are affected by number of variables, including dose and plant species. Despite any stress, NPs increase antioxidant enzymes, maintaining photosynthetic pigments and reducing growth inhibition (Jiang et al. 2021). It is necessary to develop a user-friendly, commercially viable technology that could meet the growing global population's demand for nutrient-rich food supplies with adequate levels of Fe. In the current situation, nanotechnological agricultural intervention provides a workable substitute to assure a healthier and nutrient-rich food production for people. A sustainable solution to malnutrition would be to increase the natural nutritional value of basic crops, through biofortification.

9.3 Iron Nanoparticles

One mineral that is necessary for the body to grow properly is Fe. Fe is a crucial micronutrient for plants and key component of cytochromes and hemoglobin (Claessens et al. 2013). A diverse and crucial component of plant metabolism is Fe. To enhance the Fe content in plants of different crops, Fe nanoparticles have been introduced. Its biological processes are principally supported by the reversible redox reaction between the Fe ions, i.e., ferrous (Fe^{2+}) and ferric (Fe^{3+}). Iron is involved in transferring electrons, iron–sulfur (Fe–S) clusters are utilized in electron transfer chain of photosynthesis and respiration (Hell and Stephan 2003). Chloroplasts serve as the primary source of Fe in cells. Root growth was accelerated by pre-treating different leguminous seeds with α-Fe_2O_3 nanoparticles (Palchoudhury et al. 2018). Iron oxide nanoparticles can increase the amount of Fe in chloroplast by 80–90% (Marschner 2011). Iron sulphate heptahydrate ($FeSO_4.7H_2O$) foliar spraying was successful in increasing the Fe percentage in leaves and sink organs (Niyigaba et al. 2019). Vegetables with significantly better profiles can result from biofortification of Fe to increase the accumulation of secondary metabolites such chlorogenic acid, violaxanthin, or neoxanthin (Giordano et al. 2019). In rice, nanoparticles boost nitrogen absorption, reductive sugars, activity of α-amylase and root growth (Farooq et al. 2010). The use of Fe nanoparticles increased the nutritional content, physiological metabolism and the supplementation of Fe to peanuts (Rui et al. 2016). Iron nanoparticles regulate enzyme for heme protein and ultimately enhance functioning of cytochrome also enhanced (Zahra et al. 2015). Fe-chelated nano fertilizers were used to increase nutrient availability (particularly NPK) and productivity in basil and rice (Peyvandi et al. 2011; Fakharzadeh et al. 2020). The crop yield of wheat, harvest index, kernel weight, and branches per plant were all strongly impacted by Fe fertilizers (Harsini et al. 2014). Additionally, it has been discovered that the impacts of nano-FeO, and nano-ZnCuFe oxide nanoparticles on the shoot development of mung (*Vigna radiata*) seedlings (Dhoke et al. 2013).

9.4 Zinc Nanoparticles

A well-known issue with plants is the deficiency of Zn, which results in lower productivity and lower nutritional content in crops. Among the most fundamental trace elements in living things is Zn, which plays significant roles as a structural component, catalyst in many metabolic processes and regulatory ion (Chasapis et al. 2012). Micronutrient Zn is important for both humans and plants. Zinc is crucial for numerous physiological processes that occur in living things, including the protein synthesis and production of carbohydrates as well as the breakdown of these substances (Tarafdar et al. 2014; Khan et al. 2021). Insufficient amount of Zn is a persistent problem that has a negative impact on both agricultural output and human health (Farooq et al. 2018). Zinc is needed by plants for a variety of biological processes include photosynthesis, respiration, and the formation of chlorophyll (Nishizawa 2005). During Zn deficiency, plants have developed intricate regulatory systems for accumulating and transporting Zn (Henriques et al. 2012). Gene expression in plants can be controlled by the application of Zn, a sufficient Zn supply increases agricultural yields, while too much Zn might harm plants (Cakmak 2008). Zinc fertilization is a useful way to increase the amount of Zn in grains that are grown. As a replacement for existing traditional Zn fertilizers, it is important to investigate a new Zn fertilizer with excellent performance and minimal environmental risk. Plants have the tendency to maintain growth and output in soils with reduced Zn availability is known as Zn efficiency, a temporary approach is to use Zn-containing fertilizers (also known as agronomic biofortification). Raising rhizospheric Zn levels by application of fertilizer is also employed by using this strategy. Nanoparticles can enhance the dietary value and growth and development of plant (Kah et al. 2018). ZnO nanoparticles (ZnO NPs) could be a more effective supply of Zn for maize grains, these particles enhanced biomass productivity and biofortification of crops. One of the most popular nanofertilizers is ZnO NPs, which has drawn increased attention from researchers. Numerous research has assessed how ZnO NPs affect wheat growth, nutrient intake, and Zn uptake and use efficiency (Dimkpa et al. 2020). Various development stages of plant have been affected by ZnO NPs. The use of ZnO NPs has the potential to significantly raise the Zn content of grains in cereal crops (Du et al. 2019). Amino acids (AAs) are a crucial measure of a grain's nutritional quality (Wang et al. 2019a, b, c). Zinc concentration in shoots and grains was enhanced by the use of ZnO nanoparticles. Rise in Zn concentration in grains was seen with foliar application of ZnO NPs. Zn-EDTA foliar spray has improved the nutritional content of rice grains (Wei et al. 2012). Foliar application of nanoforms of Zn and Fe in the chickpeas improves the Zn and Fe content in the grains of chickpeas (Dhaliwal et al. 2021).

9.5 Selenium and Iodine Nanoparticles

Selenium and I have great significance, these two microelements are helpful to plants and necessary for humans. Although Se is necessary for lower plants, its requirement for higher plants is currently being researched (El-Ramady et al. 2020). Biofortification helps the plant to increase the percentage of Se in roots, stems, leaves, fruits, or seeds (Abedi et al. 2021). By influencing with a number of physiological systems, Se may operate as a crucial factor (Iqra Bano et al. 2021). Microbes, animals and humans required Se to have adequate and healthy lives (El-Ramady et al. 2015). When organic Se is used, it may be preferable to use plants like broccoli and garlic, these species work best for fortified crops (Hsu et al. 2011). It is important to avoid adding extra sulphur to the production of Se-fortified crops because sulphur inhibits the uptake of Se (Broadley et al. 2006). They are especially important for their role in promoting the growth and development of all living things, their antioxidant qualities, the enhancement of immunity, and their capacity to resist biotic and abiotic stresses (Golubkina et al. 2021). Selenium biofortification of crops is utilized in Se-deficient regions of China in order to stop the terrible Keshan disorder, which is still widespread and characterized by cardiomyopathy brought on by Se shortage (Banuelos et al. 2013).

These elements promote the creation of combined agrochemical biofortification in plants. Salicylic acid and phytohormones are employed in joint Se-I biofortification to increase its effectiveness (Smoleń et al. 2019). In order to maintain human health, Se and nano-selenium (nano-Se) or (Se-NPs) have been utilized (Tan et al. 2019). Se-NPs are very stable, hydrophilic, toxic-free, and permeable (Zahedi et al. 2019). Enhancing the nutritional value of usable foods by biofortifying crops with application of Se-NPs to increase Se content, Se-NPs also reduce abiotic stress of the plant by producing antioxidants (Sarwar et al. 2020). Selenium might promote productivity and plant development and defend crops against biotic and abiotic stresses. Selenium is also used to prevent the plant from herbivores by producing some volatile compounds (dimethyl selenide) (Wang et al. 2019a, b, c). Se-NPs (particularly 5–200 nm) exhibit great solubility, and high biological activity in plants and they are nontoxic. They also improve the activities of several enzymes including GSH-Px, TrxR, and GST and might destroy free radicals (Zahedi et al. 2019). In addition, Se is thought to be an important plant PAL enzyme activator (Astaneh et al. 2018). Se-NPs have scope in producing plants which show suitable growth (Hussein et al. 2019). By the utilization of combined foliar and soil-based fertilizers, Crops with low levels of Se can be biofortified (Lopes et al. 2017). Cereals, spinach and strawberry are among the frequently biofortified food crops (Mimmo et al. 2017; Golubkina et al. 2017; Lyons 2018). It has been demonstrated that the creation and application of Se-NPs as a nutrient and biofortifier is an innovative strategy (García Márquez et al. 2020). People are using nano-Se increasingly to fortify the crops (El-Ramady et al. 2020). As innovative and developing technologies like nanotechnology evolve, the use of Se-NPs has emerged as an alternative to traditional Se fertilizers, to enhance crops. Scientists are interested in using Se-NPs to biofortify plants. It aims to raise the amount of Se in plants' edible parts in order to alleviate human and animal Se deficiency (Schiavon et al. 2020). Se-NPs have the ability to enhance the quality, nutrient content, and quantity of Se present in the edible parts of plants, there is now an increasing interest in utilizing them to biofortify crops. More than a billion individuals around the world suffer from Se deficiency which can be reduced by producing biofortified food stuff with Se (Gupta and Gupta 2017). Nutrient delivery to plants using nanoparticles is a unique method. Selenium enrichment in rice is currently being investigated by supplementation of Se-NPs (Wang et al. 2020).

9.6 Copper Oxide Nanoparticles

Copper oxide nanoparticles (CuONPs) usage in agribusiness has proven to have favorable impacts on plants because of their resilience in different stresses (Mali et al. 2019). They show good result when they apply in low amounts (Hernández-Hernández et al. 2019). Copper oxide nanoparticles (CuONPs) enhance nutrient content in fruits like in tomato and boost enzymatic and non-enzymatic antioxidant systems, as well as have positive effect on yield (López-Vargas et al. 2018a, b). Copper is one of the most important trace elements for growth of plant. Plants require copper for several physiological activities. Copper is most abundant in plant chloroplasts because it aids in the formation of chloroplasts and other pigments, it also improves grain quality and has an impact on vegetative development in agricultural crops (Siddiqi and Husen 2020). Copper also accelerates the absorption of nutrients and activity of enzymes (Shams et al. 2019). Copper oxide nanoparticles (CuONPs) show improved growth and the mitotic index in *Allium cepa*, when they were exposed to (CuONPs) at a concentration of 20 g/ml (Nagaonkar et al. 2015). Recently, CuONPs has found its application in the field of agriculture as fertilizers (Wang et al. 2019a, b, c). Nutrient of plants can be enhanced by utilizing CuONPs in low concentration (Liu et al. 2016). Copper nanoparticles directly improve fruit quality of tomato and shoot up bioactive constituent of the fruits (López-Vargas et al. 2018a, b). Plant cells use copper as catalytic centers for proteins in their metabolism (Garcia et al. 2014). Enzymatic antioxidants include catalase (CAT), APX, and GPX, which are crucial ROS eliminators, are governed by the action of nanoparticles. Non-enzymatic antioxidant mechanism become active in the presence of Cu-NPs, enzymatic and non-enzymatic substances are biostimulated by the cellular membranes (Juárez-Maldonado et al. 2019).

9.7 Conclusion

Biofortification is a process of enhancing the nutritional quality of crops to improve human health. It involves the selection and breeding of crops with high nutritional value, as well as the use of agronomic practices to increase the uptake and availability of essential nutrients. Nanoparticles have emerged as a promising tool for improving biofortification. Nanoparticles are small particles with dimensions ranging from 1 to 100 nanometers. They have unique physical and chemical properties that make them ideal for enhancing nutrient uptake and transport in plants.

Nanoparticles can be used to enhance the efficiency of fertilizers by delivering nutrients directly to the plant roots. They can also be used to increase the availability of essential micronutrients such as Fe, Zn, and Se in crops. This is achieved by using nanoparticles to modify the physicochemical properties of the soil, which in turn enhances nutrient uptake and transport in plants. In addition, the use of nanoparticles in biofortification can also help reduce the use of chemical fertilizers needed to achieve the desired crop yields, which can have a positive impact on the environment. Nanoparticles can play a significant role in improving the biofortification of food crops by enhancing nutrient uptake and improving plant growth. Nanoparticles can improve biofortification is by increasing the solubility and bioavailability of nutrients in the soil, which makes it easier for plants to absorb them. For example, iron oxide nanoparticles can be used to increase the uptake of Fe by plants, which is crucial for combating Fe deficiency anemia in humans. Nanoparticles can also protect plants from environmental stressors, such as drought or disease, by strengthening their cell walls and improving their ability to retain moisture and nutrients.

Overall, the use of nanoparticles in improving biofortification has the potential to significantly improve the nutritional quality of crops, leading to better health outcomes for populations that are most vulnerable to nutrient deficiencies. However, the use of nanoparticles in agriculture and food production is still a relatively new field, and there are concerns about their safety and potential impacts on the environment. Therefore, it is important to carefully evaluate the risks and benefits of using nanoparticles in biofortification, and to develop guidelines and regulations to ensure their safe and sustainable use.

References

Abedi S, Iranbakhsh A, Oraghi Ardebili Z, Ebadi M (2021) Nitric oxide and selenium nanoparticles confer changes in growth, metabolism, antioxidant machinery, gene expression, and flowering in chicory (*Cichorium intybus* L.): potential benefits and risk assessment. Environ Sci Pollut Res 28:3136–3148

Aguilera J, Venegas V, Oliva JM, Sayagués MJ, De Miguel M, Sánchez-Alcázar JA, Arévalo-Rodríguez M, Zaderenko A (2016) Targeted multifunctional tannic acid nanoparticles. RSC Adv 6:7279–7287

Anand U, Jacobo-Herrera N, Altemimi A, Lakhssassi N (2019) A comprehensive review on medicinal plants as antimicrobial therapeutics: potential avenues of biocompatible drug discovery. Meta 9:258

Andrade VS, Gutierrez MF, Regaldo L, Paira AR, Repetti MR, Gagneten AM (2021) Influence of rainfall and seasonal crop practices on nutrient and pesticide runoff from soybean dominated agricultural areas in Pampean streams, Argentina. Sci Total Environ 788:147676

Astaneh RK, Bolandnazar S, Nahandi FZ, Oustan S (2018) Effect of selenium application on phenylalanine ammonia-lyase (PAL) activity, phenol leakage and total phenolic content in garlic (*Allium sativum* L.) under NaCl stress. Inf Process Agric 5:339–344

Athar T, Khan MK, Pandey A, Yilmaz FG, Hamurcu M, Hakki EE, Gezgin S (2020) Biofortification and the involved modern approaches. J Elem 25

Banuelos GS, Lin ZQ, Yin X (eds) (2013) Selenium in the environment and human health. CRC Press

Bouis HE, Hotz C, McClafferty B, Meenakshi J, Pfeiffer WH (2011) Biofortification: a new tool to reduce micronutrient malnutrition. Food Nutr Bull 32:S31–S40

Bouis HE, Saltzman A (2017) Improving nutrition through biofortification: a review of evidence from HarvestPlus, 2003 through 2016. Glob Food Secur 12:49–58

Broadley MR, White PJ, Bryson RJ, Meacham MC, Bowen HC, Johnson SE, Hawkesford MJ, McGrath SP, Zhao F-J, Breward N (2006) Biofortification of UK food crops with selenium. Proc Nutr Soc 65:169–181

Cakmak I (2008) Enrichment of cereal grains with zinc: agronomic or genetic biofortification? Plant Soil 302:1–17

Castiglione D, Platania A, Conti A, Falla M, D'Urso M, Marranzano M (2018) Dietary micronutrient and mineral intake in the mediterranean healthy eating, ageing, and lifestyle (MEAL) study. Antioxidants 7:79

Chasapis CT, Loutsidou AC, Spiliopoulou CA, Stefanidou ME (2012) Zinc and human health: an update. Arch Toxicol 86:521–534

Claessens M, Contor L, Dhonukshe-Rutten R, De Groot LC, Fairweather-Tait SJ, Gurinovic M, Koletzko B, Van Ommen B, Raats MM, Van't Veer P (2013) EURRECA—principles and future for deriving micronutrient recommendations. Crit Rev Food Sci Nutr 53:1135–1146

De Valença AW, Bake A, Brouwer ID, Giller KE (2017) Agronomic biofortification of crops to fight hidden hunger in sub-Saharan Africa. Glob Food Sec 12:8–14

Dhaliwal SS, Sharma V, Shukla AK, Verma V, Behera SK, Singh P, Alotaibi SS, Gaber A, Hossain A (2021) Comparative efficiency of mineral, chelated and nano forms of zinc and iron for improvement of zinc and iron in chickpea (*Cicer arietinum* L.) through biofortification. Agronomy 11:2436

Dhoke SK, Mahajan P, Kamble R, Khanna A (2013) Effect of nanoparticles suspension on the growth of mung (*Vigna radiata*) seedlings by foliar spray method. Nanotechnol Dev 3:e1–e1

Díaz-Gómez J, Twyman RM, Zhu C, Farré G, Serrano JC, Portero-Otin M, Muñoz P, Sandmann G, Capell T, Christou P (2017) Biofortification of crops with nutrients: factors affecting utilization and storage. Curr Opin Biotechnol 44:115–123

Dimkpa CO, Andrews J, Sanabria J, Bindraban PS, Singh U, Elmer WH, Gardea-Torresdey JL, White JC (2020) Interactive effects of drought, organic fertilizer, and zinc oxide nanoscale and bulk particles on wheat performance and grain nutrient accumulation. Sci Total Environ 722: 137808

Du W, Yang J, Peng Q, Liang X, Mao H (2019) Comparison study of zinc nanoparticles and zinc sulphate on wheat growth: from toxicity and zinc biofortification. Chemosphere 227:109–116

Ekor M (2014) The growing use of herbal medicines: issues relating to adverse reactions and challenges in monitoring safety. Front Pharmacol 4:177

Elemike EE, Uzoh IM, Onwudiwe DC, Babalola OO (2019) The role of nanotechnology in the fortification of plant nutrients and improvement of crop production. Appl Sci 9:499

El-Ramady H, Abdalla N, Alshaal T, Domokos-Szabolcsy E, Elhawat N, Prokisch J, Sztrik A, Fari M, El-Marsafawy S, Shams MS (2015) Selenium in soils under climate change, implication for human health. Environ Chem Lett 13:1–19

El-Ramady H, Faizy SE-D, Abdalla N, Taha H, Domokos-Szabolcsy É, Fari M, Elsakhawy T, Omara AE-D, Shalaby T, Bayoumi Y (2020) Selenium and nano-selenium biofortification for human health: opportunities and challenges. Soil Syst 4:57

Fakharzadeh S, Hafizi M, Baghaei MA, Etesami M, Khayamzadeh M, Kalanaky S, Akbari ME, Nazaran MH (2020) Using nanochelating technology for biofortification and yield increase in rice. Sci Rep 10:4351

Farooq M, Basra S, Wahid A, Khaliq A, Kobayashi N (2010) Rice seed invigoration: a review. In: Lichtfouse E (ed) Organic farming, Pest control and remediation of soil pollutants. Sustainable agriculture reviews, vol 1. Springer, Dordrecht, pp 137–175

Farooq M, Ullah A, Rehman A, Nawaz A, Nadeem A, Wakeel A, Nadeem F, Siddique KH (2018) Application of zinc improves the productivity and biofortification of fine grain aromatic rice grown in dry seeded and puddled transplanted production systems. Field Crops Res 216:53–62

Farre G, Twyman RM, Zhu C, Capell T, Christou P (2011) Nutritionally enhanced crops and food security: scientific achievements versus political expediency. Curr Opin Biotechnol 22:245–251

Garcia L, Welchen E, Gonzalez DH (2014) Mitochondria and copper homeostasis in plants. Mitochondrion 19:269–274

García Márquez V, Morelos Moreno Á, Benavides Mendoza A, Medrano Macías J (2020) Ionic selenium and nanoselenium as biofortifiers and stimulators of plant metabolism. Agronomy 10 (9):1399

Garg M, Sharma N, Sharma S, Kapoor P, Kumar A, Chunduri V, Arora P (2018) Biofortified crops generated by breeding, agronomy, and transgenic approaches are improving lives of millions of people around the world. Front Nutr 5:12

Giordano M, El-Nakhel C, Pannico A, Kyriacou MC, Stazi SR, De Pascale S, Rouphael Y (2019) Iron biofortification of red and green pigmented lettuce in closed soilless cultivation impacts crop performance and modulates mineral and bioactive composition. Agronomy 9:290

Golubkina N, Moldovan A, Kekina H, Kharchenko V, Sekara A, Vasileva V, Skrypnik L, Tallarita A, Caruso G (2021) Joint biofortification of plants with selenium and iodine: new field of discoveries. Plan Theory 10:1352

Golubkina NA, Kosheleva OV, Krivenkov LV, Dobrutskaya HG, Nadezhkin S, Caruso G (2017) Intersexual differences in plant growth, yield, mineral composition and antioxidants of spinach (*Spinacia oleracea* L.) as affected by selenium form. Sci Hortic 225:350–358

Gupta M, Gupta S (2017) An overview of selenium uptake, metabolism, and toxicity in plants. Front Plant Sci 7:2074

Gupta U, Gupta S (2014) Sources and deficiency diseases of mineral nutrients in human health and nutrition: a review. Pedosphere 24:13–38

Harsini M, Habibi H, Talaei G (2014) Study the effects of iron nano chelated fertilizers foliar application on yield and yield components of new line of wheat cold region of Kermanshah provence. Agric Adv 3:95–102

Hatami M, Kariman K, Ghorbanpour M (2016) Engineered nanomaterial-mediated changes in the metabolism of terrestrial plants. Sci Total Environ 571:275–291

Hell R, Stephan UW (2003) Iron uptake, trafficking and homeostasis in plants. Planta 216:541–551

Henriques AR, Chalfun-Junior A, Aarts M (2012) Strategies to increase zinc deficiency tolerance and homeostasis in plants. Braz J Plant Physiol 24:3–8

Hernández-Hernández H, Quiterio-Gutiérrez T, Cadenas-Pliego G, Ortega-Ortiz H, Hernández-Fuentes AD, Cabrera de la Fuente M, Valdés-Reyna J, Juárez-Maldonado A (2019) Impact of selenium and copper nanoparticles on yield, antioxidant system, and fruit quality of tomato plants. Plan Theory 8:355

Hossain SM, Mohiuddin A (2012) Study on biofortification of rice by targeted genetic engineering. Int J Agric Res Innov Technol 2:25–35

Hussein HAA, Darwesh OM, Mekki BB, El-Hallouty SM (2019) Evaluation of cytotoxicity, biochemical profile and yield components of groundnut plants treated with nano-selenium. Biotechnol Reports 24:e00377

Hsu FC, Wirtz M, Heppel SC, Bogs J, Krämer U, Khan MS, Bub A, Hell R, Rausch T (2011) Generation of se fortified broccoli as functional food: impact of se fertilization on S metabolism. Plant Cell Environ 34:192–207

Iqra Bano E, Hira Sajjad H, Reza R (2021) Importance of micro-nutrient supplementation for livestock a mini-review. Acta Sci Vet Sci 3:54–57

Jiang L, Wang L, Tanveer M, Tian C (2019) Lithium biofortification of medicinal tea *Apocynum venetum*. Sci Rep 9:8182

Jiang M, Dai S, Wang B, Xie Z, Li J, Wang L, Li S, Tan Y, Tian B, Shu Q (2021) Gold nanoparticles synthesized using melatonin suppress cadmium uptake and alleviate its toxicity in rice. Environ Sci Nano 8:1042–1056

Juárez-Maldonado A, Ortega-Ortíz H, Morales-Díaz AB, González-Morales S, Morelos-Moreno Á, Cabrera-De la Fuente M, Sandoval-Rangel A, Cadenas-Pliego G, Benavides-Mendoza A (2019) Nanoparticles and nanomaterials as plant biostimulants. Int J Mol Sci 20:162

Kah M, Kookana RS, Gogos A, Bucheli TD (2018) A critical evaluation of nanopesticides and nanofertilizers against their conventional analogues. Nat Nanotechnol 13:677–684

Kamran F, Salampessy J, Reddy N (2016) Application of NMR spectroscopy for structural characterization of bioactive peptides derived from food protein. Appl NMR Spectrosc Food Sci 5:3–76

Khan MT, Ahmed S, Shah AA, Noor Shah A, Tanveer M, El-Sheikh MA, Siddiqui MH (2021) Influence of zinc oxide nanoparticles to regulate the antioxidants enzymes, some osmolytes and agronomic attributes in *Coriandrum sativum* L. grown under water stress. Agronomy 11:2004

Kolesnikov S, Timoshenko A, Kazeev KS, Akimenko YV, Myasnikova M (2019) Ecotoxicity of copper, nickel, and zinc nanoparticles assessment on the basis of biological indicators of chernozems. Eurasian Soil Sci 52:982–987

Lean ME (2019) Principles of human nutrition. Medicine 47(3):140–144

Liu R, Zhang H, Lal R (2016) Effects of stabilized nanoparticles of copper, zinc, manganese, and iron oxides in low concentrations on lettuce (*Lactuca sativa*) seed germination: nanotoxicants or nanonutrients? Water Air Soil Pollut 227:1–14

Lopes G, Ávila FW, Guilherme LRG (2017) Selenium behavior in the soil environment and its implication for human health. Ciência e Agrotecnologia 41:605–615

López-Vargas ER, Ortega-Ortíz H, Cadenas-Pliego G, de Alba RK, Cabrera de la Fuente M, Benavides-Mendoza A, Juárez-Maldonado A (2018a) Foliar application of copper nanoparticles increases the fruit quality and the content of bioactive compounds in tomatoes. Appl Sci 8:1020

López-Vargas ER, Ortega-Ortíz H, Cadenas-Pliego G, de Alba Romenus K, Cabrera de la Fuente M, Benavides-Mendoza A, Juárez-Maldonado A (2018b) Foliar application of copper nanoparticles increases the fruit quality and the content of bioactive compounds in tomatoes. Appl Sci 8:1020

Lowry GV, Avellan A, Gilbertson LM (2019) Opportunities and challenges for nanotechnology in the Agri-tech revolution. Nat Nanotechnol 14:517–522

Lyons K, Scrinis G, Whelan J (2018) Nanotechnology, agriculture, and food. In: Nanotechnology and global sustainability. CRC Press, pp 146–169

Mali SC, Raj S, Trivedi R (2019) Biosynthesis of copper oxide nanoparticles using *Enicostemma axillare* (lam.) leaf extract. Biochemis Biophysics Rep 20:100699

Manjunatha S, Biradar D, Aladakatti YR (2016) Nanotechnology and its applications in agriculture: a review. J Farm Sci 29:1–13

Marschner H (ed) (2011) Marschner's mineral nutrition of higher plants. Academic

Mimmo T, Tiziani R, Valentinuzzi F, Lucini L, Nicoletto C, Sambo P, Scampicchio M, Pii Y, Cesco S (2017) Selenium biofortification in *Fragaria ananassa*: implications on strawberry fruits quality, content of bioactive health beneficial compounds and metabolomic profile. Front Plant Sci 8:1887

Nagaonkar D, Shende S, Rai M (2015) Biosynthesis of copper nanoparticles and its effect on actively dividing cells of mitosis in *Allium cepa*. Biotechnol Progress 31:557–565

Ngigi PB, Lachat C, Masinde PW, Du Laing G (2019) Agronomic biofortification of maize and beans in Kenya through selenium fertilization. Environ Geochemis Health 41:2577–2591

Nishizawa NK (2005) The uptake and translocation of minerals in rice plants. Rice is life: scientific perspectives for the 21st century IRRI, Manila, Philippines 90–93

Niyigaba E, Twizerimana A, Mugenzi I, Ngnadong WA, Ye YP, Wu BM, Hai JB (2019) Winter wheat grain quality, zinc and iron concentration affected by a combined foliar spray of zinc and iron fertilizers. Agronomy 9:250

Palchoudhury S, Jungjohann KL, Weerasena L, Arabshahi A, Gharge U, Albattah A, Miller J, Patel K, Holler RA (2018) Enhanced legume root growth with pre-soaking in α-Fe 2 O 3 nanoparticle fertilizer. RSC Adv 8:24075–24083

Pérez-Massot E, Banakar R, Gómez-Galera S, Zorrilla-López U, Sanahuja G, Arjó G, Miralpeix B, Vamvaka E, Farré G, Rivera SM (2013) The contribution of transgenic plants to better health through improved nutrition: opportunities and constraints. Genes Nutr 8:29–41

Peyvandi M, Jamkany ZK, Mirza M (2011) Effects of Nano iron chelate and chelated iron on growth and activity of antioxidant enzymes in savory herb'. Mol Cell Biotechnol Magazine 2(5)

Raj R, Yadav MP, Kumar V (2016) Productivity and profitability of rice hybrids at different nutrient management levels under semi-arid conditions of North Eastern Plains Zone. Ann Agric Res (New Series) 37:36–42

Rajput V, Minkina T, Suskova S, Mandzhieva S, Tsitsuashvili V, Chapligin V, Fedorenko A (2018) Effects of copper nanoparticles (CuO NPs) on crop plants: a mini review. BioNanoSci 8:36–42

Rajput VD, Minkina T, Feizi M, Kumari A, Khan M, Mandzhieva S, Sushkova S, El-Ramady H, Verma KK, Singh A (2021) Effects of silicon and silicon-based nanoparticles on rhizosphere microbiome, plant stress and growth. Biology 10:791

Raschke M, Boycheva S, Crèvecoeur M, Nunes-Nesi A, Witt S, Fernie AR, Amrhein N, Fitzpatrick TB (2011) Enhanced levels of vitamin B6 increase aerial organ size and positively affect stress tolerance in Arabidopsis. Plant J 66(3):414–432

Ritchie H, Rosado P, Roser M (2017) Diet compositions. Our World in Data

Roco MC (2001) International strategy for nanotechnology research. J Nanopart Res 3:353–360

Rui M, Ma C, Hao Y, Guo J, Rui Y, Tang X, Zhao Q, Fan X, Zhang Z, Hou T (2016) Iron oxide nanoparticles as a potential iron fertilizer for peanut (*Arachis hypogaea*). Front Plant Sci 7:815

Sarlak N, Taherifar A (2017) Encapsulation of nanomaterials and production of Nanofertilizers and Nanopesticides: insecticides for Agri-food production and plant disease treatment. Nanosci Plant–Soil Syst 481–498

Sarwar N, Akhtar M, Kamran MA, Imran M, Riaz MA, Kamran K, Hussain S (2020) Selenium biofortification in food crops: key mechanisms and future perspectives. J Food Compost Anal 93:103615

Schiavon M, Nardi S, Dalla Vecchia F, Ertani A (2020) Selenium biofortification in the 21st century: status and challenges for healthy human nutrition. Plant Soil 453:245–270

Seabra AB, Rai M, Durán N (2014) Nano carriers for nitric oxide delivery and its potential applications in plant physiological process: a mini review. J Plant Biochem Biotechnol 23:1–10

Shams M, Ekinci M, Turan M, Dursun A, Kul R, Yildirim E (2019) Growth, nutrient uptake and enzyme activity response of lettuce (*Lactuca sativa* L.) to excess copper. J Environ Sustain 2: 67–73

Shekari F, Mohammadi H, Pourmohammad A, Avanes A, Benam MBK (2015) Spring wheat yielding and the content of protein and zinc in its grain depending on zinc fertilisation. Electron J Pol Agric Univ 18:1

Siddiqi KS, Husen A (2020) Current status of plant metabolite-based fabrication of copper/copper oxide nanoparticles and their applications: a review. Biomater Res 24:1–15

Singh U, Praharaj C, Singh SS, Singh NP (2016) Biofortification of food crops. Springer, New Delhi, pp 3–18

Smoleń S, Kowalska I, Halka M, Ledwożyw-Smoleń I, Grzanka M, Skoczylas Ł, Czernicka M, Pitala J (2019) SelectedAspects of iodate and iodosalicylate metabolism in lettuce including the activity of vanadium dependent haloperoxidases as affected by exogenous vanadium. Agronomy 10:1

Stein AJ (2010) Global impacts of human mineral malnutrition. Plant Soil 335:133–154

Tan KH, Xu W, Stefka S, Demco DE, Kharandiuk T, Ivasiv V, Nebesnyi R, Petrovskii VS, Potemkin II, Pich A (2019) Selenium-modified microgels as bio-inspired oxidation catalysts. Angew Chem Int Ed 58(29):9791–9796

Tanveer M, Shabala S (2022) Entangling the interaction between essential and nonessential nutrients: implications for global food security. In: Plant Nutrition and Food Security in the Era of Climate Change. Elsevier, pp 1–25

Tanveer M, Shahzad B, Ashraf U (2020) Nanoparticle application and abiotic-stress tolerance in plants. In: Plant life under changing environment. Elsevier, pp 627–641

Tarafdar J, Raliya R, Mahawar H, Rathore I (2014) Development of zinc nanofertilizer to enhance crop production in pearl millet (*Pennisetum americanum*). Agric Res 3:257–262

Tardy A-L, Ballesta AA, Yilmaz GC, Dan M, Ramirez DM, Lam HY, Azais-Braesco V, Pouteau E (2019) Adult's dietary intakes of selected Vitamins & Minerals Essential for energy metabolism and cognition: a comparison across countries & genders (FS10-04-19). Curr Dev Nutr 3:nzz039. FS010-004-019

Thorne-Lyman AL, Valpiani N, Sun K, Semba RD, Klotz CL, Kraemer K, Akhter N, De Pee S, Moench-Pfanner R, Sari M (2010) Household dietary diversity and food expenditures are closely linked in rural Bangladesh, increasing the risk of malnutrition due to the financial crisis. J Nutr 140:182S–188S

Tombuloglu H, Slimani Y, Tombuloglu G, Almessiere M, Baykal A, Ercan I, Sozeri H (2019) Tracking of NiFe2O4 nanoparticles in barley (*Hordeum vulgare* L.) and their impact on plant growth, biomass, pigmentation, catalase activity, and mineral uptake. Environ Nanotechnol Monit Manag 11:100223

Tombuloglu H, Tombuloglu G, Slimani Y, Ercan I, Sozeri H, Baykal A (2018) Impact of manganese ferrite ($MnFe_2O_4$) nanoparticles on growth and magnetic character of barley (*Hordeum vulgare* L.). Environ Pollut 243:872–881

United Nations Food and Agriculture Organization (2021) The state of food and agriculture 2021: making agrifood systems more resilient to shocks and stresses

Vlaic RA, Mureşan CC, Muste S, Mureşan V, Pop A, Mureşan GPA (2019) Boletus edulis mushroom flour-based wheat bread as innovative fortified bakery product. Bull Univ Agric Sci Vet Med Cluj-Napoc Food Sci Technol 76:46–56

Wang K, Wang Y, Li K, Wan Y, Wang Q, Zhuang Z, Guo Y, Li H (2020) Uptake, translocation and biotransformation of selenium nanoparticles in rice seedlings (*Oryza sativa* L.). J Nanobiotechnol 18:1–15

Wang M, Peng Q, Zhou F, Yang W, Dinh QT, Liang D (2019c) Uptake kinetics and interaction of selenium species in tomato (*Solanum lycopersicum* L.) seedlings. Environ Sci Pollut Res 26: 9730–9738

Wang Y, Jiang F, Ma C, Rui Y, Tsang DC, Xing B (2019a) Effect of metal oxide nanoparticles on amino acids in wheat grains (*Triticum aestivum*) in a life cycle study. J Environ Manag 241: 319–327

Wang Y, Lin Y, Xu Y, Yin Y, Guo H, Du W (2019b) Divergence in response of lettuce (*var. ramosa* Hort.) to copper oxide nanoparticles/microparticles as potential agricultural fertilizer. Env Pollut Bioavail 31:80–84

Waters BM, Grusak MA (2008) Quantitative trait locus mapping for seed mineral concentrations in two *Arabidopsis thaliana* recombinant inbred populations. New Phytol 179:1033–1047

Wei Y, Shohag M, Yang X, Yibin Z (2012) Effects of foliar iron application on iron concentration in polished rice grain and its bioavailability. J Agric Food Chem 60:11433–11439

White PJ, Broadley MR (2009) Biofortification of crops with seven mineral elements often lacking in human diets–iron, zinc, copper, calcium, magnesium, selenium and iodine. New Phytol 182: 49–84

Xie H, Zhang Y, Wu Z, Lv T (2020) A bibliometric analysis on land degradation: current status, development, and future directions. Land 9:28

Yin J, Wang Y, Gilbertson LM (2018) Opportunities to advance sustainable design of nano-enabled agriculture identified through a literature review. Environ Sci 5:11–26

Younes N, Hassan HS, Elkady MF, Hamed A, Dawood MF (2020) Impact of synthesized metal oxide nanomaterials on seedlings production of three Solanaceae crops. Heliyon 6:e03188

Zahedi SM, Hosseini MS, Meybodi NDH, da Silva JAT (2019) Foliar application of selenium and nano-selenium affects pomegranate (*Punica granatum cv. Malase Saveh*) fruit yield and quality. S Afr J Bot 124:350–358

Zahra Z, Arshad M, Rafique R, Mahmood A, Habib A, Qazi IA, Khan SA (2015) Metallic nanoparticle (TiO_2 and Fe_3O_4) application modifies rhizosphere phosphorus availability and uptake by *Lactuca sativa*. J Agric Food Chem 63:6876–6882

Chapter 10
Role of Nanoparticles in Improving Biofortification of Zinc and Iron in Vegetables

Musarrat Ramzan, Naheed Kauser, Touqeer Ahmad, Misbah Parveen, and Mohammad Safdar Baloch

Abstract Malnutrition affects about 1.8 billion people worldwide and is significantly attributed to micronutrient deficiencies, notably in developing countries. Humans receive their nutrients mainly from agriculture, but this sector is under pressure due to population growth and reducing arable land areas, especially in developing and underdeveloped regions. These countries are turning to farming practices to boost crop yields and address food and nutrient shortage issues. The vital microelements zinc and iron are required for both plant and human growth, however, it is commonly known that approximately one-third of the world's population suffers from a shortage of these nutrients. Crops must either have better nutritional properties or be biofortified to counteract micronutrient deficiencies. In order to provide critical nutrients, a number of biofortification projects using conventional breeding, biotechnological, and agronomic approaches have been launched with varying degrees of success, particularly in developing and underdeveloped countries. Utilizing nanoparticle fertilizers is an agronomic biofortification technique to boost crop production and nutrients, soil micronutrient availability, and plant parts assimilation. It is important to use eco-friendly and secure technologies like nanobiofortification to reduce the reliance on large quantities of chemical fertilizers. A viable strategy for enhancing the sustainability of present agricultural methods and for biofortifying food crop production by key micronutrients, resulting

M. Ramzan (✉) · M. Parveen
Department of Botany, Faculty of Chemical and Biological Sciences, The Islamia University of Bahawalpur, Bahawalpur, Pakistan
e-mail: musarrat.ramzan@iub.edu.pk

N. Kauser
Department of Biotechnology, Jinnah University of Women Karachi Pakistan, Karachi City, Sindh, Pakistan

T. Ahmad
Vegetable Section, Regional Agricultural Research Institute Bahawalpur, Bahawalpur, Pakistan

M. S. Baloch
Department of Agronomy, Faculty of Agriculture, Gomal University, Dera Ismail Khan, Pakistan

M. Hasanuzzaman et al. (eds.), *Mineral Biofortification in Crop Plants for Ensuring Food Security*, https://doi.org/10.1007/978-981-99-4090-5_10

in better nutritional quality, is the use of nanoparticles as nano fertilizers. In order to maximize the sustainable application, this chapter evaluates the existing usage of Zn and Fe-NPs for biofortification in various food crops while addressing valuable information gaps and challenges.

Keywords Nanoparticles · Biofortification · Zinc · Iron

10.1 Introduction

Health is a major concern around the globe. Developed countries and people spend a large amount of their budget for wellbeing. Almost all the environmental elements are contributors to health such as air, water, plants, soil and various microbes (Van Bruggen et al. 2019). Biofortification is a process of adding the nutrients to edible plants through fertilizers (Tiozon et al. 2021), rice (de Lima Lessa et al. 2020), sweet potato (Siwela et al. 2020), cassava (Okwuonu et al. 2021) and maize (Cheah et al. 2020) are important biofortified food crops. There are various fruits which are biofortified like strawberry (Budke et al. 2020) and some pulse crops (Jha and Warkentin 2020) too. For biofortification various macro and micro nutrients can be used such as iodine (I) (Dobosy et al. 2020), and boron (B), iron (Fe) (Okwuonu et al. 2021), copper (Cu) (Grujcic et al. 2021), selenium (Se) (Gonz'alez-García et al. 2021).

This chapter is an effort to give an overview of various factors responsible for Zn and Fe deficiencies as well as toxicities for humans and other forms of life. It also briefly sheds light on various phyto-technologies and strategies used for biofortification of Zn and Fe. Like agronomic biofortification approaches which include soil and foliar application of Zn and Fe-based fertilizers specially employing enriched organic materials and nano-sized Zn/ Fe (Zn/ Fe nanoparticles).

10.2 Role of Zinc and Iron in Human Health

Micronutrients are essential vitamins and minerals, used to perform various metabolic functions of living body and generally obtained from diet (Tripathi et al. 2015; Rehman et al. 2018). Malnutrition is a phenomenon developed by micronutrient deficiencies (MNDs) it has adverse effects which might lead to death if not treated. A study reported that around 2 billion individuals of the world are affected by MNDs (Bhowmik et al. 2010). Majority of malnourished individuals are suffering due to poor, inadequate diet and belong to developing countries.

Many human ailments originate from poor quality and vitamin-deficient diet. For instance, the most micronutrient sicknesses originate from Zn, vitamin A, Fe, and I deficiencies (Cakmak and Kutman 2018). One of the very important and essential micronutrients for life is Zn (Salgueiro et al. 2000; Khan et al. 2021). It has an important structural role in a number of proteins and transcription factors (Tripathi et al. 2015). Zinc is utilized as co factors for more than three hundred enzymes (White and Broadley 2005). The Zn dosage intake is generally based on the developmental stage and gender of animals. Usually, adult men need 12 mg Zn d^{-1} whereas adult women require 10 mg Zn d^{-1} for and for performing the routine activities. However, during pregnancy and lactation, this requirement increased up to 14 mg Zn d^{-1}. In the early 1960s, first time, effects of Zn deficiency were documented on human health (Bhowmik et al. 2010). Its importance for the human body can be determined by the fact that Zn is the most abundant component in human body after Iron Fe (Cakmak and Kutman 2018). In recent years, severe health impacts of Zn deficiency have observed on humans especially in children, women (pregnant and breastfeeding) and elderly persons. Zinc deficiency is one of the five topmost challenges related to micronutrients in developing countries and eleventh in the world (WHO (World Health Organization) 2012).

Iron is an essential element for life and constitute about 4.2% of the earth's crust (Frey and Reed 2012). It plays a vital role as cofactors in various enzymatic reactions and helps in oxygen transportation. The recommended daily allowances (RDA) of Fe (ranging from 19–50 years of age) for adult males is 8 mg/day and for females is 18 mg day^{-1}. Iron deficiency could be developed by inadequate daily intake (Miller 2013). Globally around 3 billion people are suffering from Fe deficiency. Preschool children and women in Africa and South East Asia are most affected (WHO (World Health Organization) 2012). According to few studies, Fe (Joy et al. 2015) deficiency is affecting one-third of the world population. Iron deficiency results in the reduction of mental and physical developments in children of all ages ranging from infants to school aged kids. It also a leading cause of stillbirth and maternal mortality.

The effects of Zn deficiency on human health were firstly recorded in the early 1960s (Joy et al. 2015). Its importance for the human body can be determined by the fact that it is the second-most abundant component in the body, after Fe (Semba 2012). Micronutrients are vital vitamins and minerals that must be consumed in sufficient amounts to sustain all metabolic processes.

If not treated in a timely manner, micronutrient deficiencies (MNDs) can result in malnutrition, have a severely detrimental effect on health, and ultimately result in mortality. According to a recent estimate, 2 billion individuals worldwide suffer from MNDs, with the majority living in underdeveloped nations where the disease is primarily brought on by a poor diet. Inadequate diets are a direct cause of inadequate diet intakes, which is another cause of MNDs (Saunders and Smith 2010; Chugh et al. 2022). A diet lacking in vitamins and of poor quality is the root cause of many human illnesses. For instance, deficiencies in Zn, vitamin A, Fe, and I are the main causes of micronutrient diseases. Nevertheless, zinc deficiency has become more problematic in recent year due to its worsening effects on people, particularly on young children, pregnant women, nursing mothers, old people, and flora (Burchi et al. 2011; Clemens 2014; Datta and Vitolins 2016). In a report released by the World Health Organization (WHO), Zn deficiency is ranked ninth globally and among the top five micronutrient issues in developing nations. Similarly, it is believed that one-third of the world's population is negatively impacted by iron.

10.3 Why Nano-Biofortification over Conventional Fertilization?

Large amounts of conventional fertilizers are used to increase crop output, yet about half of those fertilizers end up in rivers where they are inaccessible to plants. In addition, the majority of nutrients are not accessible to plants since they are insoluble in soil. Therefore, applying chemical fertilizer might increase productivity in the short term but be detrimental to soil health in the long run (Tanveer and Shabala 2022) because it disturbs the homeostasis of plant nutrients and nutritional status, which reduces plant development and output (Wan et al. 2021). For effective plant nutrient use and sustained crop production, a substitute for synthetic fertilizers is therefore necessary (Al-Suhaibani et al. 2021).

Due to its high use efficiency, nanotechnology can be used to add nanonutrients and carriers into fertilizers with enhanced and tailored formulations that reduce nutrient loss (Chugh et al. 2022). Due to their large surface area, size, shape, high surface mass ratio, and surface modification, nanoparticles can be employed as smart delivery devices for targeted, controlled-release kinetics (Khan et al. 2019). These characteristics help with nutrient retention and enable nutrient release that is delayed and controlled, increase crop output and nutrient usage efficiency. Additionally, biological materials can be used to make nano fertilizers as reducing, capping, and stabilizing agents, which is an advantage over chemically made nano fertilizers in terms of biocompatibility and toxicity issues. Consequently, nanoparticles provide a foundation for innovative and sustainable nanonutrient delivery methods to plants using nano porous plant surfaces with enhanced nutrient accessibility and availability.

10.4 Nano-Biofortification

An approach to supplement the human diet with micronutrients to cure malnutrition by enriching the crops and cereals with essential nutrients is known as nano-biofortification. There are various approaches for bio-fortification like nano-fertilizers or nanoparticles-based nutrients (El-Ramady et al. 2020a, 2020b, 2020c). Like other nanomaterials, these nano nutrients might have advantages and disadvantages for natural ecosystem and human life (Malakar et al. 2021; Silva et al. 2021). The fortified crops and nano remediation of water and soil could be prospective advantages, whereas, nano pollution and toxicity could be major concern for researchers (Martínez et al. 2021; Rizwan et al. 2021). The adverse effects of nano particles could be addressed by employing the green synthesis of nanoparticles (NPs) using living systems such as plant extracts (i.e., leaves, roots, flowers, and seeds), microbes (e.g., bacteria, yeast, fungi, and algae) and biomolecules (enzymes, proteins, and carbohydrates), to produce NPs instead of chemical, solvents and stabilizing agents (Bandeira et al. 2020; Abinaya et al. 2021). The biosynthesis of nanoparticles is more convenient, economical and environmentally friendly process compared to physical and chemical methods (Tanveer et al. 2020; Stephen et al. 2021).

10.5 Agronomic Mineral Biofortification

10.5.1 Fertilizer

An emerging approach is working on to provide the nutrients-based nanoparticles as fertilizers to cultivated plants for enhancement of their nutritional value which could be beneficial for human health. Nanoparticles could also be administered directly to humans or used as food additives (Deng et al. 2021) such as artificial sweeteners, emulsifiers, flavor enhancers, colorants, foaming or anti-foaming agents etc. (Medina-Reyes et al. 2020). Various nanoparticles in the form of metal oxides could be used as antimicrobial agents in packaged food which includes zinc oxide (ZnO-NPs), silver (Ag-NPs), and titanium oxide (TiO_2-NPs) (Deng et al. 2020). A number of nanoparticles are being used as nano-fertilizers or nano-pesticides to increase the crop production but its overdose is reported to cause problems related to soil plant interface (Ragab and Saad-Allah 2020). A number of research articles reported the use of engineered-NPs as nano-fertilizers (Guo et al. 2018; Farshchi et al. 2021; Madzokere et al. 2021) which improved the crop production despite various stress factors (Ye et al. 2019; Landa 2021) such as salinity (Manzoor et al. 2021; Zulfiqar and Ashraf 2021), drought (Sreelakshmi et al. 2020; Ahmed et al. 2021; Ali et al. 2021), various biotic stress (Tauseef et al. 2021a, 2021b) and heavy metal. Deposition (Noman et al. 2020; Xin et al. 2020; Manzoor et al. 2021). The nanobiotechnology has developed various advancements to utilize the intelligent nano fertilizers which would be aimed to release the desired amount of nutrients according to the need and requirements of plants, which prevents excessive fertilizers from leaking down (Yadav et al. 2023).

According to the nature/application of nano fertilizers, they can be classified into three categories: (1) micronutrient nano fertilizers, (2) macronutrient nano fertilizers and (3) nanoparticulate nano fertilizers. The nutrients which required in a minute or trace amount are known as micronutrients such as Ni, Mo, Mn, Fe and Zn in the form of nanoparticles are considered micronutrient nano fertilizers. Studies revealed that biosynthesized iron oxide nano fertilizers showed improved growth of *Zea mays* plants as compared to the chemically synthesized NPs. Delfani et al. reported that *Vigna unguiculata* showed the increase in seed weight and photosynthesis after foliar application of Mg and Fe NPs. When chickpea (*Cicer arietinum*) was treated with ZnO NPs during the seed germination and root growth stages the phytostimulatory actions of indole acetic acid (IAA) was observed, here ZnO NPs increased the level of IAA in roots (sprouts), which in turn increased the growth of plants (Asztemborska et al. 2015). The effect of various micronutrient NPs on hydroponically grown mung bean (*Vigna radiata*) and chickpea (*C. arietinum*) seedlings was studied. For this experiment various concentrations of nano-ZnO, nano-FeO and nano-Zn-Cu-Feoxide in suspension was sprayed on seedlings and effect on the shoot growth observed. It was found that the plantlets showed good growth over control, the best results were given by Zn/CuFe-Oxide NPs followed by FeO NPs and ZnO NPs proved least effective.

Iron based nano fertilizers applications improved the solubility and mobilization of Fe in the soil resulting enhance the bioavailability of the Fe for plants. Soil fortification by Fe fertilizer in Fe deficient regions improve the available Fe content. Ferrous sulfate heptahydrate ($FeSO_4.7H_2$ O) is common fertilizer for soil fortification, which is applied at 2–2.50 kg ha^{-1} at the time of land preparation. Generally common Fe-rich fertilizers for soil are not much effective and rapidly become unavailable for plants. To resolve this issue generally, chelated-Fe containing fertilizers such as Fe-ethylenediaminetetraacetate (EDTA) are advised (5.0 kg ha^{-1}). Foliar application of soluble Fe -rich fertilizers such as Fe-EDTA (Chelated iron1 g/lit water) or Ferrous sulfate heptahydrate ($FeSO_{4.}7H_2O$- 0.5 g L^{-1} water) is also used to improve the Fe availability by enhancing the Fe content of the leaves. Nano fertilizers improve the overall growth of plant by help the root to grow, enhance water content in the chlorophyll and increased seed germination. Hu et al. (2017) observed and reported an increase in the Fe concentration of *Citrus maxima* shoots when the plants were exposed to both Fe_2O_3NPs and Fe^{3+} treatment compared to controls and Fe (II)-EDTA treated plants. Furthermore, the roots assessments, showed no significant changes in the Fe levels among all Fe-treated and control plants, which indicated that the Fe had been transported to other parts from roots. Researchers believe that nutrients reach to other parts of the plant through the root; though, it has also been reported that nutrients can be translocate from the leaves to various parts of the plant tissues. Corredor et al. (2009) observed that carbon-coated iron NPs sprinkled on the leaves migrated from the leaves to other parts of the plant.

10.6 Scope of Iron and Zinc Fortification in Leafy Vegetables

In daily diet green leafy vegetables are considered very important for their health benefits, flavor, and color. These foliage vegetables are rich in iron, calcium, magnesium, and zinc, etc. These are most economical source for vitamins, antioxidants, and dietary fibers. Generally, these vegetables are of short duration and has high biomass. These are used as natural source of iron (Chatterjee et al. 2016; Chiplonkar et al. 1999). However, in some cases, the presence of certain inhibitors reduced the bioavailability of Fe. To treat iron deficiency, there is a need to characterize and promote the use of green leafy vegetables based on their Fe availability.

As a result of this strategy, micronutrient malnutrition can be reduced, resulting in improved health. In particular, the poor will consume more micronutrients as a result

Fig. 10.1 Micronutrient deficiencies and influence of application of nano fertilizers

of the addition of vitamins and micronutrients to basic foods and vegetables, which will reduce malnutrition. A plant's amounts of phytonutrients and vitamins can be easily enhanced in a vegetable, which has a shorter life cycle and a shorter juvenile time to flowering. With tree fruits and nuts, it will be challenging because the juvenile period lasts a lot longer (O'Hare 2015).

An experiment was performed using various concentrations (0.0, 10, 20, 30 and 40 g ml L^{-1}) of ZnO NPs to treat the eggplant to study the effect on seed germination, cell division, and early seedling growth. An increase in length, width and surface area of leaf was observed with increased protein content and dietary fiber (Semida et al. 2021). The foliar applications of Zn NP also improved the nutritional quality of green pea, besides increased yield has also been reported (Skiba et al. 2020). In a study conducted on *Capsicum annuum* L. the increased percentage of germination, shoot growth and root growth were observed while the concentration of ZnO NPs was increased from 0-2000 ppm (Liu et al. 2016) (Figs. 10.1, 10.2, and 10.3).

10.7 Conclusion

Nano-biofortification could be a promising tool to fight malnutrition. It has several advantages like efficiency and lower amount of application. This approach is still infant and needs more experiments and studies to be applied on the global level.

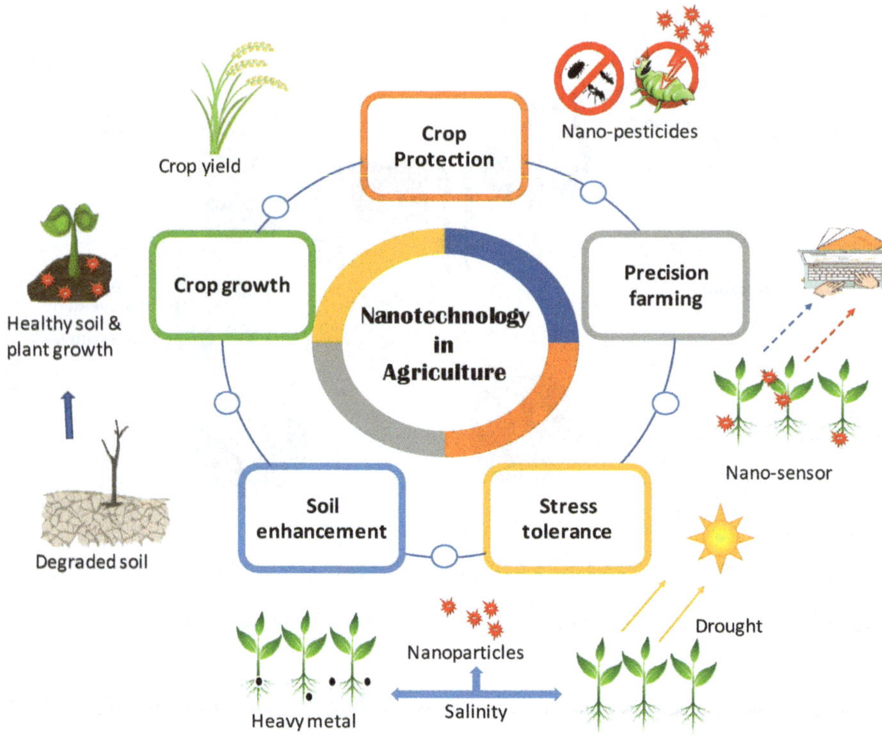

Fig. 10.2 Applications of nanotechnology in agriculture

Various strategies for biofortification have been recognized to improve human and animal health by increasing dietary Zn/Fe intake. There is an utmost need to assess the various components to understand the impact of these approaches on health, economy, environment and social acceptance perspectives. There is a need to select a most cost-effective biofortification method with improved quality, high crop yield and farmers' inclination as a function of crop growth conditions and the socio-economic environment. Nanoparticle-assisted strategies can help in optimization and improved crop Zn/Fe accumulation and crop yield. Conversely, each biofortification practice must be carefully evaluated to avoid the toxicity of plant-derived food products. Furthermore, various other issues related to nano- biofortification further needed to be addressed. High quantities of Zn/Fe fertilizers as soil or foliar application may not be the most sustainable strategy, as this can result in the leaching of excessive micronutrient, thus requiring regular applications, which can make this approach more costly. The widespread use of Zn/Fe might cause environmental (e.g., water, soil and plant) contamination, which in might have potential threats for human and animal health. The selection of crops with a high ability to accumulate greater Se concentrations is needed in the conventional breeding approach, however, the crop biodiversity and dietary diversity should also not be neglected. Besides,

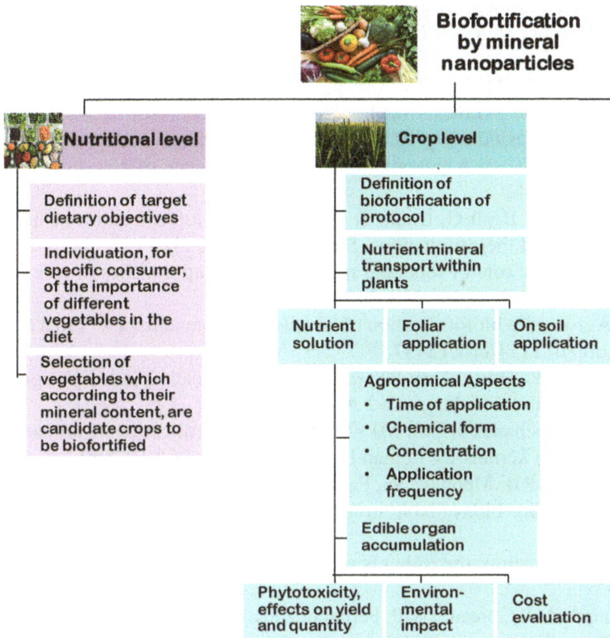

Fig. 10.3 Key aspects to be considered in agronomic biofortification by mineral nanoparticles

more studies are still needed on the application of Zn/Fe NPs as an emerging technology. The mechanism of Zn/Fe NPs uptake from soil, and its translocation and transformation in vegetable plants need to be further unraveled.

References

Abinaya S, Helen Kavitha P, Prakash M, Muthukrishnaraj A (2021) Green synthesis of magnesium oxide nanoparticles and its applications: a review. Sustain Chem Pharm 19:100368

Ahmed T, Noman M, Manzoor N, Shahid M, Abdullah M, Ali L, Wang G, Hashem A, Al-Arjani AF, Alqarawi AA, Abd-Allah EF, Li B (2021) Nanoparticle-based amelioration of drought stress and cadmium toxicity in rice via triggering the stress responsive genetic mechanisms and nutrient acquisition. Ecotoxicol Environ Saf 209:11829. https://doi.org/10.1016/j.ecoenv.2020.111829

Al-Suhaibani N, Seleiman MF, El-Hendawy S, Abdella K, Alotaibi M, Alderfasi A (2021) Integrative effects of treated wastewater and synthetic fertilizers on productivity, energy characteristics, and elements uptake of potential energy crops in an arid agro-ecosystem. Agronomy 11:2250. https://doi.org/10.3390/agronomy11112250

Ali EF, El-Shehawi AM, Ibrahim OHM, Abdul-Hafeez EY, Moussa MM, Hassan FAS (2021) A vital role of chitosan nanoparticles in improvisation the drought stress tolerance in *Catharanthus roseus* (L.) through biochemical and gene expression modulation. Plant Physiol Biochem 161:166–175. https://doi.org/10.1016/j.plaphy.2021.02.008

Asztemborska M, Steborowski R, Kowalska J, Bystrzejewska-Piotrowska G (2015) Accumulation of platinum nanoparticles by *Sinapis alba* and *Lepidium sativum* plants. Water Air Soil Pollut 226:126

Bandeira M, Giovanela M, Roesch-Ely M, Devine DM, Crespo JDS (2020) Green synthesis of zinc oxide nanoparticles: a review of the synthesis methodology and mechanism of formation. Sustain Chem Pharm 15:100223. https://doi.org/10.1016/j.scp.2020.100223

Bhowmik D, Chiranjib KP, Kumar S (2010) A potential medicinal importance of zinc in human health and chronic. Int J Pharm 1:5–11

Budke C, Thor Straten S, Mühling KH, Broll G, Daum D (2020) Iodine biofortification of field-grown strawberries – approaches and their limitations. Sci Hortic 269:109317

Burchi F, Fanzao J, Frison E (2011) The role of food and nutrition system approaches in tackling hidden hunger. Int J Environ Res Pub Health 8:358–373

Cakmak I, Kutman UB (2018) Agronomic biofortification of cereals with zinc: a review. Eur J Soil Sci 69:172–180. https://doi.org/10.1111/ejss.12437

Chatterjee R, Chowdhury RS, Dukpa P, Thirumdasu RK (2016) Iron fortification in leafy vegetables: present status and future possibilities. Innovare J Agri Sci 4(4):1–3

Cheah ZX, O'Hare TJ, Harper SM, Kochanek J, Bell MJ (2020) Zinc biofortification of immature maize and sweetcorn (*Zea mays* L.) kernels for human health. Sci Hortic 272:109559

Chiplonkar SA, Tarwadi KV, Kavedia RB, Mengale SS, Paknikar KM, Agte VV (1999) Fortification of vegetarian diets for increasing bioavailable iron density using green leafy vegetables. Food Res Int 32:169–174

Clemens S (2014) Zn and Fe biofortification: the right chemical environment for human bioavailability. Plant Sci 225:52–57

Corredor E, Testillano PS, Coronado MJ, Gonzalez MP, Fernandez PR, Marquina C, Ibarra MR, de la Fuente JM, Rubiales D, Perez de-Luque A (2009) Nanoparticle penetration and transport in living pumpkin plants: in situ subcellular identification. BMC Plant Biol 9:45

Datta M, Vitolins MZ (2016) Food fortification and supplement use-are there health implications? Crit Rev Food Sci Nutr 56:2149–2159

de Lima Lessa JH, Raymundo JF, Branco Corguinha AP, Dias Martins FA, Araujo AM, Melo Santiago FE, Pereira de Carvalho HW, Guimaraes Guilherme LR, Lopes G (2020) Strategies for applying selenium for biofortification of rice in tropical soils and their effect on element accumulation and distribution in grains. J Cereal Sci 96:103125

Deng C, Wang Y, Cota-Ruiz K, Reyes A, Sun Y, Peralta-Videa J, Hernandez- Viezcas JA, Turley RS, Niu G, Li C, Gardea-Torresdey J (2020) Bok choy (*Brassica rapa*) grown in copper oxide nanoparticles-amended soils exhibits toxicity in a phenotype-dependent manner: translocation, biodistribution and nutritional disturbance. J Hazard Mater 398:122978

Deng J, Ding QM, Jia MX, Li W, Zuberi Z, Wang JH, Ren JL, Fu D, Zeng XX, Luo JF (2021) Biosafety risk assessment of nanoparticles: evidence from food case studies. Environ Pollut 275: 116662. https://doi.org/10.1016/j.envpol.2021.116662

Dobosy P, Kröpfl K, Ov'ari M, Sandil S, N'emeth K, Engloner A, Tak'acs T, Z'aray G (2020) Biofortification of green bean (*Phaseolus vulgaris* L.) and lettuce (*Lactuca sativa* L.) with iodine in a plant-calcareous sandy soil system irrigated with water containing KI. J Food Compos Anal 88:103434

El-Ramady H, Brevik EC, Amer M, Elsakhawy T, Omara AED, Elbasiouny H, Elbehiry F, Mosa AA, El-Ghamry AM, Bayoumi Y, Shalaby TA (2020a) Soil and air pollution in the era of COVID-19: a global issue. Egypt J Soil Sci 60:437–448. https://doi.org/10.21608/ejss.2020.49996.1411

El-Ramady H, Eid Y, Brevik EC (2020b) New pollution challenges in groundwater and wastewater due to COVID-19. J Sus Agric Sci 46:61–73. https://doi.org/10.21608/jsas.2020.51353.1257

El-Ramady H, El-Mahdy S, Awad A, Nassar S, Osman O, Metwally E, Aly E, Fares E, El-Henawy (2020c) Is nano-biofortification the right approach for malnutrition in the era of COVID-19 and climate change? Egypt J Soil Sci 61:141–150. https://doi.org/10.21608/ejss.2021.75653.1445

Farshchi HK, Azizi M, Teymouri M, Nikpoor AR, Jaafari MR (2021) Synthesis and characterization of nanoliposome containing Fe^{2+} element: a superior nano-fertilizer for ferrous iron delivery to sweet basil. Sci Hortic 283:110110. https://doi.org/10.1016/j.scienta.2021.110110

Chugh G, Kadambot HMS, Zakaria MS (2022) Iron fortification of food crops through nanofertilisation. Crop Pasture Sci 73:736–748

Gonz'alez-García Y, C'ardenaś Alvarez C, Cadenas-Pliego G, Benavides-Mendoza A, Cabrera-de-la-Fuente M, Sandoval-Rangel A, Vald'es-Reyna J, Ju'arez-Maldonado A (2021) Effect of three nanoparticles (Se, Si and Cu) on the bioactive compounds of bell pepper fruits under saline stress. Plan Theory 10:217. https://doi.org/10.3390/plants10020217

Grujcic D, Yazici AM, Tutus Y, Cakmak I, Singh BR (2021) Biofortification of silage maize with zinc, iron and selenium as affected by nitrogen fertilization. Plan Theory 10:391

Guo H, White JC, Wang Z, Xing B (2018) Nano-enabled fertilizers to control the release and use efficiency of nutrients. Curr Opin Environ Sci Health 6:77–83

Frey PA, Reed GH (2012) The ubiquity of iron. ACS Chem Biol 7(9):1477–1481. https://doi.org/10.1021/cb300323q. Epub 2012

Hu J, Guo H, Li J, Wang Y, Xiao L, Xing B (2017) Interaction of -Fe_2O_3 nanoparticles with *Citrus maxima* leaves and the corresponding physiological effects via foliar application. J Nanobiotechnol 15:1–8

Jha AB, Warkentin TD (2020) Biofortification of pulse crops: status and future perspectives. Plants (Basel) 9(1):73. https://doi.org/10.3390/plants9010073

Joy EJM, Kumssa DB, Broadley MR, Watts MJ, Young SD, Chilimba ADC, Ander EL (2015) Dietary mineral supplies in Malawi: spatial and socioeconomic assessment. BMC Nutr 1:42

Khan I, Saeed K, Khan I (2019) Nanoparticles: properties, applications and toxicities. Arab J Chem 12(7):908–931

Khan MT, Ahmed S, Shah AA, Noor Shah A, Tanveer M, El-Sheikh MA, Siddiqui MH (2021) Influence of zinc oxide nanoparticles to regulate the antioxidants enzymes, some osmolytes and agronomic attributes in *Coriandrum sativum* L. grown under water stress. Agronomy 11:2004

Landa P (2021) Positive effects of metallic nanoparticles on plants: overview of involved mechanisms. Plant Physiol Biochem 161:12–24. https://doi.org/10.1016/j.plaphy.2021.01.039

Liu R, Zhang H, Lal R (2016) Effects of stabilized nanoparticles of copper, zinc, manganese, and iron oxides in low concentrations on lettuce (*Lactuca sativa*) seed germination: nanotoxicants or nanonutrients? Water Air Soil Pollut 227:1–14

Madzokere TC, Murombo LT, Chiririwa H (2021) Nano-based slow releasing fertilizers for enhanced agricultural productivity. Mater Today Proc 45:3709–3715. https://doi.org/10.1016/j.matpr.2020.12.674

Malakar A, Kanel SR, Ray C, Snow DD, Nadagouda MN (2021) Nanomaterials in the environment, human exposure pathway, and health effects: a review. Sci Total Environ 759:143470. https://doi.org/10.1016/j.scitotenv.2020.143470

Manzoor N, Ahmed T, Noman M, Shahid M, Nazir MM, Ali L, Alnusaire TS, Li B, Schulin R, Wang G (2021) Iron oxide nanoparticles ameliorated the cadmium and salinity stresses in wheat plants, facilitating photosynthetic pigments and restricting cadmium uptake. Sci Total Environ 769:145221. https://doi.org/10.1016/j.scitotenv.2021.145221

Martínez G, Merinero M, P'erez-Aranda M, P'erez-Soriano EM, Ortiz T, Villamor E, Begines B, Alcudia A (2021) Environmental impact of nanoparticles' application as an emerging technology: a review. Materials 14:166. https://doi.org/10.3390/ma14010166

Medina-Reyes EI, Rodríguez-Ibarra C, Deciga-Alcaraz A, Díaz-Urbina D, Chirino YI, Pedraza-Chaverri J (2020) Food additives containing nanoparticles induce gastrotoxicity, hepatotoxicity and alterations in animal behavior: the unknown role of oxidative stress. Food Chem Toxicol 146:111814. https://doi.org/10.1016/j.fct.2020.111814

Miller JL (2013) Iron deficiency anemia: a common and curable disease. Cold Spring Harb Perspect Med 3(7):a011866. https://doi.org/10.1101/cshperspect.a011866

Noman M, Ahmed T, Hussain S, Niazi MBK, Shahid M, Song F (2020) Biogenic copper nanoparticles synthesized by using a copper-resistant strain *Shigella flexneri* SNT22 reduced the translocation of cadmium from soil to wheat plants. J Hazard Mater 398:123175. https://doi.org/10.1016/j.jhazmat.2020.123175

O'Hare TJ (2015) Biofortification of vegetables for the developed world. Acta Hortic 1106:1–8. https://doi.org/10.17660/ActaHortic.2015.1106.1

Okwuonu IC, Narayanan NN, Egesi CN, Taylor NJ (2021) Opportunities and challenges for biofortification of cassava to address iron and zinc deficiency in Nigeria. Glob Food Secur 28: 100478. https://doi.org/10.1016/j.gfs.2020.100478

Ragab GA, Saad-Allah KM (2020) Green synthesis of sulfur nanoparticles using *Ocimum basilicum* leaves and its prospective effect on manganese-stressed *Helianthus annuus* (L.) seedlings. Ecotoxicol Environ Saf 191:110242. https://doi.org/10.1016/j.ecoenv.2020.110242

Rehman A, Farooq M, Naveed M, Nawaz A, Shahzad B (2018) Seed priming of Zn with endophytic bacteria improves the productivity and grain biofortification of bread wheat. Eur J Agron 94:98–107. https://doi.org/10.1016/j.eja.2018.01.017

Rizwan M, Ali S, Zia ur Rehman M, Riaz M, Adrees M, Hussain A, Zahir ZA, Rinklebe J (2021) Effects of nanoparticles on trace element uptake and toxicity in plants: a review. Ecotoxicol Environ Saf 221:112437. https://doi.org/10.1016/j.ecoenv.2021.11243

Salgueiro MJ, Zubillaga M, Lysionek A, Sarabia MI, Caro R, De Paoli T, Hager A, Weill R, Boccio J (2000) Zinc as an essential micronutrient: a review. Nutr Res 20(5):737–755

Saunders J, Smith T (2010) Malnutrition: causes and consequences. Clin Med (Lond) 10(6):624–627. https://doi.org/10.7861/clinmedicine.10-6-624

Semba RD (2012) The vitamin a story: lifting the shadow of death. Vitamin A Story Lift Shad Death 104:1–207

Semida WM, Abdelkhalik A, Mohamed GF, Abd El-Mageed TA, Abd El-Mageed SA, Rady MM, Ali EF (2021) Foliar application of zinc oxide nanoparticles promotes drought stress tolerance in eggplant (*Solanum melongena* L.). Plan Theory 421. https://doi.org/10.3390/plants10020421

Silva LFO, Santosh M, Schindler M, Gasparotto J, Dotto GL, Oliveira MLS, Hochella MF (2021) Nanoparticles in fossil and mineral fuel sectors and their impact on environment and human health: a review and perspective. Gondwana Res 92:184–201. https://doi.org/10.1016/j.gr.2020.12.026

Siwela M, Pillay K, Govender L, Lottering S, Mudau FN, Modi AT, Mabhaudhi T (2020) Biofortified crops for combating hidden hunger in South Africa: availability, acceptability, micronutrient retention and bioavailability. Foods 9:815

Skiba E, Michlewska S, Pietrzak M, Wolf WM (2020) Additive interactions of nanoparticulate ZnO with copper, manganese and iron in *Pisum sativum* L., a hydroponic study. Sci Rep 10:13574. https://doi.org/10.1038/s41598-020-70303-8

Sreelakshmi B, Induja S, Adarsh PP, Rahul HL, Arya SM, Aswana S, Haripriya R, Aswathy BR, Manoj PK, Vishnudasan D (2020) Drought stress amelioration in plants using green synthesised iron oxide nanoparticles. Mater Today Proc 41:723–727. https://doi.org/10.1016/j.matpr.2020.05.801

Stephen BJ, Sharma MM, Jain D, Singh A (2021) Biogenic synthesized nanoparticles a boon to human health. Mater Today Proc 43:3211–3213. https://doi.org/10.1016/j.matpr.2021.01.869

Tanveer M, Shabala S (2022) Entangling the interaction between essential and nonessential nutrients: implications for global food security. In: Srivastava AS, Suprasanna P (eds) Kumar V. Plant Nutrition and Food Security in the Era of Climate Change, Academic Press, pp 1–25

Tanveer M, Shahzad B, Ashraf U (2020) Nanoparticle application and abiotic-stress tolerance in plants. In: Singh VP, Chauhan DK, Sharma S, Prasad SM, Dubey NK, Ramawat N (eds) Tripathi DK. Plant Life Under Changing Environment, Academic Press, pp 627–641

Tauseef A, Hisamuddin Gupta J, Rehman A, Uddin I (2021a) Differential response of cowpea towards the CuO nanoparticles under *Meloidogyne incognita* stress. S Afr J Bot 139:175–182. https://doi.org/10.1016/j.sajb.2021.02.017

Tauseef A, Hisamuddin Khalilullah A, Uddin I (2021b) Role of MgO nanoparticles in the suppression of *Meloidogyne incognita*, infecting cowpea and improvement in plant growth and physiology. Exp Parasitol 220:108045. https://doi.org/10.1016/j.exppara.2020.108045

Tiozon RN, Fernie AR, Sreenivasulu N (2021) Meeting human dietary vitamin requirements in the staple rice via strategies of biofortification and post-harvest fortification. Trends Food Sci Technol 109:65–82. https://doi.org/10.1016/j.tifs.2021.01.023

Tripathi DK, Singh S, Singh S, Mishra S, Chauhan DK, Dubey NK (2015) Micronutrients and their diverse role in agricultural crops: advances and future prospective. Acta Physiol Plant 37:139. https://doi.org/10.1007/s11738-015-1870-3

Van Bruggen AHC, Goss EM, Havelaar A, van Diepeningen AD, Finckh MR, Morris JG (2019) One health - cycling of diverse microbial communities as a connecting force for soil, plant, animal, human and ecosystem health. Sci Total Environ 664:927–937

Wan L-J, Tian Y, He M, Zheng Y-Q, Lyu Q, Xie R-J, Ma Y-Y, Deng L, Yi S-L (2021) Effects of chemical fertilizer combined with organic fertilizer application on soil properties, citrus growth physiology, and yield. Agriculture 11:1207. https://doi.org/10.3390/agriculture11121207

White PJ, Broadley MR (2005) Biofortifying crops with essential mineral elements. Trends Plant Sci 10:586–593. https://doi.org/10.1016/j.tplants.2005.10.001

WHO (World Health Organization) Nutrition – Micronutrient deficiencies – Iron deficiency anemia, 2012. URL http://www.who.int/nutrition/topics/ida/en/. Accessed 25 June 2012

Xin X, Zhao F, Rho JY, Goodrich SL, Sumerlin BS, He Z (2020) Use of polymeric nanoparticles to improve seed germination and plant growth under copper stress. Sci Total Environ 745:141055. https://doi.org/10.1016/j.scitotenv.2020.141055

Yadav A, Yadav K, Abd-Elsalam KA (2023) Nanofertilizers: types, delivery and advantages in agricultural sustainability. Agrochemicals 2(2):296–336. https://doi.org/10.3390/agrochemicals2020019

Ye Y, Medina-Velo IA, Cota-Ruiz K, Moreno-Olivas F, Gardea-Torresdey JL (2019) Can abiotic stresses in plants be alleviated by manganese nanoparticles or compounds? Ecotoxicol Environ Saf 184:109671. https://doi.org/10.1016/j.ecoenv.2019.109671

Zulfiqar F, Ashraf M (2021) Nanoparticles potentially mediate salt stress tolerance in plants. Plant Physiol Biochem 160:257–268

Chapter 11
Scope of Seed Priming in Inducing Biofortification in Plants

Muhammad Talha Aslam, Muhammad Umer Chattha, Imran Khan, Muhammad Zia Ul Haq, Ayesha Mustafa, Fareeha Athar, Bisma, Muhammad Nawaz, Adnan Noor Shah, Faisal Mahmood, and Muhammad Umair Hassan

Abstract The major staple food crops are intensively being used to meet the food security demand of ever-growing world's population, though, most of the crops are less in micronutrients necessary for human nutrition. Micronutrient insufficiency poses hidden hunger and malnutrition challenges in more than two billion people globally. This may cause mortality, anemia to blondness, and under-nourished childbirth. Interestingly, it is more common in developing countries. Common approaches to addressing micronutrient deficiency include supplementation, fortification and dietary diversification. Over the last 15 years, research has demonstrated that another strategy, called biofortification, is an effective complement to these approaches in addressing micronutrient deficiency and hidden hunger. Moreover, Biofortification is an upcoming, promising, cost-effective, and sustainable technique of delivering micronutrients to a population that has limited access to diverse diets and other micronutrient interventions. Furthermore, the seed priming is potentially considered for effectively enhancing deficit micronutrient proportion in grains. Seed priming is associated with hydration process based on time duration aimed at increasing micronutrient absorption before seed plantation. Seed priming with micronutrients augmented crop performance in terms of growth, development, and grain yield. Scientist reported numerous (morphological, physiological, molecular, and biochemical) benefits of seed priming. Interestingly, the seed priming proven

M. T. Aslam · M. U. Chattha · I. Khan · M. Z. U. Haq · A. Mustafa · F. Athar · Bisma
Department of Agronomy, University of Agriculture Faisalabad, Faisalabad, Pakistan

M. Nawaz (✉) · A. N. Shah (✉)
Department of Agricultural Engineering, Khwaja Fareed University of Engineering and Information Technology, Rahim Yar Khan, Punjab, Pakistan
e-mail: dmnawaz@kfueit.edu.pk

F. Mahmood
Department of Environmental Science and Engineering, Government College University Faisalabad, Faisalabad, Pakistan

M. U. Hassan
Research Center on Ecological Sciences, Jiangxi Agricultural University, Nanchang, China

M. Hasanuzzaman et al. (eds.), *Mineral Biofortification in Crop Plants for Ensuring Food Security*, https://doi.org/10.1007/978-981-99-4090-5_11

best with improved micronutrient percentage in grains than the other methods used for seed biofortification. In addition, seed priming is cost-effective and more efficient option for seed biofortification.

Keywords Seed priming · Nutrient uptake · Micronutrient · Biofortification · Seed treatment · Agricultural practices · Yield

11.1 Introduction

Developing countries have seen continuous improvement in hunger-energy-deficits following the green revolution. One out of nine people still suffer from hunger in the world (FAO 2017). Globally, nearly 2.5 billion or one in three people remain at risk of vitamin A and other micro nutrient (iron, Fe; zinc, Zn; and iodine) deficiencies (Black et al. 2013). Most of these individuals are pregnant women and children under the age of five who are suffering from severe malnutrition (White and Broadley 2009; WHO 2012). A major focus was placed on malnutrition after the Millennium Development Goals (MDGs) were set, followed by the Sustainable Development Goals (SDGs) (Ramadas et al. 2020). Many scientists reported that 80% of the peoples living in developing countries are directly attached with agriculture for their livelihood, employment, nutritional and food security against hunger and poverty (Frison et al. 2006; Mozumdar 2012; Mugambiwa and Tirivangasi 2017). There is malnutrition all over the world. There are numerous factors responsible for malnutrition but unbalanced diet has the foremost importance among others. Consequently, meager heath and disease issues take place. The dependency on cereals for dietary requirements caused to increase in micronutrient deficiencies in people (Cakmak et al. 2010a; Bouis et al. 2011). Ramadas et al. (2020) observed 11% decrease in GDP of Asian and African countries to overcome micronutrient deficiency requirement which affect nearly 2 billion peoples in the world (Tulchinsky 2010). Additionally, the number of undernourished people is around 815 million (Zhou and Wan 2007; Das et al. 2019).

The hidden hunger index (HHI) used to confer the extent of malnutrition prevailing in Asian and African countries of the world. That can be conquer by proper utilization of vitamin A, I, Fe, Zn and other deficit micronutrients in daily diet (Ruel-Bergeron et al. 2015). The children are majorly affected by hidden hunger associated with malnutrition. It has been reported that 151 million children under the age of 5 years have stunted growth and about 51 million falls into the wasting category, meaning that there is no weight proportional to height. Moreover, 45% of children's death in the age of 5 years are related to severe malnutrition problems. Among the other continents of the world, Asia is badly affected with stunted growth in 33.3% and wasting about 15.3% (Yadava et al. 2018).

From the literature reviewed, it is noticed that low public income (poverty) in developing countries relates to the health issues (hidden hunger and malnutrition) (Ekholuenetale et al. 2020). The use of additional nutrients in the daily diet depends on the purchasing power peoples of the region. The increasing poverty rate ends with the decreased usage of expensive nutritional supplements limiting the appropriate mental and physical growth while increasing the infectious disease susceptibility among children of the developing countries. Poor health condition can impact the long-lasting effects on livelihood as it ominously reduces one's capacity to take advantage of monetary opportunities (Bryce et al. 2003; Alderman et al. 2006). Micronutrient deficiency as compared to poor nutritional quality and low energy intake observed most in developing nations (Stewart et al. 2010) and deficiency of vitamin A, Zn, Fe and I caused 20% death of children under the age of 5 years (Prentice et al. 2008). Therefore, the adaptation of multi-nutritional diet is a promising approach to combat the micronutrient deficiencies. Indispensable micronutrient (Fe, Zn, I, etc.) can be the part of human body through fortification or biofortification approach (Vakili et al. 2013).

11.2 What Is Biofortification and Strategies of Biofortification?

Biofortification is a technique used to enhance grain nutritional quality attributes through plant breeding, biotechnological approach, by adopting agronomic management practices and seed priming (SP) techniques.

11.2.1 Difference Between Fortification and Biofortification

The increase in nutritional elements with agricultural means through plant mechanism in a biological manner called biofortification whereas the nutrients added exogenously in the processed food is different and called fortification. Further, biofortification was adapted to produce deficit nutrient enriched food for low-income nations. It is a promising tool by which human food can be supplemented with I, Fe and Zn contents that are found relatively in low concentration while encompassing a great dietary benefit. Moreover, biofortification is a cost-effective technique that is concerned with increasing nutrient-rich productivity of the crop addressing hidden hunger, and malnutrition to cope the mortality rate in developing countries. Several crops with successful biofortification have been marked (Wakeel et al. 2018), among those, Fe-enriched beans, Zn-enriched rice, sweet potatoes, lentil, cowpeas, and sorghum have foremost importance (Tanumihardjo 2013; Talsma 2014). These fulfill the major food requirement of the low-income peoples residing in developing countries (Wakeel et al. 2018). Though, food supplementation is superior to the biofortification technique and is still in successfully use to mitigate the micronutrient intake gap (Bouis et al. 2011; Hotz et al. 2012). But the use of supplements is beyond the access of most of the people in developing countries (Bouis 2007; Hermann 2009; Ramadas et al. 2020).

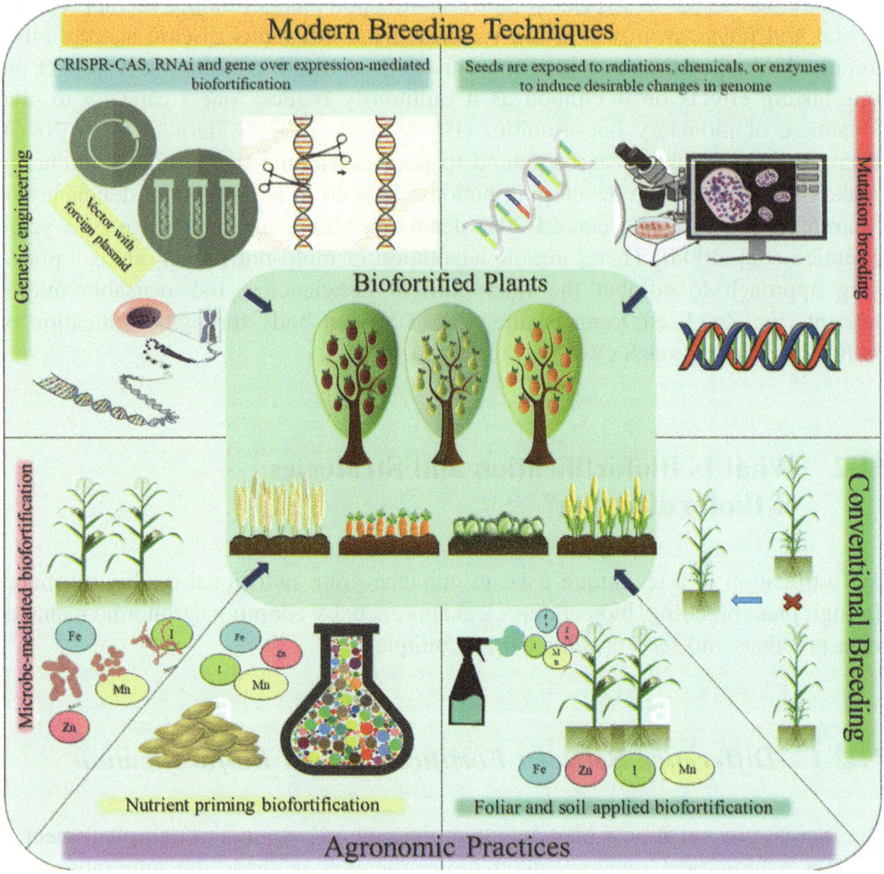

Fig. 11.1 Modern breeding and agronomic techniques used around the globe for biofortification of crops

Therefore, need of the hour is to promote and use biofortified crop that are quite cheap, accessible, and promising way to cope increasing malnutrition problems in the world. There are copious ways to produce biofortified crops which are discussed below in detail (Fig. 11.1).

11.2.2 Strategies of Biofortification

11.2.2.1 Agronomic Biofortification

A management base strategy to augment those micro and macro-nutrients in plants that are found in less than the required concentration is described as agronomic biofortification. This may be done by enhancing organic matter or by adding specific nutrients (fertilization) through soil application, foliar spray or by fertigation. Although, this strategy is considered a temporary solution to increase micronutrient, but quality attributes amongst yield of grain crops are significantly improved (Wakeel et al. 2018). Agronomic biofortification is associated with soil fertility management with combined use of organic and inorganic source of nutrients that improved micronutrient concentration in soil and in plants (Srivastava et al. 2021). Moreover, this technique gives the promising results for essential nutrients to (Zn, I, Se etc.) compared to non-essential (synthesized in plant cell) nutrients of plants (Lyons and Cakmak 2012; Bevis 2015). The concentration of trace elements significantly increased in cereals (grains) with fertilization (Jiao et al. 2012). Likewise, Broadley et al. (2006) reported that Se application enhances Se contents in Finnish foods that results in increased amount of Se present in human body with use of Se rich food. Iodine contents in food increased with iodine supply by different means of iodine application (fertigation, soil application and foliar spray) (Smolen et al. 2011; Gonzali et al. 2017; Cakmak et al. 2017; Krzepilko et al. 2019; Izydorczyk et al. 2021). Whereas, the addition of Zn via fertilizer means started from early 90s and improved Zn percentage in wheat grains (Roohani et al. 2013; Rehman et al. 2018; Hassan et al. 2019, 2021; Ali et al. 2022). In addition, Hawkesford and Zhao (2007) reported an experiment to increase Se content in wheat through biofortification. It is not possible to produce vitamin-enrich crops with agronomic measures; though, agronomic biofortification shown the tremendous potential to enhance mineral concentrations (Wakeel et al. 2018).

11.2.2.2 Genetic Engineering

The term 'genetic biofortification' states the development of essential nutrients founded in low level in field crops. It can be done either with conventional breeding or by genetic engineering techniques (Bouis 2003). Agronomic intervention of mineral augments the nutrient concentration in plant parts (fruits, seeds, tubers) which nourished form phloem tissues (White et al. 2009; Cakmak et al. 2010a, b). The transformation of these nutrients in edible plant parts depends on nutrient concentration in phloem (Broadley et al. 2007). Agronomic biofortification efficiency restricted in low quantity of nutrients present and thus, mobility decreases in phloem. In addition, Cakmak et al. (2010a) observed low buildup of Zn in grain due to less Zn mobility in phloem. Agronomic biofortification and conventional breeding improves nutrient enrichment in crops however these approaches fail to increase nutrient uptake in cultivars which are genetically not respondent to nutrients (Beyer 2010). Therefore, genetic engineering can be an alternative strategy to increase nutrient concentration in edible plant parts.

Genetic engineering and bio-supplementation aim to do the same thing and have been used primarily in products where the nutrients (e.g., Zn, Cu, Fe, I etc.) are not present at required volume. When there is not enough nutrient concentration exist genetically in a particular crop, transgenic approach offers an alternate method of achieving nutrient rich crops. This is due to the addition of new genes which are acquired from various cultivars attributed with specific traits. Though, a lot of research questions are still answerable but a significant success of nutrient (Ca, Fe, vitamin E, etc.) enriched crops (carrot, maize, rice, etc.) has been made (Hotz et al. 2012). Biofortified rice containing β-carotene developed with genetic engineering (Sautter et al. 2006). However, gene responsible for producing β-carotene naturally exist in rice plant but its expression is generally blocked in endosperm (Qamar et al. 2020). Supplementation of genes plant phytoene synthase (PSY) and bacterial carotene desaturase (CRTI) turn on the expression of blocked gene in rice plant (Al-Babili and Beyer 2005; Bai et al. 2016). The best example carrying the mentioned technique resulted in the production of golden rice comprising β-carotene in significant concentration (Rice 2016; Qamar et al. 2020). Moreover, researchers founded maximum percentage of β-carotene when PSY was inserted in rice developed from maize crop (Paine et al. 2005).

Furthermore, the Consultative Group on International Agricultural Research (CGIAR) discovered the effect of genetic variation on nutrient intake, translocation, acquisition in staple (wheat, rice, beans, cassava) crops (White and Broadley 2009). Literature revealed a major influence of genetic variability on micronutrients (Zn and Fe) availability in agronomic crops. In addition, 2.2 and >14-fold genetic variation of Zn observed in 20 and 600 germplasm of carrot and cassava respectively (Nicolle et al. 2004; Chávez et al. 2005). And level of provitamin A increased in tomato through genetic engineering (De Steur et al. 2015). Regardless the importance of genetically engineered crops for human health, problems associated with its approval hinders the genetically modified crops. Whereas, conventional breeding significantly contributes in developing biofortified crops to ameliorate mineral insufficiency in humans (Finkelstein 2015).

11.2.2.3 Conventional Breeding

Conventional breeding is widely adopted method to develop biofortified crop following the same practice as used for plant breeding to ensure the improved agronomic yield contributing traits. It is reliant on the identification or developing lines of desired characters that made the nutrient intake better to produce high yield (Bouis 2003; Pfeiffer and McClafferty 2007). The success of conventional breeding depends upon natural and genetic variation; including the presence of specific nutrients in soil and the genetic heredity to make in use of those nutrients to concentrate them in edible grain part, which results in high yield production (Ortiz-Monasterio et al. 2007; Acquaah 2015). Germplasm screening helps to determine the range of mineral nutrients and genetic diversity of a particular specie (Baligar et al. 2001; Griffiths and York 2020). Varying environmental conditions affects the genetic heredity as well as nutrient uptake during vegetative growth period of cereal crops (Jacoby et al. 1975; Karamanos 2013). Measuring the relative effects of genetics, environment, and their combine effect on nutrient intake is an important component in the traditional breeding process. Once, the better nutrient concentration in a cultivar is found to be genetically stable, that genotype is eventually backcrossed with new or existing varieties to achieve desired traits. A dominant influence of the environment or low natural genetic variation of nutrient content may drop the breeding feasibility, but other methods can be applied to achieve bio-enriched (biofortified) crop. Zn and Fe rich biofortified beans, millet, rice and wheat have been developed with conventional breeding (Hotz et al. 2012).

11.2.2.4 Mutational Breeding/Mutagenesis

Mutagenesis is the prime source that causes variation in genetic material that occurs naturally. Whereas, researcher reported that mutational breeding can also be performed using various chemicals or radioactive waves to achieve the target traits in plants (Novak and Brunner 1992; Brunner 1995; Oladosu et al. 2016). And, its role in conventional breeding cannot be denied (Dalla Costa et al. 2017); mutational breeding is quicker and shorten the conventional breeding program to achieve desired level of traits (Hallajian 2016). Scientist are working on this technique for decades, moreover, it is progressively being used in biofortification. A significant achievement has been recognized with this technique, amongst these, biofortified cereals (rice, maize, wheat, barley, sorghum, and soybean) developed with less phytic acid percentage.

11.3 Role of Seed Priming in Biofortification

11.3.1 Role of Seed Priming in Iron (Fe) Biofortification

Iron insufficiency causes anemia that is a major problem in developing nation; therefore, sufficient amount of Fe must be included in the daily diet through various strategies to combat Fe deficiency (Vasconcelos et al. 2017). Seed priming is a novel and user-friendly method for bio-fortifying wheat and increasing Fe absorption and accumulation in grains. Seed priming with 25 ppm iron oxide nanoparticles significantly increased the grain's Fe content (45.7%) when compared to the control (Sundaria et al. 2019).

Exogenous techniques such as seed priming, foliar spraying, and soil application can be used to achieve agronomic biofortification with Fe. However, if these micronutrients are present in excess, the plants may suffer. More than 2 mg L^{-1} Fe concentrations considerably reduce the mitotic index, increase the proportion of abnormal cells proliferating, in addition to increased germination percentage (80%), tillers, grain yield (Reis et al. 2018). Seed priming with 2 mg L^{-1} Fe has been shown to increase tillering and grain yield while ensuring a high rate of germination (80%) and normal proliferating cells (90%) (Carvalho et al. 2019). It was demonstrated in this study that priming bread wheat seeds with the appropriate amounts of Fe can increase the nutritional value of flour. Fe seed priming increased plant height, fresh and dry weights, as well as stem diameter, length, and weight. The yield characteristics of fruit and fiber were improved significantly (Mazhar et al. 2022). After seed priming, malondialdehyde and hydrogen peroxide levels decreased by 66% and 71%, respectively. Furthermore, antioxidant enzyme activity was increased by 28%, 56%, and 39% for superoxide dismutase (SOD), peroxidase (POD), and catalase (CAT), demonstrating the ability of iron oxide particles to reduce water stress (Mazhar et al. 2022). Seed priming lowers the amount of energy required by the plant to activate a stress tolerance mechanism, allowing the plant to conserve energy and prove cost-effective in a broader range of applications (Johnson and Puthur 2021). Triveni et al. (2021) determined that 50 ppm of nano iron oxide was the optimal concentration for seed priming in mungbean because it increased germination percentage, shoot length, root length, seedling vigor index, and speed of germination.

11.3.2 Role of Seed Priming in Iodine (I) Biofortification

Iodine is one of the most important micronutrients necessary for human diet, involved in the production of thyroid hormones that play a significant role in metabolism. The high rate of iodine consumption can result in hyperthyroidism. Hypothyroidism is caused by a lack of iodine in the diet, which manifests as weakness, weight gain, and an enlarged thyroid gland. According to the World Health Organization (WHO), the recommended daily allowance (RDA) ranges from 50 to 250 g/day depending on age, with adults receiving 150 g/day. Iodine deficiency endangers the health of pregnant and nursing women, who require 250 g of iodine per day (WHO 2014; Zimmermann and Andersson 2012). Iodization of table salt is one method of preventing iodine deficiency; however, excessive NaCl consumption can cause cardiovascular disease, particularly hypertension. While, the less intake of iodine (table salt) has a negative impact on the prevention of iodine deficiencies (Stimec et al. 2009). Despite global systemic measures such as iodization of table salt, I insufficiency continues to affect world's population. Iodine deficiency affects many populations in developed countries (Eveleigh et al. 2020). Therefore, it is required to present a new solution to support iodine prophylaxis by promoting alternative iodine sources. One possible solution is to use seed priming as a biofortification strategy of plants with iodine during their development and growth, which is a necessary approach to improve emergence, seedling vigour, stress tolerance, and grain quality of many field crops to meet human nutritional demands.

The application of iodine-containing compounds (as fertilizers, chelated compounds, or nanoparticles) to soil, plant, or seed is a quick and effective method of biofortification (Valença et al. 2017). Treating the seeds with deficient nutrient is an agronomic approach to addressing nutrient deficiency in plants and humans. As a result, seeds can be treated with macro- and micronutrients to reduce nutrient deficiencies (Atar et al. 2020). Iodine seed priming improves nutritional components such as N, which is a key component of protein content, while decreasing anti-nutritional factors such as heavy metals (Cd and phytate) (Cakmak et al. 2017).

Plants grown from primed seeds have a faster start and greater stress tolerance, owing to more efficient energy metabolism, osmotic adjustment, an enlarged embryo, increased enzyme activation, and faster cellular defense responses (Jisha et al. 2013). The less concentration of I promote plant growth and develop resistance against stress, while the high concentration causes ionic toxicity in plants (Incrocci et al. 2019). Moreover, the iodine application at 10^2–10^4 nM showed a positive effect on plants (Gonzali et al. 2017). On the other hand, Watts et al. (2010) recorded 2.6 mg kg^{-1} I concentration for optimum performance. Contrarily, Lawson (2014) recorded the increased I uptake by 50 mg kg^{-1} soil iodine application in soil, that increases plant biomass in strawberry plants (Li et al. 2016). It was reported that using IO_3^- at 7.88 M increased the content of ascorbic acid and total phenolic compounds in tomato (Smoleń et al. 2015).

As a result, if enough information is available to develop a proper biofortification strategy, farmers may find simultaneous iodine biofortification through seed priming to be an appealing approach (Watts et al. 2010). However, extensive field research is still required to determine the best conditions for such an approach.

11.3.3 Role of Seed Priming in Magnesium (Mg) Biofortification

To resolve the problems associated with seed germination and seedling establishment, from few years several strategies are evolving, seed priming is one of all possible strategies. It is a best pre-sowing priming approach due to its effectiveness, reliability, easy application, economic efficient and eco-friendly attributes (Paparella et al. 2015; Singhal and Bose 2020). Principally, seed controlled hydration is involved in seed priming approach, which allows to accelerate the specific metabolic processes in seed in a controlled manner and avoiding the seed transition before the seedling emergence (Bose et al. 2018). Therefore, Mg $(NO_3)_2$ seed priming depicted a prominent increment in number of roots, root length and root weight of different crops. It is also an important stress reliving character in crops which are cultivated under field conditions (Sharma and Bose 2006; Anaytullah and Bose 2007). Mg biofortification in seeds is an innovative and widely used seed priming technology which helps in fast and uniform emergence with successful seedling development followed by high seedling growth and eventually augmented the overall yield (Jisha et al. 2013; Hussain et al. 2016). In previous studies, researchers concluded that the technology of priming of seeds with $Mg(NO_3)_2$ performed very well in crops relative to seed priming done with KNO_3 and $Ca(NO_3)_2$ (Pandey and Bose 2006; Srivastava and Bose 2012). The former strategy is also useful in prompting the nitrogen use efficiency in rice (Srivastava et al. 2017).

After analyzing frequently used different categories of seed priming strategies, it has been stated that $Mg(NO_3)_2$ primed seeds responded relatively better than KNO_3 and $Ca(NO_3)_2$ priming tactics for crops (Pandey and Bose 2006; Srivastava and Bose 2012); the former is also able to induce nitrogen-using efficiency in rice (Srivastava et al. 2017). This response of plant is probably due to the existence of divalent cation Mg as the cofactor with ATP in different plant cells enzymatic reactions, so that the plant performs well in several physiological processes (Li et al. 2001). Moreover, Mg is also a very crucial structural component of chlorophyll molecules in higher crops. If plants suffer with Mg deficiency of Mg under any stress condition, it may cause yellowing of leaves or severe interveinal chlorosis. Results of the application of Mg salt in various crops represented that protein uptake, chlorophyll contents, activities of nitrate reductase and glutamate kinase are enhanced. Thus, an increase in number of roots, root length and root weight are obtained when seeds are primed with Mg $(NO_3)_2$ under saline field environments (Sharma and Bose 2006; Anaytullah and Bose 2007).

For cellular development, activation of enzymes and many other biological functions, Mg and Zn are essential nutrients. Their deficiency may impair the seed vigor and all characteristics related to seed establishment, thus significantly affected plant growth and production. Scientists in several studies concluded that the remarkable changes in crop stand establishment, leaf expansion, fresh and dry weight of stem and leaf, chlorophyll production and overall protein content can be examined with seed priming of $Mg(NO_3)_2$, $ZnSO_4$, and their combination in wheat plant.

$Mg(NO_3)_2$, $ZnSO_4$, and their combination in seed priming were also found to be ameliorated the germination associated characters i.e. germination percent, root and shoot length and seed weight. Moreover, the biofortified seeds played a significant role in membrane stability and permeability, α-amylase and other metabolism associated enzymes (Anaytullah and Bose 2007). For mustard and maize plants, sowing of seeds primed with $Mg(NO_3)_2$ and $ZnSO_4$ also promoted the sprouting percentage (Foti et al. 2008; Paltridge et al. 2012; Prasad 2013; Rehman et al. 2015; Kumar et al. 2020; Singhal and Bose 2020).

However, for the first time Choudhary et al. (2021) observed the combined impact of seed priming treatment of Mg and Zn in wheat and concluded that combined application of Mg and Zn on popular varieties of wheat significantly improved cultivars seedling growth. $Mg(NO_3)_2$ priming helps in promoting cell elongation and cell division. Similarly, an increment in vigor index of seed in response to $Mg(NO_3)_2$ seed priming in carrot and rice crops also observed. Likewise, enhancement in vigor index is also observed in rice and carrot crops with the use of $Mg(NO_3)_2$ as a seed priming approach. Recent research suggested that seed priming treatments enhanced the production of defense metabolites in response to unfavorable growing conditions, which ultimately hastened the seedling growth under stress (Singh et al. 2015). Besides this, seed priming treatment also enhanced the solubilization of stored food matter and helps in easy translocation of more food towards the young growing seedlings for their nourishment and early support (Zibai et al. 2012). The combined use of both nutrients as an effective seed priming technique surprisingly showed better results in crop growth.

11.3.4 Role of Seed Priming in Zinc (Zn) Biofortification

Globally, enhancement of Zn content in grain is a matter of great concern as it is an important humanitarian challenge due to the severe dietary Zn deficiency in less Zn enriched staple cereals foods (Cakmak and Kutman 2018). This deficiency subsequent in unavoidable health hitches and commercial pressure on the healthcare sector (Gödecke et al. 2018; Harding et al. 2018). Among several seed treatment techniques, priming with Zn has been considered as the most widely and cost-effective technique (Harris et al. 2007; Farooq et al. 2012). Seed priming with Zn effects sprouting of seed and seedling development of maize, rice, chickpea, wheat and barley (Ajouri et al. 2004; Harris et al. 2007, 2008; Prom-u-thai et al. 2012; Rehman et al. 2015; Imran et al. 2017). Rehman et al. (2018) and Farooq et al. (2018) in studies instituted that seed priming with Zn is the most commonly used approach for improving tiller count (productive) and final yield of wheat and rice. Moreover, it has been observed that seeds with Zn content can withstand the adverse climatic variability during seedling phase (Ajouri et al. 2004; Imran et al. 2017). Candan et al. (2018) findings revealed prominent increase in germination percentage and seedling development in Zn-biofortified wheat seeds under drought stress and low soil Zn concentration. Rashid et al. (2004) apparently concluded the strong influence of Zn biofortified on seedling vigor and emergence under field conditions which ultimately enhance the overall yield of crop. High grain Zn content also protect the seed during seedling stage against several negative impacts of biotic and abiotic environmental stresses and eventually increase the crop yield. In conditions, Unfortunately, low-Zn seeds are more vulnerable to environmental changes and seedling emergence is significantly harmed (Hassan et al. 2021).

In early germination phase, Low-Zn concentration in seed loses their capability to survive against abiotic stresses particularly in Zn deficit soils (Hassan et al. 2021). Total number of seedlings per unit land area increase and profuse tillering was obtained which in returns increase the plant population in Zn-biofortified wheat seed. In food, lower phytate to Zn molar ratio is the indication of Zn available through biological means, which is low in grain due to antinutrient concentration, biofortification application improves bioavailability Zn in grains (Cakmak et al. 2010b). Recently, in durum wheat, it was observed that young seedlings tolerate drought stress due to development of Zn enriched seeds either through foliar fertilization of field crops (Candan et al. 2018) or via an effective seed priming technique (Fallah et al. 2018).

Zinc primed seeds can also show protective response against salt stress in early stage of seed development (Imran et al. 2017). Additionally, Zn appears more than beneficial its nutritional effects. Zn also known as the enhancement factor for the antioxidant capacity of plants. The antioxidant potential elevated in Zn enriched seeds of cereals with the increase activity of superoxidase dismutase and catalase (Candan et al. 2018; Fallah et al. 2018). Rehman et al. (2018) reported the benefits of wheat and rice seed with Zn priming (0.5 mM $ZnSO_4$ for 12 h) on crops stand establishment (Ali et al. 2018; Farooq et al. 2018). For farmers perspectives, Zn

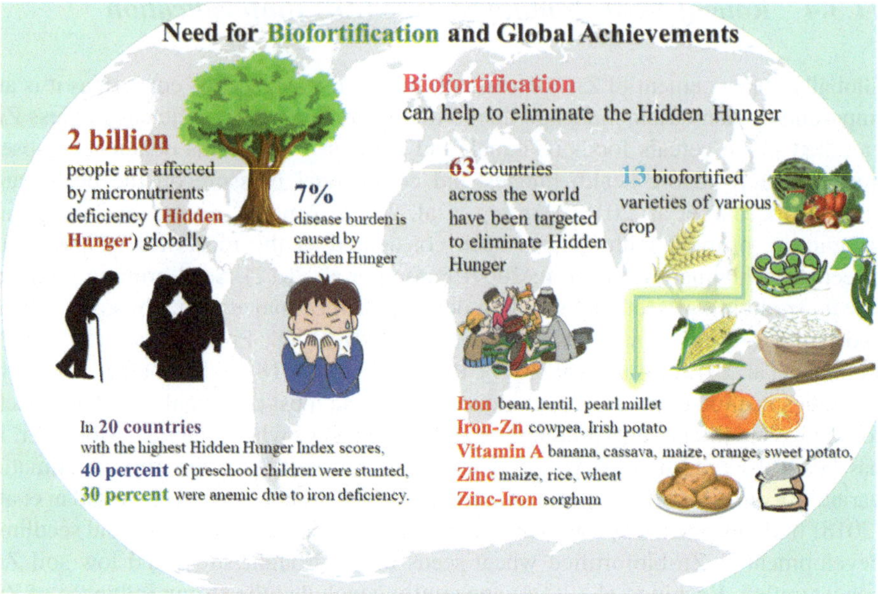

Fig. 11.2 Need of biofortification

priming is a very convenient and highly cost-efficient practice in getting high profits and yield for chickpea and wheat (Harris et al. 2008). Considering the critical requirement of development of Zn biofortified seeds in agronomic as well as human health sector, it is a dire need to introduce the approach which can address the issues related to food nutrition and security (Fig. 11.2).

Zinc as a micronutrient played many vital roles in certain plant physiological processes i.e., it acts as a cofactor for numerous enzymes such as alcohol dehydrogenase, lactic dehydrogenase and carbonic anhydrase. Phosphate-transferring enzymes such as, hexose kinase of triosephosphate dehydrogenase also contain Zn in their activation process. Moreover, Zn is also an essential component for synthesis of tryptophan, which is a precursor of very important plant hormone auxin (Prasad 2013; Kumar et al. 2020). Zn also participates in various activities of plants for example production of amino acid, protein, and chlorophyll and pollen formation. All traits related to germination in wheat are augmented by the application of seed priming with $ZnSO_4$. In fact, $ZnSO_4$ probably is considered superior for playing its role as a Zn in protein formation, cell membrane and cell elongations (Rehman et al. 2015; Singhal and Bose 2020).

Worldwide, it has been observed that range of soils are deficit in Zn content, to minimize this problem seed priming with Zn element is one of the best possible approaches.

11.4 Success Stories of Seed Priming in Biofortification in Recent Years

Priming improves seed quality, increases its germination rate under many stresses such as biotic and abiotic (Rehman et al. 2019). All these characteristics that increase competitiveness of the product are directly associated to seed vigor; a composite agronomic attribute that is influenced by numerous environmental as well as genetic factors (Rajjou et al. 2012; Jisha et al. 2013). Heydecker (1973) was the first to suggest seed priming. During a procedure known as "seed priming," seeds are moistened with the pre-sowing just enough to cause pre-germination metabolic activity without causing actual germination. In the moistened seeds (priming) metabolic processes begin but no real germination takes place. The priming technique, seed lot, vigor, and plant genotype and physiology, all contributes to making seed priming effective. Priming of seed is effective way to apply micronutrients to seeds (Johnson et al. 2005).

Seed priming activates hydrolytic enzymes that improves embryo physiology, allowing for faster germination (Bam et al. 2006). Coating of seed and priming of seed/nutria-priming are two ways or seed treatment or invigoration. Seed coating involves covering the seed with an almost constant covering of small, crushed particles or liquids containing suspended materials. Along with the dietary benefits of nutria-priming, biochemical priming successes can also be realized using this strategy (Lutts et al. 2016). It enhances emergence of seedlings, establishment of stands, yield, and grain micronutrient content (Haider et al. 2020), and tolerance to stresses (biotic and abiotic) (Haider et al. 2020).

Additionally, when needed by the brewing and cereal grain sectors, priming provides the optimal conditions to speed up seed germination (Imran et al. 2017). To increase the number of micronutrients in grain will enhance human health, the genetic engineering and breeding for biofortification procedures seem to be a viable choice (Cakmak 2008). Growing the grain's concentration of Zn with fertilization with Zn and other controlling techniques are agronomic approaches (Cakmak 2008). Zinc can be supplied using seed treatment, foliar application, or soil application (Farooq et al. 2012, 2018).

11.4.1 Benefits of Seed Priming in Different Crops

The immediate advantages of seed priming were reported by Harris et al. (2007) in grains including maize, rice, and wheat, seed priming was found to directly benefit. Greater grain output, earlier blooming, earlier harvest, better and more uniform stands, quicker emergence, more robust plants, and improved drought tolerance. Seed priming for the purpose of biofortification in different crops:

11.4.1.1 Wheat

Seed priming is a novel and user-friendly method of wheat biofortification that causes Fe to be acquired and accumulated in grains. Boron priming (0.01 M) resulted in significant increases in yield contributing characteristics, resulting in a significant increase in grain production. It could possibly be related to B role in promoting photo-assimilates' transfer from reproductive to vegetative parts (Reddy et al. 2003), which could increase wheat grain output. Boron seed priming with 0.05 and 0.01 M B solutions increased grain B content in wheat cultivars. To boost germination, early seedling growth, yield, and grain B content, wheat seedlings can be primed with 0.01 M B using borax. According to a recent study, multiwalled carbon nanotube seed priming of wheat seeds encouraged early germination and root-hair growth in addition to greater wheat productivity (Joshi et al. 2017) (Table 11.1).

Table 11.1 Effect of seed priming on grain iron (Fe) contents of different crops

Crop	Seed priming (SP) agent	Priming rate	Effect of SP on crop traits and grain Fe contents	References
Wheat	Iron oxide nanoparticles	25 ppm	SP increased 45.7% iron contents in grains	Sundaria et al. (2019)
Wheat	$FeSO_4 \cdot 7H_2O$	2 mg L^{-1}	SP increased germination by 80%, normal dividing cell (90%), grain iron contents, tillering and economic yield	Reis et al. (2018)
Wheat	$FeSO_4$	0.3%	SP with iron significantly increased the iron, protein and fat percentage in grain and yield attributes of wheat	Zulfiqar et al. (2020)
Wheat	$FeSO_4.7H_2O$	4 mg L^{-1}	Nucleolar activity in roots, total soluble protein content increased with seed priming	Carvalho et al. (2019)
Stevia	$FeSO_4.7H_2O$		Germination percentage, germination rate, mean germination time, seedling vigour index	Gorzi et al. (2017)
Wheat	$FeSO_4$	1%	Priming increased the plant height, tiller count, spike length, yield and yield contributing factors, harvest index and grain iron and protein contents	Ramzan et al. (2020)
Rice	$FeCl_3$	25 mg L^{-1}	SP increased the seed germination and seedling parameters and iron contents	Prerna et al. (2021)
Forage corn	Ferric ethylene diamine tetra acetic acid	7%	Seed priming with nano iron improved leaf chlorophyll, total dry matter, crude protein, carbohydrate and phosphorous uptake with iron contents in grains	Sharifi et al. (2016)

11.4.1.2 Rice

Boron aids seed germination by allowing seed nutrients to be remobilized (Bonilla et al. 2004). Rice (*Oryza sativa* L.) was primed, which improved Root growth, reductive sugars, α-amylase activity, and nitrogen uptake (Farooq et al. 2009). Seed priming with 0.001% B solution, for example, has been demonstrated to be an effective and practical method of B application in rice, resulting in improved yield and grain B enrichment (Rehman et al. 2012). Boron application strategies (soil, foliar, and seed priming) were examined in another study, and B seed priming was found to be the most economically viable choice for increasing rice yield (Rehman et al. 2014).

11.4.1.3 Maize

Seed priming shortens the time it takes for a plant to emerge, resulting in faster development and improved crop stand. It has been demonstrated that Zn seed priming enhances crop emergence, stand establishment, and yield. In a maize crop, Foti et al. (2008) found that Zn seed priming over 24 h improved stand establishment by 29% compared to control.

11.4.1.4 Legumes

Seed priming prior to planting has been shown to boost legume production (Musa et al. 2001; Rashid et al. 2004), owing to reduced metabolite leakage, seed healing, and higher protein and RNA synthesis (Musa et al. 2001; Rashid et al. 2004; Nasef et al. 2006).

11.4.1.5 Mungbean

On the shoot growth of mungbean seedlings (*Vigna radiata*), nano-ZnO, nano-FeO iron oxide, and nano-ZnCuFe oxide nanoparticles have also been found (Dhoke et al. 2013). Zinc oxide nanoparticles have also been shown to improve native phosphorus mobilization in rhizosphere of mungbean (Raliya et al. 2016). Zinc seeds priming therapies for mungbean, on the other hand, have yet to be perfected. We expected in this study that improved mungbean stand establishment, yield, and grain Zn biofortification will result from seed priming at a pre-optimized rate.

11.4.1.6 Canola

In another study, seed priming significantly increased canola biomass, weight, and grain yield when compared to control plants (Basra et al. 2003).

11.4.1.7 Onion

Early seedling development, a higher germination rate, improved flowering, higher fruit value, and seed weight per umbel have also been found in onion (*Allium cepa* L.) plants treated with zinc oxide ZnO nanoparticles (Rico et al. 2011) (Table 11.2).

11.4.2 Biofortification with Different Micronutrients

11.4.2.1 Zinc Biofortification

The simplest and quickest method for biofortifying pulse grains with Fe, Zn, or other desired micronutrients is agronomic biofortification. Therefore, agronomic biofortification plays a huge role in addressing hidden hunger or micronutrient deficiencies. Zinc can be provided through soil, leaf, or seed treatment to increase plant Zn uptake (Johnson et al. 2005; Farooq et al. 2012, 2018). In that way, applying Zn to the soil is the main way to provide Zn to the plants. Before planting wheat, $ZnSO_4$ is added to the soil to make up for the Zn deficit (Cakmak 2008). Zinc can also be applied to foliage, which increases wheat yield and grain Zn concentration (Johnson et al. 2005). Zinc can act as a priming agent and improves water use efficiency vigour, turgid weight, root and shoot length, leaf area and yield. Zinc in low concentration improves germination percentage, crop emergence, development of seedling, nutrient (Zn) uptake, total dry matter and yield.

Table 11.2 Effect of seed priming on grain iodine (I) contents of different crops

Crop	Seed priming agent	Priming rate	Effect of SP on crop traits and grain I Contents	References
Coriander	Karrikinolide	10–6 M	Priming improved germination rate, RWC, photosynthesis, growth and biomass production of developed seedlings	Sardar et al. (2021)
Okra	Osmopriming	5% PEG	Seed priming improved the yield parameters along with crude protein, mineral, mucilage and iodine contents in grains	Hardeep et al. (2015)
Buckwheat	Iodate (IO_3^-)	14.7 μg	Grain iodine contents was increased in seed treatment with IO_3^- as compared to I alone	Germ et al. (2019)
Safflower	Melatonin-seed-priming	0.5 mM	Seed priming increased iodine value, yield and oil contents	Akbari et al. (2020)

Zinc priming stimulates germination's metabolic processes, and the resulting sugars can be used to make proteins. This increases plant germination rates and promotes uniform growth (Rouhi et al. 2011). In compared to soil application, seeds primed with $ZnSO_4$ solutions considerably enhanced yield in a variety of conditions for diverse crops, such as maize (27%), common bean (17%), and chickpea (18%) reported by Harris et al. (2007), Kaya et al. (2007), and Harris et al. (2008), respectively.

11.4.2.2 Boron Biofortification

The most practical and economical way to apply B is by seed priming, but before using it in the field, it is crucial to evaluate and improve the concentration of the solution. In many crop plants while also reducing flowering time by 50%, grain weight, and grain yield. The use of B primed seed increases crop production. Further, the B priming improved the number of leaves and tiller count in wheat and barley (Bose et al. 2016; Mondal and Bose 2019; Misu 2021).

The application method is crucial for the plant to absorb B in an effective manner. The most common technique for B addition in the industrialized world is soil application. To apply micronutrients, seed priming can be a desirable and simple physiological technique (Farooq et al. 2009, 2012). For instance, seed priming with B significantly boosted tillering but had little effect on seed germination in oats (*Avena sativa* L.). Additionally, foliar spraying nutrients helps to quickly address deficiencies (Nadeem and Farooq 2019). Due to severe interruptions in several metabolic processes related to B, including the metabolism of carbohydrates, nucleic acids, proteins, and indole acetic acid, phenol, and integrity of membrane and purpose, crop yield is drastically reduced when there is a B deficiency (Rehman et al. 2018a).

11.4.2.3 Iron Biofortification

Although applying Fe increases wheat productivity, no comparison research has been done to look at the effectiveness of various Fe application techniques in boosting yield and wheat farmed with conventional and conservation tillage systems with grain biofortification.

11.4.2.4 Molybdenum Biofortification

Lack of molybdenum (Mo) causes plants to accumulate nitrates in their leaves rather than absorb them into proteins. Molybdenum works in symbiosis with *Rhizobium* sp. to fix atmospheric nitrogen in root nodules of leguminous. Molybdenum is a very important micronutrient that can considerably increase crop yield, nodulation, nitrogen uptake, N fixation, and dry matter accumulation when administered at extremely low concentrations for seed priming.

According to the literature, applying Mo through seed treatment is more efficient and effective than doing so through soil application (Malla et al. 2007). In soil with a deficiency in Mo, priming seeds of subterranean clover (*Trifolium subterraneum* L.) in 0.1 or 1% sodium molybdate solution were found to be identical to soil application (Kumar Rao et al. 2004). Additionally, seed priming increased the amount of Mo and N in grains compared to equal soil treatments. However, it was discovered that priming with sodium molybdate improved nitrogen fixation, yield, nodulation, and dry matter buildup in common bean seeds (Mohandas 1985).

11.4.2.5 Manganese Biofortification

Manganese (Mn) builds up charges when water is oxidized as a member of complex of oxygen-evolving connected to photosystem II. Depending on the type of crop grown, the climate, and the soil, a lack of Mn may be pervasive in various regions. Low pH of soil and higher levels of organic matter are the two main contributors to deficiency extreme chlorosis in the veins of the leaves, in addition to discoloration and deformities in the seeds of legumes are possible effects (Taiz and Zeiger 2010). Due to the multiple positive functions of Mn on plant physiology, particularly in photosynthesis and respiration, a deficit in this mineral will result in decreased yield.

11.4.2.6 Copper Biofortification

The main way that copper (Cu) works is as a cofactor for different oxidative enzymes. Additionally, lignin production requires Cu, which not only strengthens cell walls but also stops wilting (Hopkins and Huner 2008). Seeds primed with Cu sulphate resulted in increased biomass, grain production, grain weight, and number of grains (per panicle). Seed priming is easy to adopt, fast, cheap, and eco-friendly method than the most water-conserving procedures, giving farmers an extremely appealing alternative for accelerating germination and increasing crop establishment and yields. Osmotic seed priming of maize caryopses in copper sulphate, according to Maiti et al. (2006) experiments, may increase the percentage of seeds that germinate at a given time. CuEDTA solution (0.04–0.16 kg Cu ha^{-1}) was used to prime wheat seeds, which significantly boosted grain production but decreased seedling emergence (Malhi 2009). Wheat seeds that have been primed to prevent Cu deficiency at a very low concentration (0.04 kg Cu ha^{-1}) in CuEDTA solution significantly increased seed output (Malhi 2009).

11.4.2.7 Cobalt Biofortification

Cobalt (Co) is necessary for nitrogen fixation in legumes, but it is advantageous for some plants and necessary for others. Cobalt is a crucial component of many enzymes and co-enzymes and has an impact on metabolism and plant growth. Seed priming with rhizobium and a mixture of micronutrients (0.16 mg g^{-1} Mo as sodium molybdate and 0.008 mg g^{-1} Co as cobalt chloride) greatly increased nodulation, N fixation, nutrient uptake, plant development, and yield. The literature reported by Ambika et al. (2014) supports these findings regarding seed priming in various crops using cobalt as a priming agent in summer squash (*Cucurbita pepo* L.), where seed priming with cobalt sulphate significantly increased dry matter accumulation, femaleness, and fruits.

11.5 Conclusion

Seed priming is one of the easiest methods to adopt by farmers at field to enhance crop performance. In addition, it is the cheapest approach to reduce the gap between actual and potential yields. Crop produced from the soil deficient in micronutrients significantly affect human health globally. Malnutrition (micronutrient) is now becoming a severe human health and economic concern worldwide. Seed priming with micronutrients (Zn, Fe, Mg and I) produces specific proteins, sugar responsible to increase grain micronutrient contents to lowers micronutrient-malnutrition, specifically in developing nations. The increased understanding of other strategies (mutagenesis, conventional breeding, and genetic engineering), and knowing the metabolic events during priming can be helpful to increase germination and micronutrient uptake, though, these strategies failed to cope malnutrition problems among illiterate farmer's field. It is required to disseminate the knowledge of seed priming benefits to render the hidden hunger of crops and thus, malnutrition in human beings. Thus, this methodology supports to optimize yield, grain quality with less use of resources to strengthen the socio-economic conditions of farmers and global food security.

References

Acquaah G (2015) Conventional plant breeding principles and techniques. In: Al-Khayri J, Jain S, Johnson D (eds) Advances in plant breeding strategies: breeding, biotechnology and molecular tools. Springer, Cham, pp 115–158

Ajouri A, Asgedom H, Becker M (2004) Seed priming enhances germination and seedling growth of barley under conditions of P and Zn deficiency. J Plant Nutr Soil Sci 167:630–636

Akbari GA, Heshmati S, Soltani E, Amini DM (2020) Influence of seed priming on seed yield, oil content and fatty acid composition of safflower (*Carthamus tinctorius* L.) grown under water deficit. Int J Plant Prod 14:245–258

Al-Babili S, Beyer P (2005) Golden Rice–five years on the road–five years to go? Trends Plant Sci 10:565–573

Alderman H, Hoddinott J, Kinsey B (2006) Long term consequences of early childhood malnutrition. Oxf Econ Pap 58:450–474

Ali N, Farooq M, Hassan MA, Arshad MS, Saleem MK, Faran M (2018) Micronutrient seed priming improves stand establishment, grain yield and biofortification of bread wheat. Crop Pasture Sci 69:479–487

Ali J, Nawaz M, Ilyas M, Chattha MU, Khan I, Chattha MB, Hassan MU (2022) Effect of different irrigation intervals and sowing methods on growth and productivity of lentil crop grown in semi-arid conditions. J Innov Sci 8:20–28

Ambika S, Manonmani V, Deepika S (2014) Seed priming with micronutrients for quality and yield. Popular Kheti 2:35–37

Anaytullah BB, Bose B (2007) Nitrate-hardened seeds increase germination, amylase activity and proline content in wheat seedlings at low temperature. Physiol Mol Biol Plant 13:199–207

Atar B, Uygur V, Sukus UE (2020) Effects of priming with copper, zinc and phosphorus on seed and seedling composition in wheat and barley. Turk J Agric Nat Sci 7:104–111

Bai C, Capell T, Berman J, Medina V, Sandmann G, Christou P, Zhu C (2016) Bottlenecks in carotenoid biosynthesis and accumulation in rice endosperm are influenced by the precursor–product balance. Plant Biotechnol J 14:195–205

Baligar VC, Fageria NK, He ZL (2001) Nutrient use efficiency in plants. Commun Soil Sci Plant Anal 32:921–950

Bam RK, Kumaga FK, Ofori K, Asiedu EA (2006) Germination, vigour and dehydrogenase activity of naturally aged rice (*Oryza sativa* L.) seeds soaked in potassium and phosphorus salts. Asian J Plant Sci 5:948–955

Basra S, Ehsanullah MA, Warraich EA, Cheema MA, Afzal I (2003) Effect of storage on growth and yield of primed canola (*Brassica napus* L.) seeds. Int J Agric Biol 5:117–120

Bevis LE (2015) Soil-to-human mineral transmission with an emphasis on zinc, selenium, and iodine. Sci Rev 3:77–96

Beyer P (2010) Golden rice and golden crop for human nutrition. New Biotechnol 27:478–481

Black RE, Victora CG, Walker SP (2013) Maternal and child undernutrition and overweight in low-income and middle-income countries. Lancet 382:427–451

Bonilla I, El-Hamdaoui A, Bolaños L (2004) Boron and calcium increase *Pisum sativum* seed germination and seedling development under salt stress. Plant Soil 267:97–107

Bose B, Srivastava AK, Siddique A (2016) Impact of nitrate salt hardened seeds and sowing dates on seedling stand, growth, yield attributes, nitrogen and stress metabolism of rice. Int J Agric Environ Biotechnol 9:381–392

Bose B, Kumar M, Singhal RK, Mondal S (2018) Impact of seed priming on the modulation of physico-chemical and molecular processes during germination, growth, and development of crops. In: Rakshit A, Singh H (eds) . Advances in seed priming, Springer, Singapore, pp 23–24

Bouis HE (2003) Micronutrient fortification of plants through plant breeding: can it improve nutrition in man at low cost? Proc Nutr Soc 62:403–411

Bouis HE (2007) The potential of genetically modified food crops to improve human nutrition in developing countries. J Develop Stud 43:79–96

Bouis HE, Hotz C, McClafferty B (2011) Biofortification: a new tool to reduce micronutrient malnutrition. Food Nutr 32:31–40

Broadley MR, White PJ, Bryson RJ, Meacham MC, Bowen HC, Johnson SE, Hawkesford MJ, McGrath SP, Zhao FJ, Breward N (2006) Biofortification of UK food crops with selenium. Proc Nutr Soc 65:169–181

Broadley MR, White PJ, Hammond JP, Zelko I, Lux A (2007) Zinc in plants. New Phytol 173:677–702

Brunner H (1995) Radiation induced mutations for plant selection. Appl Radiat Isot 46:589–594

Bryce J, el Arifeen S, Pariyo G, Lanata CF, Gwatkin D, Habicht J (2003) The multi-country valuation of IMCI study group. "Reducing child mortality: can public health deliver?". Lancet 362:159–164

Cakmak I (2008) Enrichment of cereal grains with zinc: agronomic or genetic biofortification? Plant Soil 302:1–17

Cakmak I, Kutman UB (2018) Agronomic biofortification of cereals with zinc: a review. Eur J Soil Sci 69:172–180

Cakmak I, Pfeiffer WH, McClafferty B (2010a) Biofortification of durum wheat with zinc and iron. Cereal Chem 87:10–20

Cakmak I, Kalayaci M, Kaya Y, Torun AA, Aydin N, Wang Y, Arisoy Z, Erdem H, Yazici A, Gokmen O, Ozturk L, Horst WJ (2010b) Biofortification and localization of zinc in wheat grain. J Agric Food Chem 58:9092–9102

Cakmak I, Prom-U-Thai C, Guilherme LRG, Rashid A, Hora KH, Yazici A, Ozturk L (2017) Iodine biofortification of wheat, rice and maize through fertilizer strategy. Plant Soil 418:319–335

Candan N, Cakmak I, Ozturk L (2018) Zinc-biofortified seeds improved seedling growth under zinc deficiency and drought stress in durum wheat. J Plant Nutr Soil Sci 181:388–383

Carvalho A, Reis S, Pavia I, Lima-Brito JE (2019) Influence of seed priming with iron and/or zinc in the nucleolar activity and protein content of bread wheat. Protoplasma 256:763–775

Chávez AL, Sánchez T, Jaramillo G, Bedoya JM, Echeverry J, Bolaños EA, Ceballos H, Iglesias CA (2005) Variation of quality traits in cassava roots evaluated in landraces and improved clones. Euphytica 143:125–133

Choudhary SK, Kumar V, Singhal RK, Bose B, Chauhan J, Alamri S, Siddiqui MH, Javed T, Shabbir R, Rajendran K, Iqbal MA, Elmetwaly ZEA, Sorour S, Sabagh AE (2021) Seed priming with $Mg(NO_3)_2$ and $ZnSO_4$ salts triggers the germination and growth attributes synergistically in wheat varieties. Agronomy 11:2110

Dalla Costa L, Malnoy M, Gribaudo I (2017) Breeding next generation tree fruits: technical and legal challenges. Hortic Res 4:15–32

Das S, Chaki AK, Hossain A (2019) Breeding and agronomic approaches for the biofortification of zinc in wheat (*Triticum aestivum* L.) to combat zinc deficiency in millions of a population: a Bangladesh perspective. Acta Agrobot 72(2). https://doi.org/10.5586/aa.1770

De Steur H, Blancquaert D, Strobbe S, Lambert W, Gellynck X, Van DSD (2015) Status and market potential of transgenic biofortified crops. Nat Biotechnol 33:25–29

Dhoke SK, Mahajan P, Kamble R, Khanna A (2013) Effect of nanoparticles suspension on the growth of mung (*Vigna radiata*) seedlings by foliar spray method. Nanotech Dev 3(1):e1–e1

Ekholuenetale M, Tudeme G, Onikan A, Ekholuenetale CE (2020) Socioeconomic inequalities in hidden hunger, undernutrition, and overweight among under-five children in 35 sub-Saharan Africa countries. J Egypt Public Health Assoc 95:1–15

Eveleigh ER, Coneyworth LJ, Avery A, Welham SJM (2020) Vegans, vegetarians, and omnivores: how does dietary choice influence iodine intake? Syst Rev Nutr 12:1606

Fallah S, Malekzadeh S, Pessarakli M (2018) Seed priming improves seedling emergence and reduces oxidative stress in *Nigella sativa* under soil moisture stress. J Plant Nutr 41:29–40

FAO IFAD, UNICEF, WFP, WHO, The State of Food Security and Nutrition in the World (2017) Building resilience for peace and food security. FAO, Rome

Farooq M, Wahid A, Basra SM (2009) Improving water relations and gas exchange with brassinosteroids in rice under drought stress. J Agron Crop Sci 195:262–269

Farooq M, Wahid A, Siddique KHM (2012) Micronutrient application through seed treatments—a review. J Soil Sci Plant Nutr 12:125–142

Farooq M, Ullah A, Rehman A, Nawaz A, Nadeem A, Wakeel A, Siddique KH (2018) Application of zinc improves the productivity and biofortification of fine grain aromatic rice grown in dry seeded and puddled transplanted production systems. Field Crops Res 216:53–62

Finkelstein JL (2015) A randomized trial of iron biofortified pearl millet in school children in India. J Nutr 145:1576–1581

Foti R, Abureni K, Tigere A, Gotosa J, Gere J (2008) The efficacy of different seed priming osmotica on the establishment of maize (*Zea mays* L.) caryopses. J Arid Environ 72:1127–1130
Frison EA, Smith IF, Johns T, Cherfas J, Eyzaguirre PB (2006) Agricultural biodiversity, nutrition, and health: making a difference to hunger and nutrition in the developing world. Food Nutr Bull 27:167–179
Germ M, Stibilj V, Šircelj H, Jerše A, Kroflič A, Golob A, Maršić NK (2019) Biofortification of common buckwheat microgreens and seeds with different forms of selenium and iodine. J Sci Food Agric 99:4353–4362
Gödecke T, Stein AJ, Qaim M (2018) The global burden of chronic and hidden hunger: trends and determinants. Global Food Sec 17:21–29
Gonzali S, Kiferle C, Perata P (2017) Iodine biofortification of crops: agronomic biofortification, metabolic engineering and iodine bioavailability. Curr Opin Biotechnol 44:16–26
Gorzi A, Omidi H, Bostani AB (2017) Morpho-physiological responses of stevia (*Stevia rebaudiana* Bertoni) to various priming treatments under drought stress. Appl Ecol Environ Res 16:4753–4771
Griffiths M, York LM (2020) Targeting root ion uptake kinetics to increase plant productivity and nutrient use efficiency. Plant Physiol 182:1854–1868
Haider MU, Hussain M, Farooq M, Nawaz A (2020) Optimizing zinc seed priming for improving the growth, yield and grain biofortification of mungbean (*Vigna radiata* L. wilczek). J Plant Nutri 43:1438–1446
Hallajian MT (2016) Mutation breeding and drought stress tolerance in plants. In: Hossain M, Wani S, Bhattacharjee S, Burritt D, Tran LS (eds) Drought stress tolerance in plants. Springer, Cham, pp 359–383
Hardeep K, Neena C, Mamta P (2015) Effect of different seed priming treatments and priming duration on biochemical parameters and agronomic characters of okra (*Abelmoschus esculentus* L.). Int J Plant Physiol Biochem 7:1–11
Harding KL, Aguayo VM, Masters WA, Webb P (2018) Education and micronutrient deficiencies: an ecological study exploring interactions between women's schooling and children's micronutrient status. BMC Public Health 18
Harris D, Rashid A, Miraj G, Arif M, Shah H (2007) On-farm seed priming with zinc sulphate solution – a cost-effective way to increase the maize yields of resource-poor farmers. Field Crops Res 10:119–127
Harris D, Rashid A, Miraj G, Arif M, Yunas M (2008) On-farm' seed priming with zinc in chickpea and wheat in Pakistan. Plant Soil 306:3–10
Hassan MU, Chattha MU, Ullah A, Khan I, Qadeer A, Aamer M, Khan TA (2019) Agronomic biofortification to improve productivity and grain Zn concentration of bread wheat. Int J Agric Biol 21:615–620
Hassan MU, Aamer M, Nawaz M, Rehman A, Aslam T, Afzal U, Guoqin H (2021) Agronomic bio-fortification of wheat to combat zinc deficiency in developing countries. Pak J Agric Res 34: 201
Hawkesford MJ, Zhao FJ (2007) Strategies for increasing the selenium content of wheat. J Cereal Sci 46:282–292
Hermann M (2009) The impact of the European novel food regulation on trade and food innovation based on traditional plant foods from developing countries. Food Policy 34:499–507
Heydecker W, Higgins J, Gulliver RL (1973) Accelerated germination by osmotic seed treatment. Nature 246:42–44
Hopkins WG, Huner NPA (2008) Introduction to plant physiology, 4th edn. Wiley, Hoboken, NJ, pp 73–74
Hotz C, Loechl C, Lubowa A, Tumwine JK, Ndeezi G, Masawi AN, Baingana R, Carriquiry A, Brauw A, Meenakshi JV, Gilligan DO (2012) Introduction of β-carotene–rich orange sweet potato in rural Uganda resulted in increased vitamin A intakes among children and women and improved vitamin A status among children. J Nutr 142:1871–1880

Hussain S, Khan F, Cao W, Wu L, Geng M (2016) Seed priming alters the production and detoxification of reactive oxygen intermediates in rice seedlings grown under sub-optimal temperature and nutrient supply. Front Plant Sci 7:439

Imran M, Garbe-Schönberg D, Neumann G, Boelt B, Mühling KH (2017) Zinc distribution and localization in primed maize seeds and its translocation during early seedling development. Environ Exp Bot 143:91–98

Incrocci L, Carmassi G, Maggini R, Poli C, Saidov D, Tamburini C (2019) Iodine accumulation and tolerance in sweet basil (*Ocimum basilicum* L.) with green or purple leaves grown in floating system technique. Front Plant Sci 10:1–15

Izydorczyk G, Ligas B, Mikula K, Witek-Krowiak A, Moustakas K, Chojnacka K (2021) Biofortification of edible plants with selenium and iodine–A systematic literature review. Sci Total Environ 754:141983

Jacoby A, Altman DG, Cook J, Holland WW, Elliott A (1975) Influence of some social and environmental factors on the nutrient intake and nutritional status of schoolchildren. J Epidemiol Commun Health 29:116–120

Jiao W, Chen W, Chang AC, Page AL (2012) Environmental risks of trace elements associated with long-term phosphate fertilizers applications: a review. Environ Pollut 168:44–53

Jisha KC, Vijayakumari K, Puthur JT (2013) Seed priming for abiotic stress tolerance: an overview. Acta Physiol Plant 35:1381–1396

Johnson R, Puthur JT (2021) Seed priming as a cost effective technique for developing plants with cross tolerance to salinity stress. Plant Physiol Biochem 162:247–257

Johnson SE, Lauren JG, Welch RM, Duxbury JM (2005) A comparison of the effects of micronutrient seed priming and soil fertilization on the mineral nutrition of chickpea (*Cicer arietinum*), lentil (*Lens culinaris*), rice(*Oryza sativa*) and wheat (*Triticum aestivum*) in Nepal. Exp Agric 41: 427–448

Joshi A, Kaur S, Dharamvir K, Nayyar H, Verma G (2017) Multiwalled carbon nanotubes applied through seed-priming influence early germination, root hair, growth and yield of bread wheat (*Triticum Aestivum* L.). J Sci Food Agric 98:3148–3160

Karamanos RE (2013) Nutrient uptake and metabolism in crops. Prairie Soils Crops 6:52–63

Kaya M, Atak M, Khawar KM, Çiftçi CY, Özcan S (2007) Effect of pre-sowing seed treatment with zinc and foliar spray of humic acids on yield of common bean (*Phaseolus vulgaris* L.). Int J Agric Biol 7:875–878

Krzepilko A, Prazak R, Skwarylo-Bednarz B, Molas J (2019) Agronomic biofortification as a means of enriching plant foodstuffs with iodine. Acta Agrobot 72:150–167

Kumar Rao JVD, Harris KD, Johansen C, Musa AM (2004) Lowcost provision of molybdenum (Mo) to chickpeas grown in acid soils. Proceedings of the Zvarevashe, CARE stakeholder workshop, report IDG/00/18, Silsoe Research Institute, Bedford

Kumar V, Singhal RK, Kumar N, Bose B (2020) Micro-nutrient seed priming: a pragmatic approach towards abiotic stress management. In: Rakshit A, Singh H, Singh A, Singh U, Fraceto L (eds) New frontier stress management for durable agriculture. Springer, Singapore, pp 231–255

Lawson PG (2014) Development and evaluation of iodine biofortification strategies for vegetables. Logos Verlag Berlin GmbH, Berlin

Li L, Tutone AF, Drummond RS, Gardner RC, Luan S (2001) A novel family of magnesium transport genes in Arabidopsis. Plant Cell 13:2761–2775

Li R, Liu HP, Hong CL, Dai ZX, Liu JW, Zhou J (2016) Iodide and iodate effects on the growth and fruit quality of strawberry. J Sci Food Agric 97(1):230–235

Lutts S, Benincasa P, Wojtyla L, Kubala S, Pace R, Lechowska K, Quinet M, Garnczarska M (2016) Seed priming: new comprehensive approaches for an old empirical technique. In: Araujo S (ed) New challenges in seed biology-basic and translational research driving seed technology. IntechOpen, Rijeka, pp 1–46

Lyons G, Cakmak I (2012) Agronomic biofortification of food crops with micronutrients. In: Bruulsema TW, Heffer P, Welch RM (eds) Fertilizing crops to improve human health: a scientific review. Int Plant Nutr Institute, Paris, pp 97–122

Maiti RK, Sarkar NC, Singh VP (2006) Principles of postharvest seed physiology and technology. Agrobios, Jodhpur

Malhi SS (2009) Effectiveness of seed-soaked Cu, autumn- versus spring-applied Cu, and Cu-treated P fertilizer on seed yield of wheat and residual nitrate-N for a Cu-deficient soil. Can J Plant Sci 89:1017–1030

Malla RM, Padmaja B, Malathi S, Jalapathi RL (2007) Effects of micronutrients on growth and yield of pigeonpea. J Semi-Arid Tropics Agric Res 5:1–3

Mazhar MW, Ishtiaq M, Maqbool M, Akram R, Shahid A, Shokralla S, Elansary HO (2022) Seed priming with iron oxide nanoparticles raises biomass production and agronomic profile of water-stressed flax plants. Agronomy 12:982

Misu UM (2021) Influence of boron applications methods on morpho-physiological attributes and yield of wheat (*Triticum aestivum* L.) (Doctoral dissertation, Department of Agricultural Botany, Sher-e-Bangla Agricultural University, Dhaka)

Mohandas S (1985) Effect of presowing seed treatment with molybdenum and cobalt on growth, nitrogen and yield in bean (*Phaseolus vulgaris* L.). Plant Soil 86:283–285

Mondal S, Bose B (2019) Impact of micronutrient seed priming on germination, growth, development, nutritional status and yield aspects of plants. J Plant Nutri 42:2577–2599

Mozumdar L (2012) Agricultural productivity and food security in the developing world. Bangladesh J Agric 35:53–69

Mugambiwa SS, Tirivangasi HM (2017) Climate change: a threat towards achieving 'sustainable development goal number two (end hunger, achieve food security and improved nutrition and promote sustainable agriculture) in South Africa. Jàmbá: J Disaster Risk Stud 9:1–6

Musa AM, Harris D, Johansen C, Kumar J (2001) Short duration chickpea to replace fellow after AMAN rice: the role of on-farm seed priming in the high barind tract of Bangladesh. Exp Agric 37:509–521

Nadeem F, Farooq M (2019) Application of micronutrients in rice-wheat cropping system of South Asia. Rice Sci 26:356–371

Nasef MA, Nadia M, Amal F, El-Hamide A (2006) Response of peanut to foliar spray with boron and/or rhizobium inoculation. J Appl Sci Res 2:1330–1337

Nicolle C, Simon G, Rock E, Amouroux P, Rémésy C (2004) Genetic variability influences carotenoid, vitamin, phenolic, and mineral content in white, yellow, purple, orange, and dark-orange carrot cultivars. J Am Soc Hortic Sci 129:523–529

Novak FJ, Brunner H (1992) Plant breeding: induced mutation technology for crop improvement. IAEA Bull 4:25–33

Oladosu Y, Rafii MY, Abdullah N, Hussin G, Ramli A, Rahim HA, Usman M (2016) Principle and application of plant mutagenesis in crop improvement: a review. Biotechnol Biotechnol Equip 30:1–16

Ortiz-Monasterio JI, Palacios-Rojas N, Meng E, Pixley K, Trethowan R, Pena RJ (2007) Enhancing the mineral and vitamin content of wheat and maize through plant breeding. J Cereal Sci 46: 293–307

Paine JA, Shipton CA, Chaggar S, Howells RM, Kennedy MJ, Vernon G, Drake R (2005) Improving the nutritional value of Golden Rice through increased pro-vitamin A content. Nat Biotechnol 23:482–487

Paltridge NG, Milham PJ, Ortiz-Monasterio JI, Velu G, Yasmin Z, Palmer LJ, Guild GE, Stangoulis JC (2012) Energy-dispersive X-ray fluorescence spectrometry as a tool for zinc, iron and selenium analysis in whole grain wheat. Plant Soil 361:261–269

Pandey D, Bose B (2006) Influence of pre-sowing seed treatment with nitrate salts and different sowing dates on performance of mustard. Indian J Plant Physiol 11:261–265

Paparella S, Araújo SS, Rossi G, Wijayasinghe M, Carbonera D, Balestrazzi A (2015) Seed priming: state of the art and new prospective. Plant Cell Rep 34:1281–1293

Pfeiffer WH, McClafferty B (2007) HarvestPlus: breeding crops for better nutrition. Crop Sci 47: 88–105

Prasad AS (2013) Biochemistry of zinc, vol 11. Springer Science & Business Media, Berlin

Prentice AM, Gershwin ME, Schaible UE, Keusch GT, Victoria LG, Gordon JI (2008) New challenges in studying nutrition disease interactions in the developing world. J Clin Invest 118:1322–1329

Prerna DI, Govindaraju K, Tamilselvan S, Kannan M, Vasantharaja R, Chaturvedi S, Shkolnik D (2021) Influence of nanoscale micro-nutrient α-Fe_2O_3 on seed germination, seedling growth, translocation, physiological effects and yield of rice (*Oryza sativa*) and maize (*Zea mays*). Plant Physiol Biochem 162:564–580

Prom-u-thai C, Rerkasem B, Yazici A, Cakmak I (2012) Zinc priming promotes seed germination and seedling vigor of rice. J Plant Nutr Soil Sci 175:482–488

Qamar S, Tantray AY, Bashir SS, Zaid A, Wani SH (2020) Golden rice: genetic engineering, promises, present status and future prospects. In: Roychoudhury A (ed) Rice research for quality improvement: genomics and genetic engineering. Springer, Singapore, pp 581–604

Rajjou L, Duval M, Gallardo K, Catusse J, Bally J, Job C, Job D (2012) Seed germination and vigor. Annu Rev Plant Biol 63:507–533

Raliya R, Tarafdar JC, Biswas P (2016) Enhancing the mobilization of native phosphorus in the mung bean rhizosphere using ZnO nanoparticles synthesized by soil fungi. J Agric Food Chem 64:3111–3118

Ramadas S, Vellaichamy S, Ramasundaram P, Kumar A, Singh S (2020) Biofortification for enhancing nutritional outcomes and policy imperatives. In: Gupta OP, Pandey V, Narwal S, Sharma P, Ram S, Singh GP (eds) Wheat and barley grain biofortification. Woodhead Publishing, pp 309–327

Ramzan Y, Hafeez MB, Khan S, Nadeem M, Batool S, Ahmad J (2020) Biofortification with zinc and iron improves the grain quality and yield of wheat crop. Int J Plant Prod 14:501–510

Rashid A, Harris D, Hollington P, Ali S (2004) On-farm seed priming reduces yield losses of mungbean (*Vigna radiata* L.) associated with mungbean yellow mosaic virus in north west Frontier Province of Pakistan. Crop Prot 23:1119–1124

Reddy NYA, Shaanker RU, Prasad TG, Kumar MU (2003) Physiological approaches to improving harvest index and productivity in sunflower. Helia 26:81–90

Rehman A, Farooq M, Cheema ZA, Wahid A (2012) Role of boron in leaf elongation and tillering dynamics in fine grain aromatic rice. J Plant Nutr 2:25–37

Rehman AU, Farooq M, Nawaz A, Ahmad R (2014) Influence of boron nutrition on the rice productivity, grain quality and biofortification in different production systems. Field Crops Res 169:123–131

Rehman A, Farooq M, Ahmad R, Basra SMA (2015) Seed priming with zinc improves the germination and early seedling growth of wheat. Seed Sci Technol 43:262–268

Rehman A, Farooq M, Naveed M, Nawaz A, Shahzad B (2018) Seed priming of Zn with endophytic bacteria improves the productivity and grain biofortification of bread wheat. Eur J Agron 94:98–107

Rehman A, Farooq M, Ozturk L, Asif M, Siddique KH (2018a) Zinc nutrition in wheat-based cropping systems. Plant Soil 422:283–315

Rehman A, Shahzad B, Ullah A, Nadeem F, Tanveer M, Sharma A, Lee DJ (2019) Seed priming-mediated improvement of plant morphophysiology under salt stress. Priming and pretreatment of seeds and seedlings: implications in plant stress tolerance and enhancing productivity in crop. Plants 205–217

Reis S, Pavia I, Carvalho A, Moutinho-Pereira J, Correia C, Lima-Brito J (2018) Seed priming with iron and zinc in bread wheat: effects in germination, mitosis and grain yield. Protoplasma 255: 1179–1194

Rice G (2016) "Golden Rice" GM VitaMin-a Rice

Rico CM, Majumdar S, Duarte M, Peralta JR, Gardea JL (2011) Interaction of nanoparticles with edible plants and their possible implications in the food chain. J Agri Food Chem 59:3485–3498

Roohani N, Hurrell R, Kelishadi R, Schulin R (2013) Zinc and its importance for human health: an integrative review. J Res Medical Sci 18:144

Rouhi A, Tajbakhsh M, Bernousi A, Saeedi M (2011) Study on effect of different primings on seed germination and seedling traits of different pea cultivars. J Agric For 90:1–8

Ruel-Bergeron JC, Stevens AG, Sugimoto JD, Roos FF, Ezzati M, Black RE, Kraemer K (2015) Global update and trends of hidden hunger, 1995-2011: the hidden hunger index. PLoS One 10: e0143497

Sardar R, Ahmed S, Yasin NA (2021) Seed priming with karrikinolide improves growth and physiochemical features of *Coriandrum sativum* under cadmium stress. Environ Adv 5:100082

Sautter C, Poletti S, Zhang P, Gruissem W (2006) Biofortification of essential nutritional compounds and trace elements in rice and cassava. Proc Nutr Soc 65:153–159

Sharifi R, Mohammadi K, Rokhzadi A (2016) Effect of seed priming and foliar application with micronutrients on quality of forage corn (*Zea mays*). Environ Exp Biol 14:151–156

Sharma MK, Bose B (2006) Effect of seed hardening with nitrate salts on seedling emergence, plant growth and nitrate assimilation of wheat (*Triticum aestivum* L.). Physiol Mol Biol Plants 12: 173–176

Singh NB, Singh D, Singh A (2015) Biological seed priming mitigates the effects of water stress in sunflower seedlings. Physiol Mol Biol Plants 21:207–214

Singhal RK, Bose B (2020) Wheat seedlings as affected by Mg $(NO_3)_2$ and $ZnSO_4$ priming treatments. World Sci News 144:13–29

Smolen S, Rozek R, Ledwozyw-Smolen I, Strzetelski P (2011) Preliminary evaluation of the influence of soil fertilization and foliar nutrition with iodine on the efficiency of iodine biofortification and chemical composition of lettuce. J Elem 16:110–128

Smoleń S, Wierzbińska J, Sady W, Kołton A, Wiszniewska A, Liszka-Skoczylas M (2015) Iodine biofortification with additional application of salicylic acid affects yield and selected parameters of chemical composition of tomato fruits (*Solanum lycopersicum* L.). Sci Hortic 188:89–96

Srivastava AK, Bose B (2012) Effect of nitrate seed priming on phenology, crop growth rate and yield attributes in rice (*Oryza sativa* L.). Vegetos-An Int J Plant Res 25:174–181

Srivastava AK, Siddique A, Sharma MK, Bose B (2017) Seed priming with salts of nitrate enhances nitrogen use efficiency in rice. Vegetos 30:199–216

Srivastava AK, Wu QS, Mousavi SM, Hota D (2021) Integrated soil fertility management in fruit crops: an overview. Int J Fruit Sci 21:413–439

Stewart CP, Dewey KG, Ashoran P (2010) The undernutrition epidemic: an urgent health priority. Lancet 375:282

Stimec M, Kobe H, Smole K, Kotnik P, Sirca-Campa A, Zupancic M, Battelino T, Krzisnik C, Fidler Mis N (2009) Adequate iodine intake of slovenian adolescents is primarily attributed to excessive salt intake. Nutr Res 29:888–896

Sundaria N, Singh M, Upreti P, Chauhan RP, Jaiswal JP, Kumar A (2019) Seed priming with iron oxide nanoparticles triggers iron acquisition and biofortification in wheat (*Triticum aestivum* L.) grains. J Plant Growth Regul 38:122–131

Taiz L, Zeiger E (2010) Plant physiology, 5th edn. Sinauer Associates, Sunderland, MA

Talsma E (2014) Yellow cassava: efficacy of provitamin A rich cassava on improvement of vitamin A status in Kenyan schoolchildren. Dissertation for Wageningen University, Cham

Tanumihardjo SA (2013) Vitamin A and bone health: the balancing act. J Clin Densitom

Triveni U, Luther MM, Ramanamurthy KV, Prasad T, Mukundarao CH, Patro T (2021) Influence of seed priming with nano iron oxide on germination and seedling growth of mungbean. Int J Plan Soil Sci 33:10–14

Tulchinsky TH (2010) Micronutrient deficiency conditions: global health issues. Pub Health Rev 32:243–255

Vakili M, Abedi P, Sharifi M, Hosseini M (2013) Dietary diversity and its related factors among adolescents: a survey in Ahvaz-Iran. Glob J Health Sci 5:181–186

Valença AWD, Bake A, Brouwer ID, Giller KE (2017) Agronomic biofortification of crops to fight hidden hunger in sub-Saharan Africa. Glob Food Sec 12:8–14

Vasconcelos MW, Gruissem W, Bhullar NK (2017) Iron biofortification in the 21st century: setting realistic targets, overcoming obstacles, and new strategies for healthy nutrition. Curr Opin Biotechnol 44:8–15

Wakeel A, Farooq M, Bashir K, Ozturk L (2018) Micronutrient malnutrition and biofortification: recent advances and future perspectives. Plant Micronutrient use efficiency 225–243

Watts MJ, O'Reilly J, Maricelli A, Coleman A, Ander EL, Ward NI (2010) A snapshot of environmental iodine and selenium in La Pampa and San Juan provinces of Argentina. J Geochem Explor 107:87–93

White PJ, Broadley MR (2009) Biofortification of crops with seven mineral elements often lacking in human diets- iron, zinc, copper, calcium, magnesium, selenium and iodine. New Phytol 182: 49–84

White PJ, Bradshaw JE, Finlay M, Ramsay G, Hammond JP, Broadley MR (2009) Relationship between yield and mineral concentrations in potato tubers. HortSci 44:6–11

WHO (2012) The world health report. World Health Organization, Geneva

WHO (2014) Guideline: fortification of food-grade salt with iodine for the prevention and control of iodine deficiency disorders. WHO, Geneva

Yadava DK, Hossain F, Mohapatra T (2018) Nutritional security through crop biofortification in India: status & future prospects. Ind J Med Res 148:621–631

Zhou ZY, Wan G (2007) The public distribution systems of foodgrains and implications for food security: a comparison of the experiences of India and China. In: Guha-Khasnobis B, Acharya SS, Davis B (eds) Food insecurity, vulnerability and human rights failure. Palgrave Macmillan, London, pp 106–127

Zibai S, Rahimi A, Dashti H (2012) Effects of seed priming on growth, chlorophyll content, relative water content and dry matter distribution of safflower (*Carthamus tinctorius* cv. Gholdasht) under salinity stress. Isfahan Univ Technol J Crop Prod Process 2(5):47–59

Zimmermann MB, Andersson M (2012) Assessment of iodine nutrition in populations: past, present, and future. Nutr Rev 70:553–570

Zulfiqar U, Maqsood M, Hussain S, Anwar-ul-Haq M (2020) Iron nutrition improves productivity, profitability, and biofortification of bread wheat under conventional and conservation tillage systems. J Soil Sci Plant Nutri 20:1298–1310

Chapter 12
Biofortification Through Seed Priming in Food Crops: Potential Benefits and Future Scope

Umair Ashraf, Munazza Kiran, Muhammad Naveed Shahid, Shakeel Ahmad Anjum, and Imran Khan

Abstract Food fortification with mineral micronutrients has been recognized as an effective strategy to address micronutrient deficiencies and/or mal-nutrition problems, which can have severe consequences on human health. Biofortification involves the improvement of micronutrient content of food crops through natural means has gained attention as a sustainable approach to improve the micronutrient status of populations, particularly in resource-limited countries. Scientists have developed multiple techniques, such as agronomic practices, genetic modifications, and transgenic approaches to enhance the micronutrient content of food crops. A promising method for increasing the uptake of vital micronutrients by plants is seed priming, which will increase the amount of nutrients in the edible crop plants. Seed priming, especially nano-priming can potentially be used to grow the micro-nutrient fortified crops. This chapter discusses the value of biofortification, how seed priming enables it, and the many seed priming methods that may be employed to enhance food crop development, yield, and nutritional status. Overall, biofortification of grain crops through seed priming is a feasible option, however its optimization, lab to field scale testing and subsequent application on large scale is needed to achieve sustainable and long-term improvements in micronutrient status of the food crops.

Keywords Biofortification · Seed priming · Nano-priming · Nutrient · Malnourishment · Grains

U. Ashraf (✉) · M. Kiran · M. N. Shahid
Department of Botany, Division of Science and Technology, University of Education, Lahore, Punjab, Pakistan
e-mail: umair.ashraf@ue.edu.pk

S. A. Anjum · I. Khan
Department of Agronomy, University of Agriculture, Faisalabad, Pakistan

M. Hasanuzzaman et al. (eds.), *Mineral Biofortification in Crop Plants for Ensuring Food Security*, https://doi.org/10.1007/978-981-99-4090-5_12

12.1 Introduction

The production of nutrient-enhanced food crops developed employing cutting-edge biotechnology approaches, traditional plant breeding, and agronomic practises is referred to as biofortification and/or biological fortification. According to the United Nations Food and Agriculture Organization, the 98% of the world's malnourished population, reside in developing countries, especially in Africa and South Asia (McGuire et al. 2015). Additionally, two billion people experience 'hidden hunger' globally, owing to regular consumption of diet deficient in essential micronutrients. On the other hand, there is growing concern over nutrition as well (Hodge 2016). The lack of vitamins and minerals, sometimes known as micronutrient malnutrition and/or hidden hunger, has a negative effect on human health. Young children and pregnant women are more susceptible to malnutrition, whereas the women may suffer variety of complications with childbirth, growth and developmental abnormalities (Black et al. 2013). Despite the fact that seven minerals i.e., iodine, calcium, magnesium, selenium, zinc, and copper are necessary for our body (White and Broadley 2009). While many vitamins and minerals are lacking in human diets, iron, zinc, vitamin A, and vitamin B9 (folate) deficiencies are among the most severe (FAO 2017). As a key strategy to alleviate micronutrient deficiency, biofortification the process of increasing the levels of micronutrients in food crops require serious attention and needs to be promoted in the public interest (Beall et al. 2017).

In addition, iron and zinc deficiencies are among the most prevalent micronutrient deficiencies worldwide (Stoltzfus 2001; Wessells et al. 2012). Zinc deficiency can damage immunity and development (Hambidge 2000; Welch and Graham 2002). Micronutrient deficiency have long-lasting effects on growth, brain development, and immune system performance (White and Broadley 2009). Therefore, this "hidden hunger" has a negative impact on economic growth, human health, and attempts to reduce poverty. Food consumption is rising dramatically due to a rapidly growing population and changing habits, yet there is a limited amount of land available for cultivation, hence there is an intensive use of natural resources and food demand is dramatically rising due to a rapidly growing population and changing food demands. Especially in underdeveloped nations, agricultural interventions to increase the nutritional value of consumable and fodder crops feeding livestock could help to improve human nutrition (Kiran et al. 2022).

Right now, our agricultural system exclusively focuses increasing grain yield and crop productivity rather than promoting human health. Such agricultural practices are resulting in the deficiency of micronutrients in dietary grains and causing malnutrition problems within communities. To combat hidden hunger, it is therefore necessary to switch from producing more food crops in larger quantities to producing enough nutrient-rich food crops, especially in underdeveloped and impoverished countries where diets are primarily made up of staple foods that are low in micronutrients (Khush et al. 2012).

On the other hand, the general public has historically received vitamins and minerals through nutritional supplementation programs, however this approach fails to meet the objectives established by the international health organizations because the programs were proved not much effective as it could not fulfill the goals and objectives of international health organizations. In addition, supplementation of micronutrients through medicinal and/or pharmacological means had its own constraints e.g., inability of the poor people to purchase the multi-vitamin pills, access to healthcare and market infrastructure, and lack of knowledge about micronutrient deficient (Perez-Massot et al. 2013).

Enhancing the nutrients in food crops and creating more nutrient-rich crop varieties are two ways nutrition-sensitive agricultural intervention called biofortification seeks to improve the nutritional condition of populations with limited resources (Bouis and Welch 2010). Using conventional breeding, HarvestPlus, a global network of biofortification research and implementation organizations, increases the nutritional value of food crops in the most practical manner (Bouis and Saltzman 2016). Based on the dietary intake of targeted populations, nutritional losses during processing and storage, as well as target nutrient availability to body, micronutrient objectives for breeding biofortified crops are developed (Hotz and McClafferty 2007; Saltzman et al. 2017).

Therefore, biofortification of crops through agronomic and breeding approaches offers a long-term and sustainable strategy for providing people with crops that are high in micronutrients (Bouis and Welch 2010). Additionally, consumers have access to nutrient-dense crops with higher bioavailable concentrations of crucial micronutrients through conventional techniques used in agriculture and the food industry. With little access to a wide range of meals, supplements, and fortified foods, this provides a useful strategy for dealing with the impoverished and malnourished group families. Since fortificants no longer need to be purchased and added to food supplies during processing, the development of biofortified crops reduces the cost of biofortification to a one-time outlay that provides an affordable, long-lasting, and sustainable solution to the problem of hidden hunger (Hefferon 2016). Furthermore, a substantial rise in population in the developing world making difficult to achieve food security goals under changing climate scenario (Das et al. 2013). In order to generate high-yielding biofortified crops with increased nutrition for better human health, the World Health Organisation and the Consultative Group on International Agricultural Research (CGIAR) have this as one of their main goals.

12.2 Crop Biofortification-Need of the Day

For regular growth and development, the human body needs trace amounts of sodium (Na), potassium (K), calcium (Ca), magnesium (Mg), phosphorus (P), chlorine (Cl), and Sulphur (S), as well as vitamin A, iron (Fe), zinc (Zn), copper (Cu), manganese (Mn), iodine (I), molybdenum (Mo), cobalt (Co), and selenium (Se) (Prashanth et al. 2015). Together, these nutrients are essential to human health and determinant of physical and mental growth (White and Broadley 2005). Numerous micronutrients serve as cofactors to enable a variety of enzymes in the body for normal functioning, regulating activities and other metabolic processes (Welch and Graham 2004). Normally, agricultural crops provide the majority of human nutrition (Graham et al. 2001), but people who only ate grains like rice, wheat, cassava, and maize did not consume enough vitamin A, calcium (Ca), manganese (Mg), copper (Cu), zinc (Zn), iron (Fe), iodine (I), calcium (Ca), or selenium (Se). These nutritionally deficient agricultural goods cannot support healthy lives and may instead cause illness, increased disability and morbidity, as well as mental and physical retardation or development, diminished livelihoods, and a reduction in the socio-economic growth of the country (Chizuru et al. 2003). Micronutrient deficiency in children from prenatal development to age 4 is linked to childhood stunting, which is common in many underdeveloped nations (Branca and Ferrari 2002). Over 30% of people worldwide are malnourished and have unmet hunger, among which 43% of children and 38% of pregnant women suffer from micro-nutrient scarcities (Stevens et al. 2013). Compared to developed countries, the developing countries have a higher prevalence of anaemia especially Africa and South-East Asia whereas iron deficiency is responsible for almost half of anaemia cases (Brotanek et al. 2005). Uneven nutrient distribution among various plant sections is a crucial factor to take into account (Zhu et al. 2007a, b). For instance, polished rice grain has a low iron level compared to rice leaves. In addition to malnourishment, there is rising evidence of over nutrition, which can result in issues with a high rate of diabetes and overweight. As a result, bio fortification also aims to increase the number of required micronutrients in edible plant parts. Crops that have been biofortified should have higher mineral content, greater vitamin content, higher amounts of key amino acids, better fatty acid composition, and higher antioxidant levels (Hirschi 2009). Biofortification of food crops can deliver all the necessary nutrients for good health as well as enough calories to meet energy needs (Welch and Graham 1999).

For example, sorghum variants have been examined for high levels of protein, minerals, lutein, zeaxanthin, and beta-carotene which has been explored along with the possibility of producing sorghums high in beta-carotene and micronutrients (Guo et al. 2022). The iron (Fe) and zinc (Zn) contents of sorghum germplasm have showed significant genetic heritability. The Fe-rich sorghum lines i.e., ICSH 14002, ICSR 14001and hybrids (ICSR 196 ×ICSA 661, ICSR 94 × ICSA 318, IS 3760 ×ICSA 336) are developed by International Crops Research Institute for the Semi-Arid Tropics (ICRISAT) in India (Mansour et al. 2021). In Nigeria, new high-nutritional (Fe) varieties of sorghum i.e., 12KNICSV-22 and 12KNICSV-188 have been released and could help the malnourished inhabitants, particularly children. Interestingly, three times more iron is present in one of the novel kinds (12KNICSV-188) than in commonly produced sorghum (ICRISAT, HarvestPlus). Moreover, in India, the hybrid ICMH 1201 (Shakti-1201) and the "Dhanashakti" variety of pearl millet is biofortified with iron and zinc (Govindaraj et al. 2020) Additionally, Nirmal-7, ICMH 1202, and ICMH 1301, are presently participating in advanced farm trials. There have been reports of a number of commercial cultivars, their offspring, and hybrids with high iron and zinc contents (Suruthi et al. 2019).

HarvestPlus and International Center for Agricultural Research in the Dry Areas (ICARDA) has given the go-ahead for the bioaugmentation of zinc and iron using the breeding process and genetic variety kept in gene banks (Pfeiffer et al. 2018). Up to this point, the HarvestPlus biofortification program of ICARDA has released high Fe and Zn lentil varieties in five counties in Bangladesh. All varieties of lentil have been examined for differences in Se content in Ethiopia, L4704, and Pusa Vaibhav in India, Khajurah-1, Khajurah-2, ILL 7723, Shital, Simal and Sisir Shekhar, in Nepal (Sarker and Erskine 2006). Moreover, cowpea, commonly referred to as 'poor man's meet' owning to its high content of proteins as also been iron-biofortified through breeding techniques (Pfeiffer et al. 2018). Moreover, early maturing as well as iron and zinc enriched cowpea verities has been developed by the Govind Ballabh Pant University, Pantnagar, India in collaboration with HarvestPlus.

Furthermore, a sufficient amount of genetic variation in zinc content has been found in the gene pool of *Brassica oleracea* (Pongrac et al. 2020). The Indian Agricultural Research Institute has developed the orange colored (β-carotene rich) cauliflower variety (Garg et al. 2018). There are now several varieties of colorful cauliflower that are well-known on a global scale, with colors like purple and orange that are full in anthocyanin and beta-carotene, respectively. On the other hand, in under-developed countries, particularly in Latin America, Africa, and the Caribbean, cassava is a major vegetable. With the help of the International Institute of Tropical Agriculture and HarvestPlus, provitamin A deficiency has been targeted in Africa. Different vitamin A-fortified cassava varieties have been developed in Nigeria (Ilona et al. 2017). Since there are many genotypic variations in cassava for proteins, carotene, and minerals (Fe and Zn), a crop with a higher nutritional value has been produced (Ijaz et al. 2012).

The tomatoes, a large source of vitamins A and C, have a high commercial value. For specific traits, a wild population with a wide genetic range of tomatoes has been extensively researched and used in breeding (Delian et al. 2017). By using a traditional breeding approach, the tomato "Sun Black" a fruit with high anthocyanin concentration in the peel, anthocyanin biofortification, and rich purple fruit coloring was produced (Blando et al. 2019). Israel has reported producing another variety called "Black Galaxy" using a similar methodology (Delian et al. 2017).

In addition, banana breeding is challenging because commercial types are sterile triploids and the fertile groups may be highly cross-incompatible (Babu et al. 2018). In order to address this issue, HarvestPlus and Bioversity International (BI) conducted extensive screenings of different banana germplasm in the Democratic Republic of the Congo and Burundi for the detection of high quantities of provitamin A and developed five banana varieties (Wilson and Tenkouano 2019).

12.3 Seed Priming: A Viable Solution

Commercial seeds are regularly primed by seed technologists to increase seed vigor, including germination potential and stress tolerance. Operators of seed banks who require better ex situ conservation techniques for their collections of genetic material may also find priming useful. Different priming techniques can be used, each of which initiating the so-called 'pre-germinative metabolism' in accordance with the plant type, seed morphology, and physiology. The physiological manifestation during early stages of seed imbibition involves the seed repair response which is crucial to maintain genome integrity, ensure correct germination, and promote seedling development (Paparella et al. 2015; Chen et al. 2021).

To meet the current need for extraordinary standards in the agriculture sector, higher seed quality has become a priority. Achieving early uniform plantlet emergence is imperative for crop performance because low seed sprouting rates typically expose seedlings to unfavorable weather conditions (Osburn and Schroth 1989). A renowned method of improving seed quality is priming (Ullah et al. 2019; Rehman et al. 2019). Priming causes seeds to germinate more quickly, which increases agricultural yields and stress resistance to high levels. All of these characteristics that increase product competitiveness are directly related to seed vigor, a complex agronomic attribute that is influenced by numerous genetic and environmental factors (Jisha et al. 2013). Priming regulates seed rehydration process to start the metabolic reactions that are typically active during pre-germination period but prevents the seed from progressing to complete germination. Therefore, priming treatment needs to end before desiccation tolerance declines. It has been generally stated that seed priming has advantages. Primed seeds exhibit synchronous and quick emergence in addition to decreased photo- and thermo-dormancy, a broader range of germination temperatures, a greater ability to fend off weeds and diseases, and a more effective ability to regulate water usage and development period (Hill et al. 2008).

Effectiveness of the seed priming are linked with the plant species/genotype, seed physiology, seed vigor and the priming method (Parera and Cantliffe 1994; Cantliffe 2003). For instance, vegetable seeds such as carrot, leek, and onion seeds, celery, lettuce, endive, pepper and tomato can be primed in a short time span (Dearman et al. 1987; Parera and Cantliffe 1994; Di Girolamo and Barbanti 2012). In the flower seed industry, priming helps to improve product quality and is widely employed on premium Petunia hybrida L. Kinds (Di Girolamo and Barbanti 2012). Seed priming is also advantageous for herbs like *Salvia splendens* L. and *Rosmarinus officinalis* L (Di Girolamo and Barbanti 2012). Although the benefits of seed priming in cereals are exceptionally higher, however, it is more challenging to accomplish (Murungu et al. 2004). Additionally, priming offers ideal circumstances to fast seed germination when required by the brewing and cereal grain industries (Yaldagard et al. 2008). The established priming therapies for horticultural crops such as EasyPrime, Emergis, and PROMOTORTM, ensure enhanced seed development, quicker and more uniform seedling emergence, and a decreased frequency of aberrant plantlets. While some products (such as Advantage and Xbeet) have been developed in order to increase seed germination and seedling emergence of elite vegetables under stressful conditions, other products specifically address the issues of seed thermo- and photo-dormancy.

12.4 Techniques of Biofortification

The major techniques or methods by which crops can be biofortified are:

12.4.1 Agronomic Practices

This involves the application of fertilizers in plants that are grown in soil conditions with poor or deficient micronutrients or minerals.

12.4.2 Conventional Plant Breeding

Traditional breeding methods are used to produce desired trait in crops. This involves crosses which produce sufficient genetic variations in them.

12.4.3 Genetic Engineering

This involves inserting part of DNA containing gene of interest, into the genome of a host organism. As a result, new or different characteristics are introduced such as disease resistant in plants.

12.5 Biofortification Through Agronomic Approach

Given that conventional methods have many drawbacks, biofortification is an efficient long-term strategy for improving crop plants' nutritional status (Zhu et al. 2007a, b). Agricultural cultivars with better soil mineral element uptake capacity have been developed in an effort to increase agricultural yields (Zhu et al. 2007a, b; White and Broadley 2009). On the other hand, supplementation of micronutrients through fertilizer is an easy method to harvest mineral enriched food crops (Prasad et al. 2014). In general, agronomic biofortification is achieved by four different ways i.e., seed priming, seed coating, and soil application and foliar application of mineral fertilizers. In this chapter, we have focused the agronomic biofortification through seed priming.

12.5.1 Seed Priming

Heydecker et al. (1973) initially suggested the phrase "seed priming". It was reported as a useful method for improving the vigour and development of plants by promoting more consistent seedling emergence. Typically, seed priming is a regulated pre-sowing hydration approach in which the metabolic activities of the seed are permitted to continue before actual germination (Farooq et al. 2006a, b, c). Priming initiates pre-germination metabolic and biochemical activities without radical emergence (Rouhi et al. 2011a, b; Paparella et al. 2015). Seed priming brings uniformity in seed germination and emergence that is associated with enzyme activation, cell repairing mechanism, improved antioxidant defense mechanisms, and synthesis of proteins (Afzal et al. 2008; Jafar et al. 2012; Hussain et al. 2018). Seed priming also enhances the osmolyte accumulation e.g., proline, polyamines and glycine-betaine through altered metabolic processes (Delavari et al. 2010; Khan et al. 2018).

12.5.2 Seed Priming Techniques

Currently, different priming techniques are being used in order to improve seed characteristics, crop productivity and to induce abiotic stress tolerance. These include hydro-priming, osmo-priming, solid matrix priming (SMP), nutria-priming, chemo-priming, thermo-priming, and bio-priming, plant growth regulator (PGR)-based priming (Paparella et al. 2015; Panuccio et al. 2018; Majda et al. 2019). Each technique has its own limitations and advantages, but the efficiency of seed priming is highly dependent on seed type and the priming technique (Tarquis and Bradford 1992). In order to stimulate seed germination, a specific methodology is chosen while accounting for the chosen plant species, seed characteristics, and priming techniques (Ellis and Butcher 1988; Hill et al. 2008; Ibrahim 2016; Paparella et al. 2015).

12.5.3 Methods of Seed Priming

The manner of priming used, along with other elements like the seed type, time, and priming agent, all affect how effectively seeds are primed (Arif et al. 2008; Farooq and Wahid 2009; Nascimento 2003). Other physical and chemical factors that can influence priming and effect germination rate and timing, seedling vigor, and subsequent plant growth include osmotic potential, temperature, the presence or absence of light, aeration, and seed quality (Hussain et al. 2006; Varier et al. 2010). The general types of seed priming are as follows:

12.5.3.1 Hydro-priming

Before broadcasting, hydro-priming required air drying of seeds to a specific moisture level after a pre-determined amount of time in the water (Singh et al. 2015a, b). Seeds are allowed to imbibe water and are helped to initiate pre-germination metabolic activities (Pill and Necker 2001). However, hydropriming may cause uncontrolled water uptake and consequently radical projection, so it is essential to maintain the ideal humidity and temperature (Taylor et al. 1998). Hydro-priming can be achieved through "drum priming" (Rowse 1991). This approach consists of a drum, having seeds, and the drum is linked to a boiler which produces vapors. The vapors condense into water droplets upon entering the drum. During drum priming, changes in the relative seed mass, water volume, and amount of time needed for hydration of the seed are precisely controlled (Warren and Bennett 1997).

12.5.3.2 Halo-priming

Halo-priming, which boosts germination, involves soaking seeds in inorganic solutions including calcium chloride ($CaCl_2$), potassium nitrate (KNO_3), sodium chloride (NaCl), and calcium sulphate ($CaSO_4$). Halo-priming is critical for all phases of plant development, including seed germination, seedling emergence, and plant growth. Additionally, this method is helpful for acclimatizing plants to various stress situations (Sedghi et al. 2010).

12.5.3.3 Chemical Priming

Chemo-priming involves the use of primers such chitosan, choline, putrescine, paclobutrazol, zinc sulphate, copper sulphate, potassium hydrogen phosphate, and selenium. These substances increase stress tolerance and growth performance (Jisha et al. 2013). These substances increase the ability to withstand drought through osmoprotection generation, ROS detoxification, protein synthesis, and ionic homeostasis (Marthandan et al. 2020). Demir et al. (2012) reported improvement in seedling strength and emergence of pepper and salvia when their seeds were chemo-primed with butanolide. Drought tolerance was observed with mannose seed priming. It helped to enhance the antioxidants levels and reduce oxidative injuries (Hameed and Iqbal 2014). In addition, the yield of late-sown wheat in the semi-arid conditions was improved the seed priming with thiourea (Chattha et al. 2017) whereas growth attributes of aromatic rice were substantially reduced with seed treatment of paclobutrazol (Huang et al. 2019).

12.5.3.4 Nutrient Priming

Before planting, nutrients are primed into seeds by saturating them for a predetermined amount of time (Shivay et al. 2016). Water regulation in seeds is enhanced through micronutrient seed priming during germination (Singh 2007). For example, yield of mung bean increased when its seeds were primed with sodium molybdate dihydrate (Umair et al. 2011). Similarly, seed priming with macronutrients is also an effective technique for integrated nutrient management (Rakshit et al. 2013). Seed priming with potassium (K) enhancedstress tolerance in crops (Cakmak 2005). Nutri-priming with zinc solutions improved the grain yield of wheat and chickpea (Arif et al. 2007). Zinc priming is helpful in improving plant morphological traits and its tolerance to several environmental stresses (Marschner 1995; Harris et al. 2008; Shivay et al. 2016). Additionally, ascorbic acid-based seed priming increased the germination of *Agropyron elongatum* under salt stress (Tavili et al. 2009).

12.5.3.5 Osmo-priming

Osmo-priming is the process of soaking seeds in known-concentration of osmotic solutions (Heydecker et al. 1973). It involves controlled hydration of seed which allows certain pre-germinative physiological and biochemical activities to proceed without radicle emergence (Halmer 2004). Osmotica such as polyethylene glycol, sugar, mannitol, and sorbitol are generally used for lowering the rate of water uptake. Various salts such as sodium nitrate, magnesium chloride, sodium chloride, and potassium nitrate are also used as osmo-priming agents (Rehman et al. 2020). During osmo-priming, seeds are imbibed to a regulated amount. Increased water ingress causes ROS buildup and oxidative damage to cellular processes in seeds. Osmo-priming shields the cell from oxidative damage by delaying the entry of water into the seed, which lowers the formation of ROS. Furthermore, osmo-priming of wheat seeds resulted in early emergence and improved the competitive ability against weeds, as compared with non-primed seeds (Ullah et al. 2018).

12.5.3.6 Hormonal-Priming

This technique uses hormone-based seed priming. Numerous phytohormones that directly affect seed metabolism include abscisic acid, salicylic acid, ascorbic acid, cytokinin, auxin, gibberellin, kinetin, ethylene, and polyamines (Rehman et al. 2020).

12.5.3.7 Solid Matrix Priming

Matri-conditioning, also known as solid matrix priming (SMP), is a pre-sowing procedure in which seeds are combined with predetermined quantities of a solid substance and water (Taylor et al. 1994). Till equilibrium, when priming is most likely to take place, both the seed and matrix used in SMP compete for the available water. In order to complete the pre-germination processes within the seed, seeds are removed from the matrix and thoroughly washed before drying. Typically, solid matrix priming involves mixing seeds with solid organic or inorganic ingredients like vermiculite, calcined clay, calcium silicate, and specified amounts of water in a sealed container that allows air circulation and prevents excessive evaporation (Harman and Nelson 1994; Rogis et al. 2004; Hacisalihoglu 2007; Ermis et al. 2016).

12.5.3.8 Thermo-priming

This method improves seed germination compared to placing seeds at a constant temperature by subjecting the seed to varying temperatures at various time intervals before planting (Shin et al. 2006; Markovskaya et al. 2007). The thermo-sensory pathway is stimulated by changes in ambient temperature, which causes changes in blooming time (Franklin 2009). The kind of temperature, including cold (Runkle et al. 1999; Garner and Armitage 2008) and hot (Khalil and Rasmussen 1983), can also affect and/or alter the period of blooming. Thermo-priming is an effective strategy to increase germination efficiency in challenging climatic conditions because it can disrupt seed dormancy (Huang et al. 2002). Additionally, alternating the temperature of pre-treated cucumber and melon seed increased both crops' productivity (Markovskaya et al. 2007).

12.5.3.9 Plant Extract Priming

Certain allelochemicals, such as phenolic compounds, alkaloids, flavonoids, terpenoids, steroids, and saponins, can stimulate or inhibit plant growth (Narwal 1994). Alkaloids, saponins and phenolic compounds provide plant protection against pathogens by improving antioxidant defense (Satish et al. 2007). Application of certain physiologically active substances activate the embryo and other associated structures that result in higher water uptake with improved vigor index (Rangaswamy et al. 1993). *Azadirachta* leaves, for example, are abundant in terpenoids, steroids, flavonoids, and antiquinone, whereas the leaves of *Chlorophytum* are high in saponin and alkaloids (Raphael 2012; Chakraborthy et al. 2014) that may be employed as priming substrates. Seed priming with *Azadirachta*, *Chlorophytum*, and *Vinca* extracts reduced the mortality rate and increased the seedling vigor in tomato seeds (Prabha et al. 2016).

12.5.3.10 Bio-priming

The controlled hydration method known as "bio-priming" uses advantageous microorganisms to speed up the pre-germination preparation steps (Sukanya et al. 2018). Callan et al. (1990) were the ones to initially introduce the bio-priming process. The process of bio-priming involves pre-soaking seeds in bio-agent formulations in certain ratios. The seeds were then incubated at 25–35 °C for 48 h while being covered with a damp jute bag (sack) to maintain a high humidity level. As the bioagent adheres to the seed surface, a shield forms. This seed priming treatment method combines bacterial inoculations with seed imbibition (Callan et al. 1990; Lutts et al. 2016) and thus protects the seed from soil and seed borne pathogens (Reddy 2012). Additionally, bio-priming increases crop productivity, resistance to biotic and abiotic stresses, seed quality, seedling vigour, and seedling vigour (Rakshit et al. 2015; Sukanya et al. 2018; Bisen et al. 2015). The greatest solution to the issue of seed microbial infection that may arise during seed priming due to moisture is bio-priming using hostile bacteria (Reddy 2012).

12.5.3.11 Nano-priming

Less than 100 nm-sized nanoparticles are the focus of nanotechnology. The future of nanotechnology in several industries, including food and agriculture, seems bright (Fraceto et al. 2016). According to Upadhyaya et al. (2017), nano-materials in agriculture offer an alternative method for reducing the overuse of chemical fertilizers. Numerous crops benefit from improved seed germination, seedling development, and vigor when seeds are primed with calcium-phosphate, silica, zinc oxide, and silver nanoparticles (Ghafari and Razmjoo 2013). The effectiveness of the seed's ability to absorb nutrients and water rises because to nano-priming (Dutta 2018). Similarly, early growth and associated biochemical mechanisms of indica rice were substantially improved with ZnO nanopriming (Li et al. 2021). The impacts of nano-priming on different crops have been summarized in Table 12.1.

12.5.3.12 Seed Priming with Physical Agents

Magnetic fields, UV radiation, X-rays, microwaves and gamma radiation, are physical agents that can prime seeds (Bilalis et al. 2012). According to several studies, magnetic seed priming increases germination rate, seedling vigour, biomass, and tolerance to various environmental conditions. ROSs have been shown to decrease when antioxidant enzyme activity increases (Bhardwaj et al. 2012; Araujo et al. 2016). The properties of ionizing radiation, on the other hand, depend on the dose and intensity. Low doses of gamma radiation (less than 10 Gy) are effective. Early dormancy breakdown and improved seed germination are the results of certain alterations in the hormonal network of plant cells (Qi et al. 2015). Mechanical

Table 12.1 The impact of nano-priming in different crops

Crop	Nanoparticles	Concentration	Experiment type	Remarks	References
Wheat	ZnO nanoparticles (NPs)	0, 25, 50, 75, and 100 mg L^{-1}	Pot	Substantial increase in elemental Zn was found in root, shoot and grains of wheat applied with ZnO NPs than control (without ZnO NPs)	Munir et al. (2018)
Wheat	Silicon nanoparticles	0, 300, 600, 900, and 1200 mg L^{-1}	Pot	Seed priming with silicon NPs improved the growth and photosynthetic pigments and antioxidant defense but reduced the Cd contents by 11–60% (in roots), 10–52% (in shoot), and 12–75% (in grains) of wheat	Hussain et al. (2019)
Maize	TiO_2 nanoparticles	0, 40, 60, and 80 ppm	Pot	Seed priming with TiO_2 nanoparticles enhanced germination percentage and early growth traits of maize, biomass and antioxidant defense	Shah et al. (2021)
Rice	Na-selinite, Na-selenate, and ZnO nanoparticles	50 μmol Na-selenite, 50 μmol Na-selenate, each and 10 μmol ZnO NPs (individual applications and in combinations)	Pot and field	Seed priming with combined selenium and Zn nanoparticles improved early growth, seedling vigor, chlorophyll and protein contents, leaf area, crop growth,	Adhikary et al. (2022)

(continued)

Table 12.1 (continued)

Crop	Nanoparticles	Concentration	Experiment type	Remarks	References
				nutrient (N, P, K, B, Zn) acquisition i.e., uptake of nutrients (N, P, K, B, Zn and Si), and yield of rice	
Rice	ZnO nanoparticles	0, 25, 50, and 100 mg L^{-1}	Petri dish culture	ZnO nanoparticle-based seed priming improved by the seedling growth and associated physio-biochemical traits in in rice under Cd stress	Li et al. (2021)
Wheat	ZnO nanoparticles	0, 50, 100 and 500 mg L^{-1}	Field	ZnO nano priming improved the growth biomarkers and induced alterations in photosynthesis under salt stress conditions	Abou-Zeid et al. (2021)
Sun flower	Greenly Synthesized Sulfur Nanoparticles	0.0, 12.5, 25, 50, 100, and 200 μM	Pot	Seed Priming with Greenly Synthesized Sulfur Nanoparticles enhance the sunflower seedlings' tolerance	Ragab and Saad-Allah (2021)
Corn	Nanoscale Synthetic Zinc Oxide	0, 20, 40, 80, 160 mg L^{-1}	Field	Nanoscale synthetic zinc oxide improved by 17%, 25% and 12% higher values than control for germination, root length, and dry biomass production	Esper Neto et al. (2020)
Mustard	Carbon nanotubes		Field	Seed priming with carbon	Dhingra et al. (2022)

(continued)

Table 12.1 (continued)

Crop	Nanoparticles	Concentration	Experiment type	Remarks	References
	and silicon dioxide nanoparticles	0, 25, 50, 75, 100 and 125 μg ml^{-1}		nanotubes and silicon dioxide nanoparticles improved by 17.26% (leaf petiole length), 5.69% (number of seeds) 11.89% (leaf length) and 10.58% (length for main inflorescence)	
Lemin Balm (*Melissa officinalis* L.)	Silicon nanoparticle	0, 100, and 500 mg L^{-1}	Pot	Silicon nanoparticle-mediated seed priming improved by 57.8% (plant height) and 51.1% (tissue height)	Hatami et al. (2021)
Chickpea	MgO nanoparticles	0, 10, 50, 100, and 150 μg ml^{-1}	Pot	MgO nanoparticles priming promoted 171, 150, 132, and 142% (in shoot length) and 217, 236, 219 and 328% (in root length)	Sharma et al. (2022)
Wheat	Zinc oxide nanoparticles	0, 5, 10, 15 and 20 mg L^{-1}	Pot	Priming with zinc oxide nanoparticles improved by 30% (in root length) and 54% (in shoot length)	Rai-Kalal Jajoo (2021)
Wheat	Gibberellic acid and titanium dioxide nanoparticles	0, 100, 200, 400, 600 mg kg^{-1}	Pot	Gibberellic acid and titanium dioxide nanoparticles improved by 17%, 34%, 40%, 37% and 42% (plants height, spike length, roots,	Alharby et al. (2021)

(continued)

Table 12.1 (continued)

Crop	Nanoparticles	Concentration	Experiment type	Remarks	References
				grains and shoots)	
Pea	Titanium dioxide nanoparticles	0 and 50 mg L^{-1}	Pot	Titanium dioxide nanoparticles improved by 10% (embryonic axis biomass)	Basahi (2021)
Summer savory (*Satureja hortensis* L.)	silver nanoparticles	0, 40, 60, and 80 ppm	Pot	Effect of silver nanoparticles improved by 1% level of stem length and root length	Nejatzadeh (2021)
Wheat	Zinc oxide nanoparticles	0, 2500, 5000, 7500, 10,000, and 15,000 ppm	Pot	Effect of zinc oxide nanoparticles on *Triticum aestivum* L. improved by shoot height (23.7%), root length (66.8%)	Alsuwayyid et al. (2022)
Wheat	Fe nanoparticles	0, 100, 300, and 500 mg L^{-1}	Pot	Iron nanoparticles in *Triticum aestivum* (shoot length, root length, number of seeds, spike length) improved by 37.5, 68, 81.25, 50, 100, 33, and 14.28%	Javad et al. (2023)

waves (ultrasound) with a frequency range of 20–100 kHz, are also used as physical agents for seed priming (Bao et al. 2022; El-Sattar and Tawfik 2022). Seed priming with ultrasonic waves accelerates and enhances seed germination due to an increase in the seed porosity (Huang et al. 2021). Recent studies found that ultrasonic seed treatment improved the seed germination indices and tolerance to abiotic stresses owing to regulations in the metabolic profiles of different crops such as peanut (Li et al. 2021), fragrant rice (Mo et al. 2020), rice (Rao et al. 2018), Brassica (Rao et al. 2019), and sugarcane (Zeng et al. 2023).

12.6 Elements Affecting Seed Priming Practices

Seed priming is influenced by a number of abiotic and biotic variables, including time, temperature, aeration and seed quality. Aeration is the most significant of them. It interferes with crucial seed functions including respiration and germination (Bujalski and Nienow 1991). PEG solution seed priming treatment under aerated conditions enhanced germination percentage of onion as compared to non-aerated solution (Heydecker and Coolbear 1977; Bujalski et al. 1989). However, the influence of aeration varies with plant species as no impact was documented in case of lettuce under the aerated and non-aerated conditions (Cantliffe 1987). Temperature is another important factor that influences the germination of seeds. Majority of plant species show improved seed performance by priming at 15 °C whereas prime temperature ranges from 15 to 30 °C (Basra et al. 2005). Seeds kept at lower priming temperatures than this range have shown slower germination (McDonald, 2000). For seed priming, a different temperature range of 15–20 °C for 8–14 days was recommended, depending on the kind of plant, the osmotic solution, and the temperature (Wahid et al. 2008; Finch-Savage et al. 1991). Light influences seed priming by affecting seed dormancy variably in different plant species. Dormancy was reduced by illumination at the time of celery seeds priming (Khan et al. 1977), whereas lettuce seeds showed enhanced germination rate in the dark (Cantliffe 1987). Another crucial element impacting seed germination is seed quality (Cantliffe 1987; Powell et al. 2000; Podlaski et al. 2003; Halmer 2004; Nascimento and De Aragão 2004).

The process of seed priming and germination is significantly influenced by additional seed properties. For sorghum seedlings, osmo-priming with PEG solution had no beneficial effects (Patanè et al. 2008). Lower seed germination in sorghum could be attributed to the tannins present in their seeds. Alkaline solution priming (KNO_3 and K_3PO_4) is effective for such seeds e.g., pepper rather than mannitol and PEG solutions (Passam et al. 1989; O'Sullivan and Bouw 1984).

Seed priming does not always produce the beneficial effects and in some cases, it is even disadvantageous. Seed priming can cause inhibition of seed germination (Farooq et al. 2005a, b, 2006a, b, c). It can reduce rate of germination as reported in seeds of barley, cowpea, wheat, maize, palisade grass and tomato (Ogbonna and Abraham 1989; Giri and Schillinger 2003; Abdulrahmani et al. 2007; Ghodrat and Rousta 2012; Asgharipour et al. 2014; Amorim et al. 2015). Plant species or variations, the kind and water potential of the priming chemical, and the amount of time the seeds are soaked are some of the variables that affect seed priming (Geetha et al. 1997; Afzal et al. 2004; Ashraf and Foolad 2005; Abdulrahmani et al. 2007).

12.6.1 Effect of Priming Agents

The dosage of the priming agents has an impact on seed priming, but their effectiveness varies since a particular agent may be favoured by a particular species and/or cultivar, making it difficult to generalize the recommendations. Wheat growth was shown to be less negatively affected by salinity when calcium chloride or calcium sulphate were used as halopriming agents rather than sodium chloride (Afzal et al. 2008). When compared to unprimed tomato seeds, potassium nitrate increased final germination, germination rate, and seedling vigor more effectively than sodium chloride and PEG (Farooq et al. 2005a, b). The most successful agent for improving the salt tolerance of two rice varieties was calcium chloride. Additionally, it was found that hydrogen peroxide had a negative effect on both cultivars' seedling growth and germination rates (Afzal et al. 2012). On *Stevia rebuadiana* seeds, the effects of different nutria-priming agents (selenium, iron, and boron) and their combinations were investigated for salt tolerance. The highest values for ultimate germination rate and seedling vigor were demonstrated by seeds treated with a mixture of iron and boron (Shahverdi et al. 2017). Hydropriming, as opposed to halo-priming or osmo-priming, enhanced the rate of germination of two wheat cultivars (Yari et al. 2010). Different priming media, such as halopriming with 1% KNO_3 or $CaCl_2$ and hormopriming with 100 ppm GA_3, SA, or AA, significantly increased maize seed germination and vigour indices. However, GA_3 produced the most fruitful outcomes (Kumari et al. 2017). When sesame seeds were treated with -2 bar of NaCl, or PEG 6000, under both normal and drought, osmo-priming with PEG (16 h) was the most successful remedy (Morteza et al. 2017). In comparison to unprimed seeds, hydropriming and halopriming both improved the melon seed sprouting and early seedling growth indices under two levels of salt stress. But using 5 g L^{-1} KNO_3, hydropriming outperformed halopriming (Oliveira et al. 2019). In a different investigation, 2% KNO_3 significantly outperformed 500 ppm GA_3 or 1% NaCl in terms of seed germination and fresh weight (FW) and dry weight (DW) of tomato (Mavi et al. 2006). When compared to PEG, KNO_3 significantly increased the germination rate of tomato seeds while reducing the mean germination time (Lara et al. 2014). The best treatments for increasing the germination ability and yield of two bread wheat cultivars were PEG-6000, 2.5% KCl, and hydropriming (Toklu et al. 2015). Compared to NaCl and $CaCl_2$, 10 mmol KCl was the most successful seed treatment for *Acacia cyanophylla* seeds' germination characteristics (Eshkab et al. 2015). When compared to unprimed seeds, hydro-priming significantly increased the germination rate and index, and vigor of maize. While osmopriming with urea (1.2 MPa) significantly reduced the germination and vigor (Dezfuli et al. 2008). Improvement of salt stress tolerance was noted with the seed priming (Akter et al. 2018; Haghighat et al. 2012).

12.6.2 Effect of the Priming Agent Concentration

The priming agent's chemical concentration directly affects osmotic potential. There have been a variety of concentrations employed, whether the priming agent was halo or osmo. Various NaCl concentrations, such as 35 mM, 40 mM, 50 mM, 4 g L^{-1} (68.5 mM), 100 mM, 1% (171 mM), 197 mM, 50, 150, or 250 mM NaCl, and 280 mM, have been used to treat seeds of various species (Sivritepe et al. 2003, 2005; Farooq et al. 2005a, b; Mavi et al. 2006; Amjad et al. 2007; Patade et al. 2009; Haghighat et al. 2012; Tian et al. 2014; Jisha and Puthur 2014; Ben Fredj et al. 2014). Likewise, several osmo-priming substances with varied concentrations have been used. For example, investigated concentrations of PEG (the most often used medium) include: PEG-800 (–1.0 MPa) (Szopińska et al. 2014), PEG-8000 (–1.25 MPa) (Amjad et al. 2007), PEG-8000 (–0.3 to –1.8 MPa) (Lemmens et al. 2019), PEG-8000 (–0.1 to –1.8 MPa) (Khalil et al. 2001), PEG-6000 (–0.2 MPa) (Morteza et al. 2017), PEG-6000 (–0.4, –0.8, or –1.2 MPa) (Asgharipour et al. 2014), PEG-6000 (–1.0 and –1.5 MPa) (Kuppusamy and Ranganathan 2014), PEG-6000 (10%) (Toklu et al. 2015), and PEG-6000 (5, 10, 15 or 20%) (Tian et al. 2014; Faijunnahar et al. 2017). In order to optimize the concentration for better seed development and growth, different amounts of hormones have also been used. Different scientists applied different concentrations of GA_3 ranged from 1.5 to 500 mg L^{-1} on various crops and concluded that the effects of seed priming are associated with duration of seed priming, concentration and type of seed (Naeem and Muhammad 2006; Mavi et al. 2006; Audi et al. 2009; Ghodrat and Rousta 2012; Iqbal and Ashraf 2013; Younesi and Moradi 2015; Kumari et al. 2017).

Even though the necessary effects of the priming media may be attained at a particular concentration or within a range of concentrations, some concentrations may have unfavourable effects, like germination inhibition or even reduced germination. For instance, Ahmad et al. (2017) investigated the effects of $CaCl_2$, KNO_3, KCl and hydro-priming on *Gerbera jamesonii* and *Zinnia elegans*. They found that priming expressively improved germination, root-shoot length, seedling fresh-dry weight, and decreased mean germination time and T50%, with the exception of *Z. elegans* seeds. However, $CaCl_2$ (25 mM) was the most effective treatment for *G. jamesonii*, whereas $CaCl_2$ (50 mM) was the best treatment for *Z. elegans*. Additionally, it was noted that, under both normal and salt-stress conditions, 300 mM NaCl was more effective than 5 mM as a priming agent for promoting the growth of *Crithmum maritimum* L. (rock samphire) (Atia et al. 2006). Similar results were reported for 30 mM $CaSO_4$ compared to 10 mM in terms of improving the ability of barley cultivated under salt-stress to germinate (Naeem and Muhammad 2006). Additionally, it was discovered that GA_3 with 200 mg L^{-1} concentration was more effective than 100 mg L^{-1} for increasing the germination of salt stressed sugar beetroot compared to the normal conditions (Jamil and Rha 2007).

When compared to unprimed seedlings, the final germination, plant height, and shoot biomass of *Leymus chinensis* (Chinese lyme grass) seeds treated with GA_3 (0.005, 0.01, 0.05, 0.1, or 0.2 mg L^{-1}) were found to be significantly improved.

However, the most efficient concentration was 0.05 mg L^{-1} (Ma et al. 2018). Similarly, Iqbal and Ashraf (2013) demonstrated that priming with GA_3 (150 mg L^{-1}) was more successful than other concentrations to reduce the salinity impacts on the grain yield of wheat. In addition, Gill and Al-Shankiti (2015) found that *Prosopis cineraria* seeds primed with fulvic acid (0.5 and 1%) had significantly greater germination percentages, coefficients of germination rate, cumulative germination speed, and leaf numbers, shoot-root biomass than untreated seeds. Similar findings were made by Mahdavi and Rahimi (2013), who discovered that chitosan (0.05, 0.1, 0.2, and 0.5%) significantly improved the germination indices of ajowan (*Carum copticum*) compared to un-primed seeds, 0.2% being the highest. It has been noted that increasing the salt tolerance of wheat with sodium nitroprusside at 0.1 mM rather than 0.2 mM was more effective (Qasim et al. 2017). Additionally, it was found that some doses of priming agents efficiently produced the desired effects, whereas other doses of the same priming material were not as effective. For example, Afzal et al. (2007) investigated how halopriming with varying concentrations of NaCl or $CaCl_2.H_2O$ controlled germination indices and vigour under salt-stress; however, seed priming with $CaCl_2.H_2O$ (25 or 50 mM) not only significantly increased the shoot length and seedling biomass but also decreased mean germination time. Shahri et al. (2015) discovered that soaking perennial rye seeds in NaCl (125 mM for 24 h) significantly boosted the germination under stress condition compared with untreated seeds. According to Tian et al. (2014), 10 or 15% PEG significantly increased maize seed germination rates while 20% PEG diminished it comparable to un-primed seeds.

Even though priming has been shown to have positive effects on a variety of crops and tree species, it has also produced unintended outcomes. For instance, Abdulrahmani et al. (2007) found that seed priming of barley seeds with Zn at 50 or 100 mM reduced their ability to germinate and slowed their rate of germination. Barsa et al. (2005) also documented a decrease in the germination rate of wheat seeds primed with 25 mM $Ca(NO_3)_2$, 50 mM NaCl, or 100 mM $CaCl_2$. Similar results were seen with perennial rye seeds primed with 125, 150, or 250 mM NaCl. According to Ghodrat and Rousta (2012), in both normal and saline circumstances, the germination percentage and germination rate of maize seeds primed with 1.5, 2.5, or 5 mg L^{-1} GA_3 were considerably lower than those of unprimed seeds. Additionally, they discovered that both indices were significantly reduced as the priming agent's concentration increased. Shahverdi et al. (2017) also shown that although the germination capacity of untreated *Stevia rebuadiana* seeds was statistically equivalent to untreated seeds at 0, 30, and 60 mM NaCl, it was considerably decreased at 90 mM NaCl compared to unprimed seeds. Similar findings were made by Arin et al. (2011) who discovered that halopriming of onion cv. Alix seeds with 2 or 4% KNO_3, KH_2PO_4, or 0.5, 0.1, or 1.5 MPa PEG significantly reduced the percentage of emergence at 20 °C in comparison to unprimed seeds. Additionally, they noticed that all concentrations of PEG reduced the emergence rate compared to untreated seeds, while no halopriming treatments affected the emergence rate of this cultivar. Furthermore, although priming tomato seeds with −0.4, −0.8, or −0.12 MPa PEG significantly increased germination rates and shoot and root lengths compared to unprimed seeds, doing osmotic potential above −0.4 Mpa significantly decreased root length (Asgharipour et al. 2014). Additionally, it has been noted that seed priming can inhibit germination. For instance, Farooq et al. (2006a, b, c) clarified the impact of ethanol aerated solutions at 1, 5, 10, and 15% (v/v) on the potential revivification of fine rice (Super-Basmati) seed. The seeds were dried back to almost their original weight after being steeped in the priming solution for 42 h. The findings indicated that while none of the concentrations examined increased the ability of seeds to germinate when compared to seeds that had not been primed, concentrations of 10 and 15% completely prevented rice from germinating and emerging. The germination of durum wheat (*Triticum durum*) seeds osmoprimed with −1.2 and −1.8 MPa PEG was completely inhibited, as opposed to 96% of un-primed seeds (Lemmens et al. 2019). Similar to this, Farooq et al. (2011) demonstrated that 0.5% boron completely reduced the germination and development of rice (Super Basmati) seeds, but that 0.001 or 0.1% boron had a good effect on the germination capacity, germination energy, germination rate, and germination index. Farooq et al. (2005a, b) also showed that osmo-priming with 5% urea inhibited the germination and emergence of both fine and coarse rice. Furthermore, Iqbal et al. (2012) came to the conclusion that 0.1% of boron negatively impacted germination and seedling growth while 0.001% of it improved stand uniformity and early seedling growth of two wheat cultivars.

12.6.3 Effect of Soaking Duration

It has been suggested that the length of time seeds soak in priming chemicals and their osmotic potential interact considerably. In order to achieve the required priming effects, one of the critical criteria is the soaking time (Vanangmudi et al. 2000; Giri and Schillinger 2003; Afzal et al. 2005; Subedi and Ma 2005; Farooq et al. 2006a, b, c; Atrip and O'Reilly 2007; Dezfuli et al. 2008; Govinden-Soulange and Levantard 2008). Furthermore, from the two wheat cultivars i.e., Azar-2 and Sardari 101, the 'Azar-2' seeds primed with 20% PEG for 12 h showed the highest germination rate, whereas seeds primed with 10% PEG for 24 h produced the longest stems (Yari et al. 2010). Yari et al. (2012) examined the effects of seed soaking times in wheat at 0.5 and 1% $CaCl_2$ solutions for 12, 24, or 36 h and reported that 24 h was the optimal amount of time to soak the seeds. Dezfuli et al. (2008) investigated the effect of hydropriming duration on the seed vigour of two maize inbred lines, namely "B73" and "MO17," and found that seed soaking for 36 h was most effective in boosting the germination percentage, germination index, vigour index, and radical and coleoptile length, as well as in reducing mean germination time for both inbred lines. Tall *Festuca arundinacea* seeds were soaked in distilled water for different time intervals at 15 or 20 °C in complete darkness before being dried to their original weight. In comparison to unprimed seedlings, all treatments significantly improved germination and growth characteristics; however, soaking for 12 h at 15 °C was found to be the most successful treatment (Rouhi et al. 2011).

According to Asgharipour et al. (2014), seed soaking time in PEG 6000 had a significant impact on tomato's germination ability, germination rate, and shoot length. According to Akter et al. (2018), hydropriming maize seeds for 24 or 48 h successfully reduced the salt stress compared to unprimed seeds, however 48 h was found to be more beneficial than 24 h. Similar to this, although hydro-priming for 8 or 16 h or halo-priming with –0.2 MPa PEG 6000 or –0.2 MPa NaCl improved sesame seed germination indices under both normal and drought stress conditions, osmo-priming with PEG for 16 h was the most successful treatment (Morteza et al. 2017). Additionally, the optimal soaking time for seeds of various crops was found to be 12 and 16 h for barley seeds (Rashid et al. 2006), 6–8 h for mungbean (*Vigna radiata*) (Rashid et al. 2004), 24 h for maize, 10 h for chickpea (Harris et al. 1999), 7 h for pinto bean (*Phaseolus vulgaris*) (Ghassemi-Golezanik et al. 2010). Similar to this, the safe limit for soaking was less than 12 h when the seeds of two wheat cultivars were soaked for 12, 24 and 36 h in water, or in 2 and 4% KCl, 0.5 and 1% KH_2PO_4, or 10 or 20% PEG (Giri and Schillinger 2003). Additionally, chicory (*Cichorium intybus*) and endive (*Cichorium endivia*) seeds primed for 15 min as opposed to 30 min with $NaHClO_3$, methyl jasmonate, and dictamus essential oil had longer radicles (Tzortzakis 2009). The safe duration for soaking seeds of several sorghum varieties, including Abshir, Dekeba, Meko-1, Melkam, and Teshale, 0.2 g L^{-1} $ZnSO_4$, and 9 g L^{-1} urea solutions for 10 h was also reported by Teshome et al. (2018) and found that soaking for more than 10 h inhibited the germination owing to ion leakage and/or ion buildup. To prevent further harm to seedling growth, it is crucial to establish the ideal time period for which seeds can be soaked. To improve seed germination and stand establishment in this respect, safe limits and/or the ideal time period for seed priming must be established or optimised for each plant species (Harris et al. 1999, 2001).

12.6.4 Responsivity of Plant Species and Varities to Priming

Different plant species and cultivars react differently to pre-sowing seed treatments. Increased growth and yield were seen in all crops after hydropriming with 3% KNO_3 and −1.0 MPa PEG on a variety of crops, including tomato, onion, and watermelon, however there was heterogeneity in how each crop responded to the priming treatments (Maiti et al. 2011). Two wheat cultivars were halo-primed with 50 mM of $CaCl_2$, NaCl, or $CaSO_4$ and the responses were variable (Afzal et al. 2008). Different varieties of wheat respond differently to priming treatments (Afzal et al. 2006, 2008; Yari et al. 2010, 2012; Nawaz et al. 2012). In two separate maize cultivars, hydro-priming, osmo-priming, and PEG-6000 all responded differently (Dezfuli et al. 2008). Different rice, beetroot, parsley, and green gram cultivars responded differently to priming medium (Sliwinska and Babinska 1999; Podlaski et al. 2003; Farooq et al. 2005a, b; Jisha and Puthur 2014). Different varieties of wheat responded differently to priming treatments (Afzal et al. 2006, 2008; Yari et al. 2010, 2012; Nawaz et al. 2012). Furthermore, three sugarcane cultivars with different salt tolerance capacities were haloprimed with 100 mM NaCl, and the tolerant and moderately tolerant types' seed germination and germination rate significantly improved as compared to the sensitive variety in salty circumstances (Patade et al. 2009). Four coriander cultivars were haloprimed with NaCl to reduce the severe effects of salinity, although one of the cultivars responded more significantly than the others (Ben Fredj et al. 2014). Cumin and lemon balm seeds both responded differently to halo-priming (Younesi and Moradi 2015).

Pre-sowing seed treatments are more beneficial for low vigour seeds than high vigour seeds thus enhancing their germination capacity and germination rate (Halmer 2004; Afzal et al. 2004). Furthermore, hydropriming resulted in the enhancement of storability of low vigour seeds e.g., cauliflower (Powell et al. 2000).

12.7 Potential Benefits and Future Scope

Biofortification through seed priming have several benefits in terms of improving crop growth, productivity and nutritional quality. For example, seed priming can improve the uptake and accumulation of essential micronutrients i.e., Zn, Fe, and vitamin A in edible plant parts, leading to increased nutrient content which potentially could help to address malnutrition and nutrient deficiencies in vulnerable populations. Seed priming can also improve the seed germination and seedling vigor, allowing crops to establish more quickly and compete more effectively with weeds and other stresses which lead to improvements in crop yield and quality, including nutrient contents in edible plant parts. Seed priming could potentially reduce the cost of production and improve the profitability of farming by reducing the need for synthetic fertilizers and pesticides, making agriculture more sustainable and environmentally friendly. Moreover, seed priming can help to reduce the incidence of seed-borne diseases, which otherwise could result in severe loss in crop yield and quality. Seed treatment with a solution containing fungicides or other biocontrol agents, can help to prevent the transmission of diseases from seed to seedlings. In addition, seed priming can improve crop resilience to environmental stresses such as drought, salinity, heavy metals, and temperature fluctuations that can help to ensure food security and stability in regions that are vulnerable to climate change.

The future scope of biofortification through seed priming includes:

Development of novel priming techniques: Scientists are exploring new priming techniques/strategies that can further improve crop productivity and nutritional quality. For instance, priming with nanoparticles or beneficial microorganisms could provide novel ways to improve nutrient uptake and plant growth.

Breeding crops for biofortification: Seed priming can be used in combination with plant breeding techniques to develop new crop varieties with high nutritional quality. This could involve selecting for traits that promote nutrient uptake and accumulation, as well as other desirable agronomic characteristics.

Scaling up production: The widespread adoption of biofortified crops will require large-scale production and distribution systems. Seed priming can be used to improve the productivity and efficiency of these systems, making biofortified crops more accessible to farmers and consumers.

Overall, biofortification through seed priming has the potential to address global nutrition challenges and improve the sustainability of agriculture. Further research and development will be needed to fully realize the benefits of this promising technology to overcome the malnutrition problem on global scale.

References

Abdulrahmani K, Ghassemi-Golezani K, Valizadeh M, Asl VF (2007) Seed priming and seedling establishment of barley (*Hordeum vulgare* L.). J Food Agric Environ 5(3–4):179–184

Abou-Zeid HM, Ismail GSM, Abdel-Latif SA (2021) Influence of seed priming with ZnO nanoparticles on the salt-induced damages in wheat (*Triticum aestivum* L.) plants. J Plant Nutr 44(5):629–643

Adhikary S, Biswas B, Chakraborty D, Timsina J, Pal S, Chandra Tarafdar J, Banerjee S, Hossain A, Roy S (2022) Seed priming with selenium and zinc nanoparticles modifies germination, growth, and yield of direct-seeded rice (*Oryza sativa* L.). Sci Rep 12(1):7103. https://doi.org/10.1038/s41598-022-11307-4

Afzal I, Aslam N, Hameed FMA, Ahmead SIG (2004) Enhancement of germination and emergence of canola seeds by different priming techniques. Caderno de Pesquisa Série Biologica 16(1): 19–34

Afzal I, Barsa SMA, Iqbal A (2005) The effects of seed soaking with plant growth regulators on seedling vigour of wheat under salinity. J Stress Physiol Biochem 1(1):6–14

Afzal I, Barsa SMA, Hameed A, Farooq M (2006) Physiological enhancements for alleviation of salt stress in wheat. Pak J Bot 38(5):1649–1659

Afzal I, Barsa SMA, Ahmad N, Lodhi TE (2007) Counteraction of salinity stress on wheat plants by pre-sowing treatments. Pak J Agric Sci 44(1):50–58

Afzal I, Rauf S, Barsa SMA, Murtaza G (2008) Halopriming improves vigour, metabolism of reserves and ionic contents in wheat seedlings under salt stress. Plant Soil Environ 54(9): 382–388

Afzal I, Butt A, Ur Rehman H, Basra MA, Afzal A (2012) Alleviation of salt stress in fine aromatic rice by seed priming. Aust J Crop Sci 6(10):1401–1407

Ahmad I, Saleem AM, Mustafa G, Ziaf K, Afzal I, Qasim M (2017) Seed halopriming enhances germination performance and seedling vigor of *Gerbera jamesonii* and *Zinnia elegans* Sarhad. J Agri 33:199–205

Akter L, Fakir OA, Alam MK, Ul Islam M, Chakraborti P, Alam MJ, Rashid MH, Begum M, Kader MA (2018) Amelioration of salinity stress in maize seed germination and seedling growth attributes through seed priming. Open J Soil Sci 8:137–146

Alharby HF, Rizwan M, Iftikhar A, Hussaini KM, Rehman MZ, Bamagoos AA, Ali S (2021) Effect of gibberellic acid and titanium dioxide nanoparticles on growth, antioxidant defense system and mineral nutrient uptake in wheat. Ecotoxicol Environ Saf 221:112436

Alsuwayyid AA, Alslimah AS, Perveen K, Bukhari NA, Al-Humaid LA (2022) Effect of zinc oxide nanoparticles on *Triticum aestivum* L. and bioaccumulation assessment using ICP-MS and SEM analysis. J King Saud Univ Sci 34(4):101944

Amjad M, Ziaf K, Iqbal AI, Riaz MA, Saqib ZA (2007) Effect of seed priming on seed vigor and salt tolerance in hot pepper. Pak J Agric Sci 44(3):408–416

Amorim MM, Vieir HD, Amorim IM, Dobbss LB, Deminici BB, Xavier PB (2015) Effects of the humic acid extracted from vermin compost on the germination and initial growth of *Brachiaria brizantha* cv. MG5 AJB 14(18):1576–1583. https://doi.org/10.5897/AJB2015144443

Araujo SDS, Paparella S, Dondi D, Bentivoglio A, Carbonera D, Balestrazzi A (2016) Physical methods for seed invigoration: advantages and challenges in seed technology. Front Plant Sci 7: 646

Arif M, Waqas M, Nawab K, Shahid M (2007) Effect of seed priming in Zn solutions on chickpea and wheat. Afr Crop Sci 8:237–240

Arif M, Jan MT, Marwat KB, Khan MA (2008) Seed priming improves emergence and yield of soybean. Pak J Bot 40(3):1169–1177

Arin L, Polat S, Deveci M, Salk A (2011) Effects of different osmotic solutions on onion seed emergence. Afr J Agric Res 6(4):986–991. https://doi.org/10.5897/AJAR10.1046

Asgharipour MR, Pasandi R, Pooresmail H (2014) Effect of osmopriming on the germination of tomato seeds. Int J Biosci 5(1):359–364

Ashraf M, Foolad MR (2005) Pre-sowing seed treatment- a shotgun approach to improve germination, plant growth, and crop yield under saline and non-saline conditions. Adv Agron 88:223–271

Atia A, Debez A, Rabhi M, Athar H, Abdelly C (2006) Alleviation of salt-induced seed dormancy in the perennial halophyte *Crithmum maritimum* L. (Apiaceae). Pak J Bot 38(5):1367–1372

Atrip N, O'Reilly C (2007) Germination response of alder and brich seeds to applied gibberelic acid and priming treatments in combination with chilling. Ann For Sci 64(4):385–394. https://doi.org/10.1051/forest:2007015

Audi AH, Mukhtar F, Bayero B (2009) Effect of pre-sowing hardening treatments using various plant growth substances on cowpea germination and seedling establishment. Bayero J Pure Appl Sci 2(2):44–48. https://doi.org/10.4314/bajopas.v2i2.63762

Babu AG, Prabhuling G, Karani RS, Patil RK, Mulla SR, Raghavendra G, Jagadeesha RC (2018) Genetic diversity analysis among banana cultivars through ISSR markers. J Pharmacogn Phytochem 7(6):1576–1580

Bao G, Zhou Q, Li S, Ashraf U, Huang S, Miao A, Cheng Z, Wan X, Zheng Y (2022) Transcriptome analysis revealed the mechanisms involved in ultrasonic seed treatment-induced aluminum tolerance in peanut. Front Plant Sci 12:3314. https://doi.org/10.3389/fpls.2021.807021

Barsa SMA, Afzal I, Ali R, Farooq RM (2005) Pre-sowing seed treatments to improve germination and seedling growth in wheat (*Triticum aestivum* L.). Caderno de Pesquisa Sêrie Biologica 17(1):155–164. http://hdl.handle.net/1807/5392

Basahi M (2021) Seed germination with titanium dioxide nanoparticles enhances water supply, reserve mobilization, oxidative stress and antioxidant enzyme activities in pea. Saudi J Biol Sci 28(11):6500–6507

Basra SMA, Farooq M, Tabassam R, Ahmad N (2005) Physiological and biochemical aspects of pre-sowing seed treatments in fine rice (*Oryza sativa* L.). Seed Sci Technol 33(3):623–628

Beall T, Massiot E, Arsenault JE, Smith MR, Hijmans RJ (2017) Global trends in dietary micronutrient supplies and estimated prevalence of inadequate intakes. PLoS One 12:e0175554

Ben Fredj M, Kaouther Z, Chérif H, Tijani M, André B (2014) Effect of priming on growth, biochemical parameters and mineral composition of different cultivars of coriander (*Coriandrum sativum* L.) under salt stress. J Stress Physiol Biochem 10(3):84–109

Bhardwaj J, Anand A, Nagarajan S (2012) Biochemical and biophysical changes associated with magnetopriming in germinating cucumber seeds. Plant Physiol Biochem 57:67–73

Bilalis DJ, Katsenios N, Efthimiadou A, Karkanis A, Efthimiadis P (2012) Investigation of pulsed electromagnetic field as a novel organic pre-sowing method on germination and initial growth stages of cotton. Electromagn Biol Med 31(2):143–150

Bisen K, Keswani C, Mishra S, Saxena A, Rakshit A, Singh HB (2015) Unrealized potential of seed biopriming for versatile agriculture. In: Rakshit A, Singh HB, Sen A (eds) Nutrient use efficiency: from basics to advances, 1st edn. Springer, New Delhi, pp 193–206

Black RE, Victora CG, Walker SP, Bhutta ZA, Christian P, De Onis M, Ezzati M, Grantham-McGregor S, Katz J, Martorell R et al (2013) Maternal and child undernutrition and overweight in low income and middle-income countries. Lancet 382:427–451

Blando F, Berland H, Maiorano G, Durante M, Mazzucato A, Picarella ME, Nicoletti I, Gerardi C, Mita G, Andersen ØM (2019) Nutraceutical characterization of anthocyanin-rich fruits produced by "Sun Black" tomato line. Front Nutr 6:133. https://doi.org/10.3389/fnut.2019.00133

Bouis H, Saltzman A (2016) Improving nutrition through biofortification: a review of evidence from HarvestPlus, 2003 through 2016. Glob Food Sec 12:49–58

Bouis HE, Welch RM (2010) Biofortification-A sustainable agricultural strategy for reducing micronutrient malnutrition in the global south. Crop Sci 50:20–32

Branca F, Ferrari M (2002) Impact of micronutrient deficiencies on growth: the stunting syndrome. Ann Nutr Metab 46:8–17

Brotanek JM, Halterman JS, Auinger P, Flores G, Weitzman M (2005) Iron deficiency, prolonged bottle-feeding, and racial/ethnic disparities in young children. Arch Pediatr Adolesc Med 159: 1038–1042

Bujalski W, Nienow AW (1991) Large-scale osmotic priming of onion seeds: a comparison of different strategies for oxygenation. Sci Hortic 46(1–2):13–24

Bujalski W, Nienow AW, Gray D (1989) Establishing the large-scale osmotic priming of onion seeds by using enriched air. Ann Appl Biol 115(1):171–176

Cakmak I (2005) The role of potassium in alleviating detrimental effects of abiotic stresses in plants. J Plant Nutr Soil Sci 168(4):521–530

Callan NW, Mathre D, Miller JB (1990) Biopriming seed treatment for biological control of *Pythium ultimum* pre-emergence damping-off in sh-2 sweet corn. Plant Dis 74:368–372

Cantliffe DJ (1987) Priming of lettuce for early and uniform emergence under conditions of environmental stress. Acta Hortic 122:29–38

Cantliffe DJ (2003) Seed enhancements. Acta Hortic 607:53–59

Chakraborthy GS, Aeri V, Verma P, Singh S (2014) Phytochemical and antimicrobial studies of *Chlorophytum borivilianum*. Pharmacophore 5:258–261

Chattha MB, Khan I, Mahmood A, Chattha MU, Anjum SA, Ashraf U, Zain M, Bilal U (2017) Seed priming with thiourea improves the performance of late sown wheat. J Agric Res 55(1):29–39

Chen J, Zeng X, Yang W, Xie H, Ashraf U, Mo Z, Liu J, Li G, Li W (2021) Seed priming with multiwall carbon nanotubes (MWCNTs) modulates seed germination and early growth of maize under cadmium (Cd) toxicity. J Soil Sci Plant Nutr 21:1793–1805

Chizuru N, Ricardo U, Shiriki K, Prakash S (2003) The joint WHO/FAO expert consultation on diet, nutrition and the prevention of chronic diseases: process, product and policy implications. Public Health Nutr 7(1a):245–250

Das JK, Kumar R, Salam RA, Bhutta ZA (2013) Systematic review of zinc fortification trials. Annu Nutr Metab 62(1):44–56

Dearman J, Brocklehust PA, Drew RLK (1987) Effect of osmotic priming and aging on the germination and emergence of carrot and leek seed. Ann Appl Biol 111:717–722

Delavari PM, Baghizadeh A, Enteshari SH, Kalantari KM, Yazdanpanah A, Mousavi EA (2010) The Effects of salicylic acid on some of biochemical and morphological characteristic of *Ocimum basilicum* under salinity stress. Aust J Basic Appl Sci 4:4832–4845

Delian EL, Bădulescu LI, Dobrescu A, Chira LE, Lagunovschi-Luchian V (2017) A brief overview of seed priming benefits in tomato. Rom Biotechnol Lett 22(3):12505–12513

Demir I, Ozuaydın I, Yasar F, Van SJ (2012) Effect of smoke-derived butenolide priming treatment on pepper and salvia seeds in relation to transplant quality and catalase activity. S Afr J Bot 78: 83–87

Dezfuli PM, Sharif-Zadeh F, Janmohammadi M (2008) Influence of priming techniques on seed germination behavior of maize inbred lines (*Zea mays* L.). J Agric Biol Sci 3(3):22–25

Dhingra P, Sharma S, Singh KH, Kushwaha HS, Barupal JK, Haq S, Kachhwaha S (2022) Seed priming with carbon nanotubes and silicon dioxide nanoparticles influence agronomic traits of Indian mustard (*Brassica juncea*) in field experiments. J King Saud Univ Sci 34(4):102067

Di Girolamo G, Barbanti L (2012) Treatment conditions and biochemical processes influencing seed priming effectiveness. IJA 7(2):25–25. https://doi.org/10.4081/ija.2012.e25

Dutta P (2018) Seed priming: new vistas and contemporary perspectives. In: Advances in seed priming. Springer, Singapore, pp 3–22

Ellis RH, Butcher PD (1988) The effects of priming and 'natural' differences in quality amongst onion seed lots on the response of the rate of germination to temperature and the identification of the characteristics under genotypic control. J Exp Bot 39:935–950

El-Sattar AMA, Tawfik E (2022) Effects of ultrasonic waves on seedling growth, biochemical constituents, genetic stability of fenugreek (*Trigonella foenum-graecum*) under salinity stress. Vegetos:1–10

Ermis S, Kara F, Özden E, Demir I (2016) Solid matrix priming of cabbage seed lots: repair of ageing and increasing seed quality. Tarım Bilimleri Dergisi 22:588–595

Eshkab IA, Shanta MB, El Waer HN (2015) Enhancement of *Acacia cyanophylla* seed germinability and vigour through salt hardening technique. Alexandria Sci Exchang J 36(3):221–225. https://doi.org/10.21608/asejaiqjsae.2015.2904

Esper Neto M, Britt DW, Lara LM, Cartwright A, dos Santos RF, Inoue TT Batista MA (2020) Initial development of corn seedlings after seed priming with nanoscale synthetic zinc oxide. Agronomy 10(2):307

Faijunnahar M, Abdullahil Baque M, Habib MA, Hossain HM (2017) Polyethylene glycol (PEG) induced changes in germination, seedling growth and water relation behavior of wheat (*Triticum aestivum* L.) genotypes. Univers J Plant Sci 5(4):49–57. https://doi.org/10.13189/ujps.2017.050402

FAO (2017) FAOSTAT. Food and Agriculture Organization of the United Nations.

Farooq MA, Wahid DJ (2009) Exogenously applied polyamines increase drought tolerance of rice by improving leaf water status, photosynthesis and membrane properties. Acta Physiol Plant 31: 937–945. https://doi.org/10.1007/s11738-009-0307-2

Farooq M, Barsa SMA, Saleem BA, Nafees M, Chishti SA (2005a) Enhancment of tomato seed germination and seedling vigour osmopriming. Pak J Agric Sci 42(3-4):3641

Farooq M, Barsa SMA, Hafeez K, Asad SA, Ahmad N (2005b) Use of commercial fertilizer as osmotica for rice priming. J Agric Sci 2:172–175

Farooq M, Barsa S, Tabassum R, Afzal I (2006a) Enhancing the performance of direct seeded fine rice by seed priming. Plant Prod Sci 9(4):446–456. https://doi.org/10.1626/pps.9.446

Farooq M, Barsa SMA, Hefeez-ue-Rehman MT (2006b) Germination and early seedling growth as affected by pre-sowing ethanol seed treatments in fine rice. Int J Agric Biol 1:19–22

Farooq M, Basra SMA, Wahid A (2006c) Priming of field-sown rice seed enhances germination, seedling establishment, allometry and yield. Plant Growth Regul 49:285–294

Farooq M, Atique-ur-Rehman AT, Habib M (2011) Boron nutripriming improves the germination and early seedling growth of rice (*Oryza sativa* L.). J Plant Nutr 34(10):1507–1515. https://doi.org/10.1080/01904167.2011.585207

Finch-Savage WE, Gray D, Dickson GM (1991) The combined effects of osmotic priming with plant growth regulator and fungicide soaks on the seed quality of five bedding plant species. Seed Sci Technol 19(2):495–503

Fraceto LF, Grillo R, de Medeiros GA, Scognamiglio V, Rea G, Bartolucci C (2016) Nanotechnology in agriculture: which innovation potential does it have? Front Environ Sci 4:20

Franklin KA (2009) Light and temperature signal crosstalk in plant development. Curr Opin Plant Biol 12:63–68

Garg M, Sharma N, Sharma S, Kapoor P, Kumar A, Chunduri V, Arora P (2018) Biofortified crops generated by breeding, agronomy, and transgenic approaches are improving lives of millions of people around the world. Front Nutr 5:12. https://doi.org/10.3389/fnut.2018.00012

Garner JM, Armitage AM (2008) Cooling and long-day lighting influences growth and flowering of *Phlox paniculata* L 'Ice Cap'used for cut flowers. Hortic Sci 43:707–709

Geetha CK, Valsalakumari PK, Rajeevan PK (1997) Effect of seed priming on germination and seedling vigour in *cassia* spp. J Appl Hortic 3(1-2):89–92

Ghafari H, Razmjoo J (2013) Effect of foliar application of nano-iron oxidase, iron chelate and iron sulphate rates on yield and quality of wheat. Int J Plant Prod 4(11):2997–3003

Ghassemi-Golezanik K, Chadordooz-Jeddi A, Nasrullahzadeh S, Moghaddam M (2010) Influence of hydro-priming duration on field performance of pinto bean (*Phaseolus vulgaris* L.) cultivars. Afr J Agric Res 5(9):893–897. https://doi.org/10.5897/AJAR.9000422

Ghodrat V, Rousta MJ (2012) Effect of priming with gibberellic acid (GA_3) on germination and growth of corn (*Zea mays* L.) under saline conditions. Int J Agric Crop Sci 4(13):882–885

Gill S, Al-Shankiti A (2015) Priming of *Prosopis cineraria* (L.) Druce and *Acacia tortilis* (Forssk) seeds with fulvic acid extracted from compost to improve germination and seedling vigor. Global J Environ Sci Manag 1(3):225–232. https://doi.org/10.7508/gjesm.2015.03.005

Giri GS, Schillinger WF (2003) Seed priming winter wheat for germination, emergence and yield. Crop Sci 43:2135–2141. https://doi.org/10.2135/cropsci2003.2135

Govindaraj M, Virk PS, Kanatti A, Cherian B, Rai KN, Anderson MS, Pfeiffer WH (2020) Biofortified pearl millet cultivars offer potential solution to tackle malnutrition in India. In: Quantitative genetics, genomics and plant breeding. CABI, Wallingford, pp 385–396

Govinden-Soulange J, Levantard M (2008) Comparative studies of seed priming and pelleting on percentage and meantime to germination of seeds of tomato (*Lycopersicon esculentum* Mill.). Afr J Agric Res 3(10):725–731. https://doi.org/10.5897/AJAR.9000117

Graham RD, Welch RM, Bouis HE (2001) Addressing micronutrient malnutrition through enhancing the nutritional quality of staple foods: principles, perspectives and knowledge gaps. Adv Agron 70:77–142

Guo X, Zhi W, Feng Y, Zhou G, Zhu G (2022) Seed priming improved salt-stressed sorghum growth by enhancing antioxidative defense. PLoS One 17(2):e0263036

Hacisalihoglu G (2007) Germination characteristics of three warm-season turfgrasses subjected to matriconditioning and aging. Hortic Technol 17:480–485

Haghighat Z, Ebadi A, Khomari S (2012) Influence of seed priming on germination and early growth of *Sorghum bicolor* (L.) Moench under Salt Stress. Plant Physiol 4:159–162

Halmer P (2004) Methods to improve seed performance in the field. In: Benech-Arnold RL, Sànchez RA (eds) Handbook of seed physiology applications to agriculture. Food Products Press, New York, pp 125–165

Hambidge M (2000) Human Zinc deficiency. J Nutr 130:1344–1349

Hameed A, Iqbal N (2014) Chemo-priming with Mannose, Mannitol and H_2O_2 mitigate drought stress in wheat, cereal research communications. Cereal Res Commun 42:450–462

Harman GE, Nelson EB (1994) Mechanisms of protection of seed and seedlings by biological seed treatments: implications for practical disease control. In: Seed treatment: progress and prospects, BCPC Monograph, vol 57, Thornton Heath, Surrey, pp 283–292

Harris D, Joshi A, Khan PA, Gotdhkar P, Sodhi P (1999) On-farm seed priming in semi-arid agriculture: development and evaluation in maize, rice and chickpea in India using participatory methods. Exp Agric 35:15–29. https://doi.org/10.1017/S0014479799001027

Harris D, Pathan AK, Gothkar P, Joshi A, Chivasa W, Nyamudeza P (2001) On-farm seed priming: using participatory methods to revive and refine a key technology. Agric Syst 69:151–164. https://doi.org/10.1016/S0308-521X(01)00023-3

Harris D, Rashid A, Miraj G, Arif M, Yunas M (2008) 'On-farm' seed priming with zinc in chickpea and wheat in Pakistan. Plant Soil 306(1–2):3–10

Hatami M, Khanizadeh P, Bovand F, Aghaee A (2021) Silicon nanoparticle-mediated seed priming and *Pseudomonas* spp. inoculation augment growth, physiology and antioxidant metabolic status in *Melissa officinalis* L. plants. Ind Crops Prod 162:113238

Hefferon KL (2016) Can biofortified crops help attain food security? Curr Mol Biol Rep 2(4): 180–185

Heydecker W, Coolbear P (1977) Seed treatments for improved performance-survey and attempted prognosis. Seed Sci Technol 5:353–425

Heydecker W, Higgins J, Gulliver RL (1973) Accelerated germination by osmotic seed treatment. Nature 246(5427):42–44

Hill H, Bradford KJ, Cunningham J, Taylor AG (2008) Primed lettuce seeds exhibit increased sensitivity to moisture during aging. Acta Hortic 782:135–141

Hirschi KD (2009) Nutrient biofortification of food crops. Annu Rev Nutr 29:401–421. https://doi.org/10.1146/annurev-nutr-080508-141143

Hodge J (2016) Hidden hunger: approaches to tackling micronutrient deficiencies. In: Gillespie S, Hodge J, Yosef S, Pandya-Lorch R (eds) Nourishing millions: stories of change in nutrition. IFPRI, Washington, pp 35–43

Hotz C, McClafferty B (2007) From harvest to health: challenges for developing biofortified staple foods and determining their impact on micronutrient status. Food Nutr Bull 28:271–279

Huang YM, Wang HH, Chen KH (2002) Application of seed priming treatments in spinach (*Spinacia oleracea* L) production. Chin Soc Hortic Sci 48:117–123

Huang S, Luo H, Ashraf U, Abrar M, He L, Zheng A, Wang Z, Zhang T, Tang X (2019) Seed treatment with paclobutrazol affects early growth, photosynthesis, chlorophyll fluorescence and physiology of rice. Appl Ecol Environ Res 17(1):999–1012

Huang S, Rao G, Ashraf U, Deng Q, Dong H, Zhang H, Tang X (2021) Ultrasonic seed treatment improved morpho-physiological and yield traits and reduced grain Cd concentrations in rice. Ecotoxicol Environ Saf 214:112119

Hussain M, Farooq M, Basra SMA, Ahmad N (2006) Influence of seed priming techniques on the seedling establishment, yield and quality of hybrid sunflower. Int J Agric Biol 8:14–18

Hussain S, Khaliq A, Tanveer M, Matloob A, Hussain HA (2018) Aspirin priming circumvents the salinity-induced effects on wheat emergence and seedling growth by regulating starch metabolism and antioxidant enzyme activities. Acta Physiol Plant 40:1–12

Hussain A, Rizwan M, Ali Q, Ali S (2019) Seed priming with silicon nanoparticles improved the biomass and yield while reduced the oxidative stress and cadmium concentration in wheat grains. Environ Sci Pollut Res 26:7579–7588

Ibrahim EA (2016) Seed priming to alleviate salinity stress in germinating seeds. J Plant Physiol 192:38–46

Ijaz A, Tasneem KAA, Shahzad MB, Zuhair H, Amjed A (2012) Effect of seed priming with ascorbic acid, salicylic acid and hydrogen peroxide on emergence, vigor and antioxidant activities of maize. Afr J Biotechnol 11(5):1127–1137

Ilona P, Bouis HE, Palenberg M, Moursi M, Oparinde A (2017) Vitamin A cassava in Nigeria: crop development and delivery. Afr J Food Agri Nutr Develop 17(2):12000–12025

Iqbal M, Ashraf M (2013) Gibberellic acid mediated induction of salt tolerance in wheat plants: growth, ionic partitioning, photosynthesis, yield and hormonal homeostasis. Environ Exp Bot 86:76–85. https://doi.org/10.1016/j.envexpbot.2010.06.002

Iqbal S, Farooq M, Nawaz A, Atique-Ur-Rehman AR (2012) Optimizing boron seed priming treatments for improving the germination and early seedling growth of wheat. J Agri Social Sci 8(2):57–61

Jafar MZ, Farooq M, Cheema MA, Afzal I, Basra SMA, Wahid MA, Aziz T, Shahid M (2012) Improving the performance of wheat by seed priming under saline conditions. J Agron Crop Sci 198:38–45

Jamil M, Rha ES (2007) Gibberellic acid (GA_3) enhance seed water uptake, germination and early seedling growth in sugar beet under salt stress. Pak J Biol Sci 10(4):654–658. https://doi.org/10.3923/pjbs.2007.654.658

Javad S, Maqsood S, Shah AA, Singh A, Shah AN, Nawaz M, Mosa WF (2023) Iron nanoparticles mitigates cadmium toxicity in *Triticum aestivum*; Modulation of antioxidative defense system and physiochemical characteristics. J King Saud Univ Sci 35(3):102498

Jisha KC, Puthur JT (2014) Halopriming of seeds imparts tolerance to NaCl and PEG induced stress in *Vigna radiata* (L.) Wilczek varieties. Physiol. Mol Biol 20(3):303–312. https://doi.org/10.1007/s12298-014-0234-6

Jisha KC, Vijayakumari K, Puthu JT (2013) Seed priming for abiotic stress tolerance: an overview. Acta Physiol Plant 35:1381–1396

Khalil MAK, Rasmussen RA (1983) Sources, sinks, and seasonal cycles of atmospheric methane. J Geophys Res 88:5131–5144

Khalil SK, Mexal JG, Murray LW (2001) Germination of soybean seed primed in aerated solution of polyethylene glycol (8000). J Biol Sci 1(3):105–107. https://doi.org/10.3923/jbs.2001.105.107

Khan AA, Tao KL, Knypl JS, Borkowska B, Powell LE (1977) Osmotic conditioning of seeds: physiological and biochemical changes. Symp Seed Probl Horticult 83:267–278

Khan F, Hussain S, Tanveer M, Khan S, Hussain HA, Iqbal B, Geng M (2018) Coordinated effects of lead toxicity and nutrient deprivation on growth, oxidative status, and elemental composition of primed and non-primed rice seedlings. Environ Sci Pollute Res 25:21185–21194

Khush GS, Lee S, Cho JI, Jeon JS (2012) Biofortification of crops for reducing malnutrition. Plant Biotechnol Rep 6:195–202

Kiran A, Wakeel A, Mahmood K, Mubaraka R, Haefele SM (2022) Biofortification of staple crops to alleviate human malnutrition: contributions and potential in developing countries. Agronomy 12(2):452. https://doi.org/10.3390/agronomy12020452

Kumari N, Rai PK, Bara BM, Singh I (2017) Effect of halo priming and hormonal priming on seed germination and seedling vigour in maize (*Zea mays* L.) seeds. J Pharmacogn Phytochem 6(4): 27–30

Kuppusamy N, Ranganathan U (2014) Storage potential of primed seeds of okra (*Abelmoschus esculentus*) and beet root (*Beta vulgaris*). Aust J Crop Sci 8(9):1290–1297

Lara TS, Lira JMS, Rodrigues AC, Rakocevic M, Alvarenga AA (2014) Potassium nitrate priming affects the activity of nitrate reductase and antioxidant enzymes in tomato Germination. J Agric Sci 6(2):72–80. https://doi.org/10.5539/jas.v6n2p72

Lemmens E, Deleu LJ, De Brier N, De Man WL, De Proft M, Prinsen E, Delcour JA (2019) The impact of hydro-priming and osmo-priming on seedling characteristics, plant hormone concentrations, activity of selected hydrolytic enzymes, and cell wall and phytate hydrolysis in sprouted wheat (*Triticum aestivum* L.). ACS Omega 4:22089–22100. https://doi.org/10.1021/acsomega.9b03210

Li Y, Liang L, Li W, Ashraf U, Ma L, Tang X, Mo Z (2021) ZnO nanoparticle-based seed priming modulates early growth and enhances physio-biochemical and metabolic profiles of fragrant rice against cadmium toxicity. J Nanobiotechnol 19(1):1–19

Lutts S, Benincasa P, Wojtyla L, Kubala S, Pace R, Lechowska K, Quinet M, Garnczarska M (2016) Seed priming: new comprehensive approaches for an old empirical technique. new challenges in seed biology-Basic and translational research driving seed technology. InTech Open, Rijeka, Croatia. 12:1–46

Ma HY, Zhao DD, Ning Q, Wei JLY, Wang M, Liu X, Jiang C, Liang Z (2018) A multi-year beneficial effect of seed priming with gibberellic acid-$_3$ (GA_3) on plant growth and production in a perennial grass, *Leymus chinensis*. Sci Rep 8:13214. https://doi.org/10.1038/s41598-018-31471-w

Mahdavi B, Rahimi A (2013) Seed priming with chitosan improves the germination and growth performance of ajowan (*Carum copticum*) under salt stress. EurAsian J Biosci 7:69–76. https://doi.org/10.5053/ejobios.2013.7.0.9

Maiti RK, Vidyasagar P, Rajkumar D, Ramaswamy A, Rodriguez HG (2011) Seed priming improves seedling vigor and yield of few vegetable crops. Technol Dev 2(1):125–130

Majda C, Khalid D, Aziz A, Rachid B, Badr AS, Lotfi A, Mohamed B (2019) Nutri-priming as an efficient means to improve the agronomic performance of molybdenum in common bean (*Phaseolus vulgaris* L). Sci Total Environ 661:654–663

Mansour MMF, Emam MM, Salama KHA, Morsy AA (2021) Sorghum under saline conditions: responses, tolerance mechanisms, and management strategies. Planta 254(2):1–38

Markovskaya EF, Sysoeva MI, Sherudilo EG, Topchieva LV (2007) Differential gene expression in cucumber plants in response to brief daily cold treatments. Russ J Plant Physiol 54:607–611

Marschner H (1995) Mineral nutrition of higher plants, 2nd edn. Academic, San Diego, pp 330–355

Marthandan V, Geetha R, Kumutha K, Renganathan VG, Karthikeyan A, Ramalingam J (2020) Seed priming: a feasible strategy to enhance drought tolerance in crop plants. Int J Mol Sci 21(21):8258. https://doi.org/10.3390/ijms21218258
Mavi K, Ermis S, Demir I (2006) The effect of priming on tomato rootstocks seeds in relation to seedling growth. Asian J Plant Sci 5(6):940–947. https://doi.org/10.3923/ajps.2006.940.947
McDonald MB (2000) Seed priming. In: Black M, Bewley JD (eds) Seed technology and its biological basis. Sheffield Academic Press, Sheffield, pp 287–325
McGuire S, FAO, IFAD, WFP (2015) The state of food insecurity in the world 2015: meeting the 2015 international hunger targets: taking stock of uneven progress. Rome FAO Adv Nutr 6(5): 623–624
Mo Z, Liu Q, Xie W, Ashraf U, Abrar M, Pan S, Duan M, Tian H, Wang S, Tang X (2020) Ultrasonic seed treatment and Cu application modulate photosynthesis, grain quality, and Cu concentrations in aromatic rice. Photosynthetica 58(3):682–691
Morteza MM, Reza KR, Ahmad N, Reza EH (2017) Influence of priming treatments on seed germination of sesame (*Sesamum indicum* L.) under osmotic conditions. Asian. J Biol Sci 10(3): 104–109. https://doi.org/10.3923/ajbs.2017.104.109
Munir T, Rizwan M, Kashif M, Shahzad A, Ali S, Amin N, Imran M (2018) Effect of zinc oxide nanoparticles on the growth and Zn uptake in wheat (*Triticum aestivum* L.) by seed priming method. Dig J Nanomater Biostruct 13(1):1892759. https://doi.org/10.1155/2022/1892759
Murungu FS, Chiduza C, Nyamugafata P, Clark LJ, Whalley WR, Finch-Savage WE (2004) Effects of 'on-farm seed priming' on consecutive daily sowing occasions on the emergence and growth of maize in semi-arid Zimbabwe. Field Crops Res 89:49–57
Naeem MA, Muhammad S (2006) Effect of seed priming on growth of barley (*Hordeum vulgare*) by using brackish water in salt affected soils. Pak J Bot 38(3):613–622
Narwal SS (1994) Allelopathy in crop production. Scientific Publishers, Jodhpur, p 288
Nascimento WM (2003) Muskmelon seed germination and seedling development in response to seed priming. Sci Agricol 60:71–75. https://doi.org/10.1590/S0103-90162003000100011
Nascimento WM, De Aragão FAS (2004) Muskmelon seed priming in relation to seed vigor. Sci Agric 61(1):114–117. https://doi.org/10.1590/S0103-90162004000100019
Nawaz F, Ashraf MY, Ahmad R, Waraich EA (2012) Selenium (Se) seed priming induced growth and biochemical changes in wheat under water deficit conditions. Biol Trace Elem Res 151(2): 284–293
Nejatzadeh F (2021) Effect of silver nanoparticles on salt tolerance of *Satureja hortensis* L. during *in vitro* and *in vivo* germination tests. Heliyon 7(2):e05981
O'Sullivan J, Bouw WJ (1984) Pepper seed treatment for low-temperature germination. Can J Plant Sci 64(2):387–393
Ogbonna JC, Abraham PG (1989) Effect of seed-pretreatment with some plant growth regulator on germination, growth and yield of cowpea (*Vigna sinensis* Endl.). J Crop Sci 58(4):641–647. https://doi.org/10.1626/jcs.58.641
Oliveira CED, Steiner F, Zuffo AM, Zoz T, Alves CZ, De Aguiar VCB (2019) Seed priming improves the germination and growth rate of melon seedlings under saline stress. Cienc 49(7): e20180588. https://doi.org/10.1590/0103-8478cr20180588
Osburn RM, Schroth MN (1989) Effect of osmopriming sugar beet deed on exudation and subsequent damping-off caused by *Pythium ultimum*. Phytopathology 78:1246–1250
Panuccio MR, Chaabani S, Roula R, Muscolo A (2018) Bio-priming mitigates detrimental effects of salinity on maize improving antioxidant defense and preserving photosynthetic efficiency. Plant Physiol Biochem 132:465–474
Paparella S, Araujo SS, Rossi G, Wijaya SM, Carbonera D, Balestrazzi A (2015) Seed priming: State of the art and new perspectives. Plant Cell Rep 34:1281–1293
Parera CA, Cantliffe DJ (1994) Pre-sowing seed priming. Hortic Rev 16:109–141
Passam HC, Karavites PI, Papandreou AA, Thanos CA, Georghiou K (1989) Osmo-conditioning of seeds in relation to growth and fruit yield of aubergine, pepper, cucumber and melon in unheated greenhouse cultivation. Sci Hortic 38(3–4):207–216

Patade VY, Bhargava S, Suprasanna P (2009) Halopriming imparts tolerance to salt and PEG induced drought stress in sugarcane. Agric Ecosyst Environ 134(1-2):24–28. https://doi.org/10.1016/j.agee.2009.07.003

Patanè C, Cavallaro V, D'Agosta G, Cosentino SL (2008) Plant emergence of PEG-osmoprimed seeds under suboptimal temperatures in two cultivars of sweet sorghum differing in seed tannin content. J Agron Crop Sci 194(4):304–309

Perez-Massot E, Banakar R, Gomez-Galera S, Zorrilla-Lopez U, Sanahuja G, Arjo G et al (2013) The contribution of transgenic plants to better health through improved nutrition: opportunities and constraints. Genes Nutr 8(1):29–41

Pfeiffer WH, Andersson MS, Govindaraj M, Parminder V, Cherian B, Illona P, Mulambu J (2018) Biofortification in underutilized staple crops for nutrition in Asia and Africa. In Regional expert consultation on underutilized crops for food and nutritional security in Asia and the Pacific – Thematic, strategic papers and country status reports. Bangkok. APAARI November 13–15, pp 70–81. http://oar.icrisat.org/10746/

Pill WG, Necker AD (2001) The effects of seed treatments on germination and establishment of Kentucky bluegrass (*Poa pratense* L.). Seed Sci Technol 29:65–72

Podlaski S, Chrobak Z, Wyszkowska Z (2003) The effect of parsley seed hydration treatment and pelleting on seed vigour. Plant Soil Environ 49(3):114–118. https://doi.org/10.17221/4099-PSE

Pongrac P, Fischer S, Thompson JA, Wright G, White PJ (2020) Early responses of *Brassica oleracea* roots to zinc supply under sufficient and sub-optimal phosphorus supply. Front Plant Sci 10:1645. https://doi.org/10.3389/fpls.2019.01645

Powell AA, Yule LJ, Jing H, Groot SPC, Bino RJ, Pritchard HW (2000) The influence of aerated hydration seed treatment on seed longevity as assessed by the viability equations. J Exp Bot 51: 2031–2043. https://doi.org/10.1093/jexbot/51.353.2031

Prabha D, Negi S, Kumari P, Negi YK, Chauhan JS (2016) Effect of seed priming with some plant leaf extract on seedling growth characteristics and root rot disease in tomato. Int J Agric Syst 4(1):46–51

Prasad R, Shivay YS, Kumar D (2014) Agronomic biofortification of cereal grains with iron and zinc Adv Agron 125:55–91

Prashanth L, Kattapagari KK, Chitturi RT, Baddam VR, Prasad LK (2015) A review on role of essential trace elements in health and disease. J Dr. NTR Uni Health Sci 4:75–85

Qasim A, Daud MK, Haider MZ, Ali S, Rizwan M, Aslam N, Noman A, Iqbal N, Shahzad F, Deeba F, Ali I, Zhu SJ (2017) Seed priming by sodium nitroprusside improves salt tolerance in wheat (*Triticum aestivum* L.) by enhancing physiological and biochemical parameters. Plant Physiol Biochem 119:50–58. https://doi.org/10.1016/j.plaphy.2017.08.010

Qi W, Zhang L, Wang L, Xu H, Jin Q, Jiao Z (2015) Pretreatment with low-dose gamma irradiation enhances tolerance to the stress of cadmium and lead in *Arabidopsis thaliana* seedlings. Ecotoxicol Environ Saf 115:243–249

Ragab G, Saad-Allah K (2021) Seed priming with greenly synthesized sulfur nanoparticles enhances antioxidative defense machinery and restricts oxidative injury under manganese stress in *Helianthus annuus* (L.) seedlings. J Plant Growth Regul 40:1894–1902

Rai-Kalal P, Jajoo A (2021) Priming with zinc oxide nanoparticles improve germination and photosynthetic performance in wheat. Plant Physiol Biochem 160:341–351

Rakshit A, Pal S, Rai S, Rai A, Bhowmick MK, Singh HB (2013) Micronutrient seed priming: potential tool in integrated nutrient management. SATSA Mukhaptra Annu Tech Issue 17:77–89

Rakshit A, Sunita K, Pal S, Singh A, Singh HB (2015) Biopriming mediated nutrient use efficiency of crop species. In: Nutrient use efficiency: From basics to advances. Springer, New Delhi, pp 181–191

Rangaswamy A, Purushothaman S, Devasenapathy P (1993) Seed hardening in relation to seedling quality characters of crops. Madras Agric J 80(9):535–537

Rao G, Ashraf U, Huang S, Cheng S, Abrar M, Mo Z, Pan S, Tang X (2018) Ultrasonic seed treatment improved physiological and yield traits of rice under lead toxicity. Environ Sci Pollut Res 25:33637–33644

Rao G, Huang S, Ashraf U, Mo Z, Duan M, Pan S, Tang X (2019) Ultrasonic seed treatment improved cadmium (Cd) tolerance in *Brassica napus* L. Ecotoxicol Environ Saf 185:109659. https://doi.org/10.1016/j.ecoenv.2019.109659

Raphael E (2012) Phytochemical constituents of some leaves extract of *Aloe vera* and *Azadirachta indica* plant species. J Adv Res 1(2):014–017

Rashid A, Harris D, Hollington P, Rafiq M (2004) Improving the yield of mungbean (*Vigna radiata*) in the north west frontier province of Pakistan using on-farm seed priming. Exp Agric 40(2):233–244. https://doi.org/10.1017/S0014479703001546

Rashid A, Hollington PA, Harris D, Khan P (2006) On farm seed priming for barley on normal, saline and saline-sodic soils in north west frontier province. Pak Eur J Agron 24(3):276–281. https://doi.org/10.1016/j.eja.2005.10.006

Reddy PP (2012) Bio-priming of seeds. In: Reddy PP (ed) Recent advances in crop protection. Springer, Cham, pp 83–90

Rehman A, Shahzad B, Ullah A, Nadeem F, Tanveer M, Sharma A, Lee DJ (2019) Seed priming-mediated improvement of plant morphophysiology under salt stress. In: Hasanuzzaman M, Fotopoulos V (eds) Priming and pretreatment of seeds and seedlings. Springer, Singapore, pp 205–217. https://doi.org/10.1007/978-981-13-8625-1_10

Rehman MS, Rauf F, Tania SS, Khatun M (2020) Seed priming methods: application in field crops and future perspectives. Asian J Crop Sci 5(2):8–19

Rogis C, Gibson LR, Knapp AD, Horton R (2004) Can solid matrix priming with GA3 break seed dormancy in eastern gamagrass? J Range Manag 57:656–661

Rouhi HR, Aboutalebian MA, Sharif-Zadeh F (2011a) Effects of hydro and osmopriming on drought stress tolerance during germination in four grass species. Int J Agric Sci 1:107–114

Rouhi HR, Sharif-Zadeh F, Aboutalebian MA (2011b) Seed priming improves the germination traits of tall fescue (*Festuca arundinacea*). Not Sci Biol 3(2):57–63. https://doi.org/10.15835/nsb.3.2.5409

Rowse HR (1991) Methods of priming seeds. UK Patent 2(192):781

Runkle ES, Heins RD, Cameron AC, Carlson WH (1999) Photoperiod and cold treatment regulate flowering of *Rudbeckia fulgida* Goldsturm. Hort Sci 34:55–58

Saltzman A, Birol E, Oparinde A, Andersson MS, Asare-Marfo D, Diressie MT, Gonzalez C, Lividini K, Moursi M, Zeller M (2017) Availability, production, and consumption of crops biofortified by plant breeding: current evidence and future potential. Ann New York Acad Sci 1390:104–114

Sarker A, Erskine W (2006) Recent progress in the ancient lentil. J Agric Sci 144(1):19–29

Satish S, Mohana DC, Ranhavendra MP, Raveesha KA (2007) Antifungal activity of some plant extracts against important seed borne pathogens of *Aspergillus* sp. J Agric Technol 3(1): 109–119

Sedghi M, Nemati A, Esmaielpour B (2010) Effect of seed priming on germination and seedling growth of two medicinal plants under salinity. Emirates J Food Agri 22(2):130–139

Shah T, Latif S, Saeed F, Ali I, Ullah S, Alsahli AA, Ahmad P (2021) Seed priming with titanium dioxide nanoparticles enhances seed vigor, leaf water status, and antioxidant enzyme activities in maize (*Zea mays* L.) under salinity stress. J King Saud Univ Sci 33(1):101207

Shahri S, Tilaki GAD, Behtari B, Alizadeh MA (2015) Growth responses of *Secale cereale* and *S. ceremont* to priming treatments under salinity stress. J Rangeland Sci 5(3):202–211

Shahverdi MA, Omidi H, Tabatabaei SJ (2017) Effect of nutri-priming on germination indices and physiological characteristics of stevia seedling under salinity stress. J Seed Sci 39(4):353–362. https://doi.org/10.1590/2317-1545v39n4172539

Sharma P, Gautam A, Kumar V, Guleria P (2022) MgO nanoparticles priming promoted the growth of black chickpea. J Agric Res 10:100435

Shin JS, Raymer P, Kim W (2006) Environmental factors influencing germination in seeded seashore paspalum. Hort Sci 41:1330–1331

Shivay YS, Singh U, Prasad R, Kaur R (2016) Agronomic interventions for micronutrient biofortification of pulses. Indian J Agron 61:161–172

Singh MV (2007) Efficiency of seed treatment for ameliorating zinc deficiency in crops. Zinc Crops 24–26

Singh H, Jassal RK, Kang JS, Sandhu SS, Kang H, Grewal K (2015a) Seed priming techniques in field crops-a review. Agric Rev 36(4):251–264

Singh U, Praharaj CS, Shivay YS, Kumar L, Singh SS (2015b) Ferti-fortification: an agronomic approach for micronutrient enrichment of pulses. In: Pulses: challenges and opportunities under changing climatic scenario. In: Proceedings of the national conference on "Pulses: challenges and opportunities under changing climatic scenario", vol 29, pp 208–222

Sivritepe N, Sivritepe HO, Eriş A (2003) The effect of nacl priming on salt tolerance in melon seedlings grown under saline conditions. Sci Hortic 97(3-4):229–237. https://doi.org/10.1016/S0304-4238(02)00198-X

Sivritepe H, Sivritepe N, Eriş A, Turhan E (2005) The effect of NaCl pre-treatments on salt tolerance of melons grown under long-term salinity. Sci Hortic 106(4):568–581. https://doi.org/10.1016/j.scienta.2005.05.011

Sliwinska E, Babinska L (1999) Effect of presowing hydration treatment on DNA replication activity in the embryo of sugar beet (*Beta vulgaris* L.). Electron J Pol Agric 2(2):04. http://www.ejpau.media.pl/series/volume2/issue2/agronomy/art.-04html.

Stevens GA, Finucane MM, De-Regil L, Paciorek CJ, Flaxman SR, Branca F et al (2013) Global, regional, and national trends in haemoglobin concentration and prevalence of total and severe anaemia in children and pregnant and non-pregnant women for 1995–2011: a systematic analysis of population-representative data. Lancet Glob Health 1(1):e16–e25

Stoltzfus R (2001) Defining iron-deficiency anemia in public health terms: a time for reflection. J Nutr 131:565–567

Subedi KD, Ma BL (2005) Seed priming does not improve corn yield in a humid temperate environment. Agron J 97(1):211–218. https://doi.org/10.2134/agronj2005.0211a

Sukanya V, Patel RM, Suthar KP, Singh D (2018) An overview: Mechanism involved in biopriming mediated plant growth promotion. Int J Pure Appl Biosci 6(5):771–783

Suruthi S, Sujatha K, Menaka C (2019) Evaluation of different seed priming methods for improvement of seedling growth of barnyard millet (*Echinochloa frumentacea* L.) cv MDU1. Elec. J Plant Breed 10(2):400–405

Szopińska D, Li W, Dorna H (2014) The effect of priming on germination and vigour of pansy (*Viola* × *Wittrockiana* Gams.) seeds. Acta Sci Pol 13(6):15–29

Tarquis AM, Bradford KJ (1992) Prehydration and priming treatments that advance germination also increase the rate of deterioration of lettuce seeds. J Exp Bot 43:307–317

Tavili A, Zare S, Enayati A (2009) Hydropriming, ascorbic and salicylic acid influence on germination of *Agropyro nelongatum* host. Seeds under salt stress. Res J Seed Sci 2(1):16–22

Taylor AG, Harman GE, Nielsen PA (1994) Biological seed treatments using *Trichoderma harzianum* for horticultural crops. Hortic Technol 4(2):105–108

Taylor AG, Allen PS, Bennett MA, Bradford KJ, Burris JS, Misra MK (1998) Seed enhancements. Seed Sci Res 8(2):245–256

Teshome W, Tana T, Dechassa N, Singh TN (2018) Effect of seed priming on germination and seedling growth of grain sorghum (Sorghum bicolor L. Moench) varieties. East Afr J Sci 12 (1):51–60

Tian Y, Guan B, Zhou D, Yu J, Li G, Lou Y (2014) Responses of seed germination, seedling growth, and seed yield traits to seed pretreatment in maize (*Zea mays* L.). Sci World J 14: 834630. https://doi.org/10.1155/2014/834630

Toklu F, Baloch F, Karaköy T, Özkan H (2015) Effects of different priming applications on seed germination and some agromorphological characteristics of bread wheat (*Triticum aestivum* L.). Turk J Agric 39(6):1005–1013

Tzortzakis GN (2009) Effect of pre-sowing treatment on seed germination and seedling vigour in endive and chicory. Hortic Sci 36(3):117–125

Ullah A, Khaliq A, Riaz A, Noor MA, Fiaz S, Waqas MA, Zain M, Ashraf U, Nawaz A (2018) Seed pre-treatment and planting geometry positively influence herbicide efficacy in wheat (*Triticum aestivum*). Planta Daninha 36:e018170372. https://doi.org/10.1590/S0100-83582018360100008

Ullah A, Shahzad B, Tanveer M, Nadeem F, Sharma A, Lee DJ, Rehman A (2019) Abiotic stress tolerance in plants through pre-sowing seed treatments with mineral elements and growth regulators. In: Hasanuzzaman M, Fotopoulos V (eds) Priming and pretreatment of seeds and seedlings. Springer, Singapore, pp 427–445. https://doi.org/10.1007/978-981-13-8625-1_21

Umair A, Ali S, Hayat R, Ansar M, Tareen MJ (2011) Evaluation of seed priming in mung bean (*Vigna radiata*) for yield, nodulation and biological nitrogen fixation under rainfed conditions. Afr J Biotechnol 10(79):1812–18129

Upadhyaya H, Begum L, Dey B, Nath PK, Panda SK (2017) Impact of calcium phosphate nanoparticles on rice plant. J Plant Sci Phytopathol 1:1–10

Vanangmudi K, Mallika V, Venkatesh A, Rai RSV, Umarani B, Balaji S (2000) Effect of osmotic priming on seed germination and vigour of neem (*Azadirachta indica*). J Tropic Forest Sci 12(1):181–184

Varier A, Vari AK, Dadlani M (2010) The subcellular basis of seed priming. Curr Sci 99:450–456

Wahid A, Noreen A, Basra SM, Gelani S, Farooq M (2008) Priming-induced metabolic changes in sunflower (*Helianthus annuus*) achenes improve germination and seedling growth. Bot Stud 49(4):343–350

Warren JE, Bennett MA (1997) Seed hydration using the drum priming system. Hortic Sci 32(7): 1220–1221

Welch RM, Graham RD (1999) A new paradigm for world agriculture: meeting human needs-productive, sustainable, nutritious. Field Crops Res 60:1–10

Welch RM, Graham RD (2002) Breeding crops for enhanced micronutrient content. Breeding crops for enhanced micronutrient content. Plant Soil 245:205–214

Welch RM, Graham RD (2004) Breeding for micronutrients in staple food crops from a human nutrition perspective. J Exp Bot 55:353–364

Wessells KR, Singh GM, Brown KH (2012) Estimating the global prevalence of inadequate zinc intake from national food balance sheets: effects of methodological assumptions. PLoS One 7: e50565

White J, Broadley MR (2005) Biofortifying crops with essential mineral elements. Trends Plant Sci 10:586–593

White PJ, Broadley MR (2009) Biofortification of crops with seven mineral elements often lacking in human diets – iron, zinc, copper, calcium, magnesium, selenium and iodine. New Phytol 182: 49–84

Wilson V, Tenkouano A (2019) Effects of storage and priming on seed emergence in soil and embryo culture of *Musa acuminata* Calcutta 4. Asian Plant Res J 2(1):1–9

Yaldagard M, Mortazavi SA, Tabatabaie F (2008) Application of ultrasonic waves as a priming technique for accelerating and enhancing the germination of barley seed: Optimization of method by the Taguchi approach. J Inst Brewing 114:14–21

Yari L, Aghaalikani M, Khazaei F (2010) Effect of seed priming duration and temperature on seed germination behavior of bread wheat (*Triticum aestivum* L.). J Agric Biol Sci 5(1):1–6

Yari L, Sheidaie S, Sadeghi H, Khazaei F (2012) Evaluation of temperature and seed priming duration on seed germination behavior of rice (*Oryza sativa* L). Int J Agric Res 1:7–11

Younesi O, Moradi A (2015) Effect of different priming methods on germination and seedling establishment of two medicinal plants under salt stress conditions. Cercetări Agronomice Moldova 3(163):43–53. https://doi.org/10.1515/cerce-2015-0040

Zeng Z, Liu X, Deng Q, Ashraf U, Chen J, Shen W (2023) Transcriptome analysis revealed mechanisms involved in improved germination and growth of sugarcane by ultrasonic treatment. Ind Crops Prod 192:116104. https://doi.org/10.1016/j.indcrop.2022.116104

Zhu C, Naqvi S, Gomez-Galera S (2007a) Transgenic strategies for the nutritional enhancement of plants. Trends Plant Sci 12(12):548–555. https://doi.org/10.1016/j.tplants.2007.09.007

Zhu C, Naqvi S, Gomez-Galera S, Pelacho AM, Capell T, Christou P (2007b) Transgenic strategies for the nutritional enhancement of plants. Trends Plant Sci 12:548–555

Chapter 13
Biochar for the Improvement of Crop Production

Jeetendra Singh, Santendra Kumar Soni, and Rajiv Ranjan

Abstract Rapid urbanization and population increase had a significant negative impact on soil health and fertility, putting additional burden on farming systems. Due to the excessive use of chemical fertilisers and pesticides in farming techniques, results in large greenhouse gas emissions. To fulfill the rising food demand has also contributed to these unsustainable farming methods. It is crucial that agricultural techniques improve soil microbiology, water retention, and fertiliser effectiveness. In recent years, biochars have gained attention for their potential as soil amendments. Biochars are produced by thermochemically processing biomass and are rich in carbon. The source feedstock and the reaction circumstances have an impact on the quality of the biochar during the pyrolysis processes, which heat up to 100 °C in oxygen-limited environments. Due to its potential to reduce greenhouse gas emissions, promote soil fertility, increase agricultural yield, and improve crop quality, biochar is attracting a lot of attention. The sequestration of stable carbon molecules in soil by biochar may also directly contribute in reducing climate change, as well as indirect effects such as enhancing tree carbon absorption. Global climate change and soil degradation are causing increasing difficulties that can be alleviated by the use of biochar. An effective technique to increase soil fertility is through making efficient use of crop residues and other farm wastes by turning them into a valuable source of soil amendment. Biochar's addition as a soil amendment has had a significant impact on the physical, chemical, and biological characteristics of the soil. When applied to soil, biochar decreases the bulk density and increases microbial activity, pH, and water holding capacity, however the effects vary depending on the texture of the soil. The addition of biochar has also increase soil porosity and infiltration rates.

Keywords Crop production · Biochar · Soil fertility · Plant growth · Pyrolysis · Thermochemical

J. Singh · S. K. Soni · R. Ranjan (✉)
Plant Molecular Biology Laboratory, Department of Botany, Dayalbagh Educational Institute, Dayalbagh, Agra, India

M. Hasanuzzaman et al. (eds.), *Mineral Biofortification in Crop Plants for Ensuring Food Security*, https://doi.org/10.1007/978-981-99-4090-5_13

13.1 Introduction

Considering the unprecedented climate change challenges, it is imperative that extensive research be conducted to significantly reduce greenhouse gas emissions (Rashid et al. 2020). Soil is the main terrestrial source for storing organic carbon (Van Der Voort et al. 2016). When land with natural or unmanaged vegetation is used for agricultural production, enormous amounts of carbon are released from standing biomass and soil (Banwart 2011). As a result, continuous international initiatives like the "4 per 1000" campaign aim to increase and better the level of organic carbon in soils (4p1000). Both the sustainability of local biogeochemical cycles and agriculture depend on healthy, high-quality soil. Reduced soil fertility and lower agricultural output, particularly in tropical areas, are the results of increasing population pressure on the agricultural sector (Lal 2015). Due to the rapid mineralization of soil organic matter and subsequent fall in soil carbon content, overuse of synthetic fertilisers can result in nutrient deficiencies, a loss of soil fertility, and global warming (Foley et al. 2005a, b). As a result, inorganic fertiliser application rates have increased to maintain crop output due to declining soil quality and a decrease in per-capita farmland.

Synthetic fertilisers are essential for the agricultural productivity. Agriculture has a variety of issues in the twenty-first century. It must supply the increasing population's food and industrial needs while also conserving the environment. By 2050, there will be 9.72 billion people on the planet (up from 7.35 billion in 2015), which means that the world's food supply must expand by about 70% from its present state (FAO 2009) in order to fulfil the world's food demand and one of the biggest issues is the over application of fertilisers, particularly N fertilisers like urea. In 2012, there was a 180 million tonne global demand for NPK fertiliser, of which 110 million tonnes (or around 61%) were for nitrogen fertiliser. The global demand for nitrogen fertiliser is to reach over 116 billion tonnes.

With a 1.3% annual growth rate, the demand for nitrogen fertiliser reached upto 116 million tonnes globally in 2016. According to the FAO (2012), the need for six million tonnes of nitrogen would increase by 60% between 2012 and 2016, 19% in America, 13% in Europe, 7% in Africa, and 1% in Oceania.

A study on, sustainable food system also supported by team of (UNEP 2012). The nutrients for plants are provided by organic material and elements in the soil. Natural soil mechanisms, such as the metabolism of nitrogen and mineral release and absorption, are disturbed through the agricultural process (Bot and Benites 2005). In order to sustain or boost agricultural productivity, the rate of synthetic fertiliser application has continuously grown from year to year due to decreasing land area per resident and declining soil quality (Srivastava 2009). It has been widely acknowledged that excessive inorganic fertiliser application, specifically N, may cause soil deterioration and other environmental concerns because of faster organic matter mineralization and ensuing losses in organic matter preserves (Palm et al. 2001; Foley et al. 2005a, b; Liu et al. 2010). Chemical fertiliser application, however, is not a sustainable strategy for improving soil fertility and maintaining yield improvements. Soil degradation is the major bio-physical barrier that much of the world's agricultural productivity is constrained by (Pender 2009; Lal 2015). Global agriculture is genuinely concerned about soil degradation, which includes declining soil fertility and increasing deleterious processes (Jianping 2006). Long-term farming can degrade soils by reducing soil organic matter, increasing soil acidity, and causing major erosion of soil (De Meyer et al. 2011). Additionally, as soil organic matter is lost, the overall stability of the soil deteriorates (Annabi et al. 2011). Consequently, it is crucial to use low-tech, environmentally friendly methods to restore the degraded soils. Pathogens, heavy metals, and medications can contaminate farmland over a lengthy period in manures and composts. Additionally, the release of ammonia and methane from manures and composts has the potential to worsen global warming and cause major nutrient pollution in streams and groundwater.

Biochar is a viable resource for managing soil fertility since it is a renewable resource with positive economic and environmental effects. Due to unsustainable extended land usage, degradation of soil micronutrients is a serious problem that is connected to poverty (Henao and Baanante 1999). According to estimates, there are 20 Tera-gram (Tg, 10^{12} g) of total nutritional deficits in the world, 75% of the deficits are found in emerging nations, 14% in nations with industrialization, and 11% in the least industrialised nations. Considering an overall the NPK deficiency for four different varieties (grains such as wheat, rice, maize, and barley), where particular nutritional deficits for N, P, and K each account for 28%, 12%, and 60% of the shortfall (Tan et al. 2005). The most common treatment for this deficiency is the use of soil amendments in the form of fertilisers that include the three major nutrients—nitrogen, phosphorus, and potassium. The most restricting ingredient for plant development is nitrogen. The formation of hormones, chlorophyll, vitamins, enzymes, and protein structures, as well as the expansion of stems and leaves, are all facilitated by nitrogen. Synthetic or commercialised chemical fertilisers were the dominant kind of soil amendment since the dawn of the industrial age. Nitrogen fertilisers are typically produced using the Haber-Bosch process, which employs nitrogen gas (N_2) from the air as the raw material and natural gas (CH_4) as the source of hydrogen, and the product of the reaction is ammonia (NH_3). According to Jatav et al. 2021, this ammonia is utilised as a precursor for nitrogen fertilisers such as urea ($CO(NH_2)_2$) and anhydrous ammonium nitrate (NH_4NO_3). Inorganic fertilisers have consistently demonstrated to have favourable impacts on plant development since the "green revolution" (Vanlauwe et al. 2010). In the previous 50 years, inorganic fertilisers have contributed significantly to rising agricultural output and productivity. The incredibly efficient fertiliser and seeding innovations, however, may be reaching their limiting point (Gruhn et al. 2000; Rosset et al. 2000). Due to the demand from the expanding human population and the consequent decrease in the amount of cultivable land available, it is no longer possible to employ protracted periods of silence for restoring soil quality in the tropics (Lal 2009).

A viable strategy for maintaining food security for the expanding global population is sustainable agricultural modernization and increasing productivity per unit land area (Tilman et al. 2011). Global study on biochar for its potential advantages as a solution to these issues (Shackley et al. 2012; Shafer et al. 2015). According to Jeyasubramanian et al. 2021, biochar is a particular kind of carbonaceous substance with distinct physicochemical features. Due to its benefits in the more significant fields of agriculture, soil health, wastewater treatment, and climate change. In the past 10 years, it has attracted a lot of interest (Vijay et al. 2021). According to studies, biochar enhances crop production, carbon sequestration, and Greenhouse (N_2O, CO_2, and CH_4) emission decrease (Arif et al. 2020). Furthermore, according to Zhang et al. (2020), biochar that has been enriched with ammonium, or nitrate, and phosphate may act like a slow-release fertiliser to increase the fertility of soil. The by-product of pyrolyzing biomass in an atmosphere with little oxygen is biochar. A variety of functional groups and porous carbonaceous structure are present (Nidheesh et al. 2021). Humic and fluvic compounds may be extracted from biochar due to its very porous structure (Nath et al. 2022).

Carbon, nitrogen, hydrogen, and a few less-nutritional substances such as K, and Mg, Ca, Na are the core constituents of biochar (Zhang et al. 2015). Typically, when the pyrolysis temperature rose from 300 to 800 °C, the carbon content dropped while the nitrogen and hydrogen contents increased.

According to Kammann et al. (2015) and Schmidt et al. (2015), biochar contains a large specific surface area and a variety of polar or nonpolar molecules that bind to chemical ions like phosphate, calcium nitrates and heavy metal ions. It has been proposed that biochar can improve soil's chemical, physical, and microbiological properties. Biochar can increase the electrical conductivity of soil by 124.6%, cation exchange ability by 20%, and acidity of soil by 31.9%, according to Oguntunde et al. (2004), Laird et al. (2010a, b) and Oguntunde et al. (2004). Biochar has additionally been shown to increase microbial biomass by 125% and the biological community makeup of soil (Grossman et al. 2010; Liang et al. 2010). According to Steiner et al. (2008a, b), basal respiration rose 35 h later, with the addition of substrate and the application of biochar, the CO_2 concentration increased by 30.1%.

13.2 History of Biochar

The name "biochar" word coined in the latter part of the twentieth decade and is derived from the Greek words "bios," and stands for existence, and "char," which refers to the biochar produced by the pyrolysis of biomass. Evidence from archaeology suggests that biochar was created and utilised by humans for the first time approximately 2500 years ago (Gabhane et al. 2020).

The Terra Preta, commonly referred to as "Indian black earth," was the first instance of biochar used to improve soil. Western Amazonia is where the Terra Preta soil type identified (Glaser and Birk 2012). Indicators of its uniqueness include its dark colour, high aggregate stability brought on by the presence of more carbon, which include substances and a significant nutrient content associated with an increase in microbial presence (Glaser and Birk 2012). The Terra Preta's intentional introduction of biochar to improve Its most notable characteristic is the soil profile, according to prehistoric Indigenous cultures (Glaser and Birk 2012). The tropical rainforest has been able to thrive and grow because of the application using biochar to improve soil (Lima et al. 2002). The Amazon rainforest's Terra Preta provides an example of how deficient soils can modified to enhance health and biomass fertility (Alling et al. 2014). Tropical sandy soils lack natural fertility; life depends primarily on the organic matter that the forest canopy provides (Alling et al. 2014).

When Lima et al. (2002) examined the Terra Preta's various features, they found mineral mica fragments from different types that had once been used in pottery in soil sub-layers. It is evident that the char mostly produced in clay ovens, and that both the clay particles and the resulting biochar into the soil content (Lima et al. 2002). According to Novotny et al. (2015), this technique is thought to have existed for close to 2000 years. Terra Preta soils include a diverse range of microbial communities (Kim et al. 2007).

Acido-bacteria species are particularly prevalent, and the Terra Preta has a 25% higher variety of bacterial species than other soils (Kim et al. 2007). This is significant because diverse types of bacteria mostly required in soils to give growing plants with a source of nitrogen (Kim et al. 2007). The Terra Preta finding demonstrates that the indigenous people of the Amazon basin either knew how to produce biochar or consciously put this substance to the soil to boost fertility (Lima et al. 2002).

13.3 Feed Stock for Production of Biochar

Switchgrass (*Panicum virgatum* L.), hardwoods, peanut hulls (*Arachis hypogaea* L.), maize husk, pecan (*Carya illinoinensis*) the shells, tree bark, sugarcane (*Saccharum officinarum* L.), sorghum (*Oryza sativa* L.), animal excrement, sewage effluent, urban backyard garbage, commercial leftovers, and the aquaculture sector waste in general, waste (Diatta et al. 2020; Hassan et al. 2020).

Non-woody biomass made up of garbage such as animal and poultry manures, agricultural and crop residues, urban and commercial garbage, and non-woody biomass has characteristics such as a significant mass density, lower void age, excessive calorie value, minimal moisture and debris material, and a significant calorific value, according to Jafri et al. (2018). Biochar formed from seaweed, agricultural leftovers, and animal manures is more nutrient-rich, has an elevated pH, yet contains less durable carbon than woody biochar (Gul et al. 2015; Aller 2016; El-Naggar et al. 2019; Diatta et al. 2020).

According to Omotade et al. (2020), Biochar was created by pyrolyzing corn cobs, nut shells, cow manure, and chicken litter at 300–600 °C for 3 h. According to the study's findings, cow dung had significant concentrations of N (0.62%) and K (16.2 mg g1), P (66.4 mg g1), Mg (0.28%), In addition to Ca (4.21 mg g1), peanut shell contains sulphur (0.28%). Maize cob had the lowest concentrations of N, Mg, and S. Thus, biochar's made from wood are useful when the end goal is to store carbon on ground soil (Wang et al. 2017; Ippolito et al. 2020), However, biochar made from animal excrement and grassy plants is used when the goal is to increase the soil's N, P, K, magnesium, and calcium, and sulphur content (Amoah-Antwi et al. 2020; Ippolito et al. 2020).

Ippolito et al. (2020) Biochar formed from hardwood provides 0.002, 2.2, and 17% of total nitrogen, phosphorus, and potassium, respectively, but biochar created from softwoods may provide 27% total P and 6% K. The phrase "feedstock" pertains to the material that is pyrolyzed in order to produce biochar. Thermal degradation of fibre and its lignin occurs at degrees ranging from 240 to 350 °C and 280 to 500°C, respectively. Biochar having a high mineral content, including grasses, grains husks, and straw-like remnants, is often used to make ash-rich biochar (Demirbas 2004). Because of the rigid ligninolytic nature of the parent material maintained in the biochar residue, pyrolysis of wood-based feedstocks generates brittle, more durable biochar having carbon percentages of up to 80% (Winsley 2007).

13.4 Production Process

Despite its older origins in the Amazon region, where it is known as "terra preta" soil, biochar has received significant attention in the current organic farming system over the last 20 to 25 years (Gabhane et al. 2020). The Amazonians have relied on these rich, black soils for millennia to meet their needs for food and grazing. There are many ways to produce biochar which are as follows:

13.4.1 Traditional Method

According to archaeological evidence, biochar production and use began in the South American Amazon basin thousands years ago (Gabhane et al. 2020). Previously, there were two methods for producing biochar: (1) pile the hardwood in soil mines, conceal the pits with layers of soil, and slowly burn the wood with no or little oxygen, or (2) burn the wood in the open air and immediately cover the half portion-burned hardwood with sections of soil (Gabhane et al. 2020; Thines et al. 2017). Clay burners, firebrick pits, and brick kilns are traditional methods for producing biochar from a pit surrounded by clay, bricks, or metal (Gabhane et al. 2020). In these operations, a fixed-bed reactor system is employed. Organic biomass is heaped and fired in earthen or metal kilns without air for several hours to days. In these approaches, the rate of heating is always quite low, and there might be no ventilation for a few hours to days. Because of limited heat transmission on the inside reactors, the temperature rise in these techniques is always very little and may not be uniform for all wood particulates (Gholizadeh et al. 2020).

13.4.2 Pyrolysis

The term "pyrolysis" is a combination of the words "lysis" (meaning "destroying breakdown") and "pyro" (heat). Biochar, bio-oil, and synthetic gas (syngas) are produced by thermally degrading biomass around 350 °C to 1000 °C inside kilns, reacts, and additional specialised machinery while being partially or entirely O_2depleted (Yadav and Khare 2020). Each product's recovery % is affected by several factors, including feedstock supply, pyrolysis heat, and warming rate. Biological oil is a concentrated blend of cyclic chemicals that requires minimal reformation and hence has a wide range of uses. Syngas is formed as a by-product when plastic is used as an initial material or substrate during pyrolysis (Jouhara et al. 2018; Amini et al. 2015).

Pyrolysis is categorised into three forms depending on criteria such as the temperature, heating rate, residential time, and operational parameters: slow pyrolysis, rapid pyrolysis, and intermediate pyrolysis (Choi et al. 2017; Diatta et al. 2020).

13.4.2.1 Slow Pyrolysis

This procedure is carried out under the environment's pressure and at relatively low temperatures (300–500 °C). There are minimal temperature rises ranging from 0.01–2.0 °C S^{-1}. Biochar preparation typically takes 5 to 30 min, and the biochar output rate is around 35%.

13.4.2.2 Fast Pyrolysis

This technique operates at a high pressure and a temperature between 501 and 8000 ° C. There are rapid heating rates of 10 to 50 °C per s. The preparation of biochar often takes time.

13.4.2.3 Intermediate Pyrolysis

It has a temperature of 500 °C and combines slow and quick pyrolysis. The manufacture of biochar typically takes 10 to 20 s, and around 25% of biochar is produced.

13.4.3 Gasification

Hot temperatures (700–1000 °C) are required for the thermochemical process known as gasification (Diatta et al. 2020; Zhang et al. 2017). The frequency at which biochar gets produced as a byproduct of the gasification process is affected by the burning agents, oxygenation equivalency ratio, atmospheric pressure, and feedstock types used in the method (Benedetti et al. 2018). According to You et al. (2017), the biochar produced by the process of gasification is typically much lower (200 g kg^{-1}) than the pyrolysis method (200–500 g kg^{-1}). This process involves the gasification of biomass fuel, which is to some part oxidised within the combustion chambers temperatures >800 °C and or ambient reduced pressure. In this technique, the biochar synthesis period ranges between 10 and 20 s, and the biochar manufacturing rate ranges between 1 and 10%.

13.4.4 Torrefaction

Torrefaction, sometimes referred to as moderate pyrolysis, is a novel thermochemical biochar production technique (Barskov et al. 2019). In the first step of the torrefaction, the feedstock is heated between 160 and 180 °C to eliminate moisture, and then the temperature is raised to 200 °C. A partial process of biomass decomposition begins when the temperature is between 250 and 270 °C. Torrefaction produces far more charcoal than pyrolysis due to the lower temperature employed in the process (Liu et al. 2015) found that heating biomass at 290 °C for 10 and 40 min generated 80 and 62% of biochar, respectively.

13.5 Characteristics of Biochar

The ratios of the labile and stable carbon pools in biochar, which have a biphasic nature, are influenced by amounts of hemicellulose, the feedstock contains cellulose and lignin (Joseph et al. 2013; Sohi et al. 2009). Temperatures of pyrolysis are the major cause of produced biochar become more arromatic (non-volatile, high C, and low O). These chars undergo slower oxidation and create surface groups that operate with oxygen (Jung et al. 2016). In contrast, biochar's created at a lower warmth have more volatile, and mostly aliphatic components with low Carbon concentration and high Oxygen concentrations (Novak et al. 2010; Zimmerman 2010; Fang et al. 2013). Biochar is a byproduct made primarily of carbon-product derived from plant biomass pyrolysis that is purposely contribute in soil improvement for agronomic and sustainable benefits agronomic as well as sustainability benefits. Organic materials are pyrolyzed under low oxygen conditions to produce biochar (Sohi et al. 2010; Lehmann and Joseph 2009; Jeffery et al. 2011; Kookana et al. 2011; Enders et al. 2012; Biederman and Harpole 2013; Hass and Gonzalez 2014; Thomas and Gale 2015; Wang et al. 2015). Biochar's intended application as a soil addition (Lehmann and Rondon 2006) as well as for an extended period carbon storage approach (Mašek et al. 2013) set it apart from similar materials and are two of its important properties that are like those of biochar (Zimmerman 2010).

As an agricultural soil type supplement, biochar be able to boost fertility of the soil (Kloss et al. 2013; Lehmann et al. 2003) and ensure the efficiency of crops (Asai et al. 2009; Lehmann and Rondon 2006) by making nutrients more readily available while also reducing leaching losses. It might potentially boost nutritional availability to plants and decrease the need for fertilizer (Glaser et al. 2002; Laird et al. 2010a, b; Woolf et al. 2010; Yao et al. 2012). Biochar can also promote microbial growth variety and activity (Steiner et al. 2008a, b; Lehmann et al. 2011; Gomez et al. 2013). Furthermore, Biochar has been scientifically shown to increase soil water retention capacity (Karhu et al. 2011; Sun and Lu 2013; Wang et al. 2013), lowering carbon gas emissions (Spokas et al. 2009; Singh et al. 2010; Woolf et al. 2010), and limit pollutants' accessibility, the bioavailability and toxicity (Beesley et al. 2010; Hale et al. 2011; Uchimiya et al. 2011; Ahmad et al. 2014). Furthermore, using biochar may enhance soil's capacity to retain carbon, thus helping in the fight against global warming (Lehmann and Rondon 2006; Barrow 2012; Sohi 2012).

13.6 Effect of Biochar on Soil Properties

The physiochemical properties of the soil have a substantial influence on crop development and production, microbial activity, and nutrient retention and absorption (Kumar et al. 2019; Seleiman et al. 2020). According to research, biochar can enhance the physiological, chemical, and biological characteristics of soil, hence fostering a favourable Root health, absorption of nutrients, and development of plants environment (Blanco-Canqui 2017; Diatta et al. 2020). In addition to other attributes, biochar utilisation influences Soil infiltration of water, WHC, aggregate durability, soil drainage and permeability, bulk thickness, soil strength, pH, CEC, are all important factors to consider (Kavitha et al. 2018; Purakayastha et al. 2019; Adekiya et al. 2020; Wang et al. 2020).

13.6.1 Physiochemical Properties of Soil

13.6.1.1 Soil Porosity

The term "porosity" refers to the interstitial area among dirt particles and has an impact on aeration of the soil, retained nutrients, and water retention transport (Blanco-Canqui 2017). The porosity of different soil types of ranges, with soils rich in clay having significant porosity and sandy soils having lesser. Increasing soil aggregation and decreasing bulk density of the soil are two ways that biochar might promote soil porosity (Blanco-Canqui 2017). In comparison to the control, Adekiya et al. (2020) observed that applying biochar increased soil porosity by 65%.

13.6.1.2 Bulk Density

Bulk density serves as a measure of soil compaction and overall soil health, influencing water intrusion, plant roots depth, and nutrition transport (Alghamdi 2018). As reported by Omondi et al. (2016), introducing biochar prevented soil erosion bulk density in 19 out of 22 soils ranging between 3 and 31%. After addition of biochar, coarse-textured soils have shown a greater decrement in bulk density as compared to fine-textured soils (Alghamdi 2018). Maximum reductions in soil bulk density of up to 31% were observed in sand following the application of biochar in a series of short-term lab experiments, after applying biochar, the soil bulk density could fall by as much as 31% in sand, 14.2% in coarse soil, and 9.2% in fine soil (Liu et al. 2017).

13.6.1.3 Soil Aggregation

Biochar has a significant power to expand soil when formed at low temperatures, the emergence of aggregates is caused by its intense contact with soil substance particles (Brodowski et al. 2005). Furthermore, Biochar carbon combines with both organic and inorganic soil constituents, affecting soil consolidation and other related phenomena (Briggs et al. 2012). Biochar has the potential to enhance wet combine which is more resilient in sand soils rather than in clayey ones. In compared to fine-textured soils, biochar organic materials may increase particle interaction and soil cohesiveness (Burrell et al. 2016).

13.6.2 Chemical Properties of soil

13.6.2.1 Soil pH

The transportation as well as the accessibility of numerous nutrients and chemical components in the soil are influenced by soil pH. Biochar soil application generally raises the soil's pH, even though variances are heavily impacted by soil kind, substrate material, and biochar precipitation value (DeLuca et al. 2015; Tian et al. 2017). According to Martinsen et al. (2015), Biochar rates from palm kernel (*Elaeis guineensis*) shell, cocoa bean shells (*Theobroma cacao* L.), and husks of rice (*Oryza sativa* L.) were applied to 31 acidic soils. After cocoa shell biochar was applied, the pH of the soil increased from 4.7 to 5.0, was greater than that of oil palm husk and husk of rice. El-Naggar et al. (2018) discovered that adding umbrellas plant (*Maesopsis eminii*) remains, silvergrass, and rice husks to unprocessed sandy soil led to a considerable rise in pH (Chathurika et al. 2016). After 70 days of biochar application, soil pH increased substantially more than quasi-biochar-amended soils.

13.6.2.2 Cation Exchange Capacity (CEC)

Cation exchange capacity is a significant soil property that regulates soil fertility as well as mineral elements and ion such as Na^+,Ca^{2+}, Mg^{2+}, K^+, and NH^{4+} are absorbed through adsorption. Even though cation exchange capacity is a fundamental property of the soil. The occurrence of strong carboxyl and phenol functional chains having deleterious consequences charges on the surface of biochar particles may be the cause of the various studies that have been suggested the value of adding Biochar has the potential to enhance soil CEC (Chathurika et al. 2016; Gao et al. 2017; Pandit et al. 2018; El-Naggar et al. 2018; Karimi et al. 2020). According to Zhang et al. (2017), using biochar increased the soil CEC by 21% when compared to unapplied soil.

El-Naggar et al. (2018) used biochar manufactured from silvergrass by-products, umbrella tree leftovers, and rice straw to sandy soils and discovered that the CEC raised by 906, 180, and 130%, in comparison to unaltered soil.

13.6.2.3 Water Holding Capacity (WHC)

WHC, or water retention in the soil refers to the greatest possible amount of water which a soil's pores can keep or hold. According to several field investigations, adding biochar improved soil water retention through having a favourable impact on the soil's porosity—the area between its particles—as well as other structural and textural characteristics (Razzaghi et al. 2020; Mclennon et al. 2020). Furthermore, regarding the space between each biochar particle, which acts as a porosity, biochar particles also contain intrapore space, which offers additional room for water storage (Liu et al. 2017).

Kätterer et al. (2019) discovered that continual biochar application for 10 years significantly enhanced soil Water Holding Capacity when compared to un-amended soil. The quantity of biochar used affects the soil's moisture content. Ndor et al. (2015) showed a 10.8% raise in soil water content after treating with sawdust that originate and rice husk biochar at 5–10 Mg ha^{-1} when compared with no biochar treatment.

13.6.2.4 Increase Nutrient Availability

Biochar application can also boost soil carbon stocks, which can contribute to soil nutrient retention, Short-term conversion of native organic carbon (positive priming), enhanced soil fertility, carbon in the soil imprisonment, and better crop production and development (Ouyang et al. 2014; Song et al. 2018). Xie et al. (2022) found that the addition of biochar to soil has risen total nitrogen material created by 27.5%, carbon content by 75.5%, and peanut production by 50.5% after 4 years of testing. Multiple research investigations indicate that incorporating biochar into the soil can reduce nitrogen leaching and increase nitrogen retention as well as the overall nitrogen concentration (Major et al. 2012). Due to its porous characteristics, biochar has a significant capacity to absorb nitrate and ammonium. As a result, it can improve nitrate and ammonium storage while also increasing nitrogen immobilisation. On the other hand, it lowers ammonium volatilization and promotes nitrogen bioavailability to increase the yield and development of crops (Rondon et al. 2006). Biochar applied at the right rates to saline soil lowers nitrogen leaching, preserves soil nitrogen, and prevents ammonium combustion (Rondon et al. 2006). When phosphorus resources reduced, biochar application may become more important and help in recycling phosphorus (P). Because of the significant phosphorus levels in its litter components, adding biochar to soil increases phosphorous accessibility (Zhai et al. 2015).

13.6.3 Biological Properties of Soil

Biochar, with its porous structure, high CEC, and high sorption capacity, can boost the variety and activity of soil microbial communities (Lehmann et al. 2011; Zheng et al. 2013). These characteristics also aid in the physical and chemical characteristics of soil. The inherent properties the use of biochar may improve preservation of nutrients and accessibility to microorganisms while additionally having an impact on the interactions of soil, plant, and intestinal microbes (Lehmann et al. 2011; Colombo et al. 2013). Small beneficial soil creatures encapsulated in the structure of biochar, including Symbiotic mycorrhizal fungus that may infiltrate deeper into the cavity of charcoal and increase the number of enzymes in the soil.

ten Berge et al. (2022) studied the effect of biochar on enzyme activity in soil and came to a verdict that application of biochar increased combining urea and b-glucosidase enzymes' activity in comparison to control. In the presence of biochar, symbionts (such as *Rhizobium*) become activated, nodulation and the activity of nitrogenase should be enhanced. Additionally, *Azotobactor* sp. and *Azospirillum* colonise and develop in soil that has been treated with biochar due to the abundance of habitat as well as necessary oxygen (Gabhane et al. 2020).

13.7 Biochar Micro-organisms and Soil Fertility

Biochar has been reported to enhance soil biological properties as well as improve soil physicochemical parameters in research by Pietikäinen et al. (2000), Lehmann et al. (2006), Kim et al. (2007), O'Neill et al. (2009), Grossman et al. (2010), and Liang et al. (2010). These changes could strengthen the composition of soil by increasing plant based/mineral combinations (aggregates) and pore dimensions (Rillig and Mummey 2006), or they may improve the flow of nutrients by increasing nutrient absorption and encapsulation and decreasing nutrient leaching (Steiner et al.

2008a, b), and thereby foster the growth of plants (Warnock et al. 2007). Additionally, certain microbes, bacteria and fungi in the rhizosphere may directly promote plant development (Schwartz et al. 2006; Compant et al. 2010). Changes in microbial community composition or activity caused by biochar may have an influence on organic matter in the soil cycling, the movement of nutrients, and plant development (Wardle et al. 2008; Kuzyakov et al. 2009; Liang et al. 2010).

13.8 Effect of Biochar on Improvement of Crop Productivity

It was previously established that applying biochar to the soil is beneficial in agricultural fields will encourage plant growth. Extraordinary agronomic gains for crop yield are provided by long-life biochar in a long-term manner (Chan et al. 2008). In general, the fundamental source of the increased crop productivity and nutrient absorption may be direct nutritional additions from sprinkled biochar comprising various nutrients. In terms of the impact of biochar on the growth of plants in the world's dry or semi arid areas applying Biochar is being shown to boost above-ground plant growth by between 10 and 25% (Biederman and Harpole 2013; Liu et al. 2016).

When biochar is applied, the physical, chemical, and biologically characteristics of soil are changed, which causes an increase in plant growth and agricultural yields. Humid and mesic regions with higher potential output. Because these extensively abused soils had improved soil qualities, biochar application in conjunction with fertilisers promoted higher agricultural yields in Amazonia's humid tropics (Steiner et al. 2008a, b; Lehmann et al. 2003). According to Agegnehu et al. (2021), Sub-Saharan Africa, as well as parts of South the United States, the southeast Asia region, and southern North America, are affected should respond favourably in terms of agricultural yield response to the use of biochar.

The improvement in the soil aggregation, retention of nutrients and absorption of water brought through the addition of biochar to the surface. According to Murtaza et al. (2021), The utilisation of biochar boosted the yields of corn, cowpea, as well as a peanut when applied as a fertilizer because it raised pH, cation exchange capacity, and nutrition levels availability of the soil. Major et al. (2010) applied 15 and 20 t/ha biochar, respectively where maize was cultivated and it was found that the yield had increased by 150 and 98%, respectively.

13.9 Conclusion

Biochar originates from recyclable materials, and not only solves the agricultural waste management problems but also improves soil qualities and encourages the growth of beneficial microbial variety. It has demonstrated through studies that applying biochar into the soils has the potential to significantly enhance agricultural production and soil fertility. This is accomplished by improving soil properties and increasing the accessibility of plant nutrients etc. Consequently, using biochar manufactured from organic waste to increase the soil's fertility and crop yield is an effective method.

Acknowledgments We are thankful to the Director, Dayalbagh Educational Institute, Dayalbagh, Agra for providing kind support and infrastructure.

References

Adekiya AO, Agbede TM, Olayanju A, Ejue WS, Adekanye TA, Adenusi TT, Ayeni JF (2020) Effect of biochar on soil properties, soil loss, and cocoyam yield on a tropical sandy loam Alfisol. Sci World J 2020:1–9

Agegnehu G, Amede T, Erkossa T, Yirga C, Henry C, Tyler R, Nosworthy MG, Beyene S, Sileshi GW (2021) Extent and management of acid soils for sustainable crop production system in the tropical agroecosystems: a review. Acta Agriculturae Scandinavica Sect B—Soil Plant Sci 71 (9):852–869

Ahmad M, Rajapaksha AU, Lim JE, Zhang M, Bolan N, Mohan D, Vithanage M, Lee SS, Ok YS (2014) Biochar as a sorbent for contaminant management in soil and water: a review. Chemosphere 99:19–33

Alghamdi AG (2018) Biochar as a potential soil additive for improving soil physical properties—a review. Arab J Geosci 24:76

Aller MF (2016) Biochar properties: transport, fate, and impact. Crit Rev Environ Sci Technol 46: 1183–1296

Alling V, Hale SE, Martinsen V, Mulder J, Smebye A, Breedveld GD, Cornelissen G (2014) The role of biochar in retaining nutrients in amended tropical soils. J Plant Nutr Soil Sci 177:671–680

Amini S, Ghadiri H, Chen C, Marschner P (2015) Salt-affected soils, reclamation, carbon dynamics, and biochar: a review. J Soils Sediments 16:939–953

Amoah-Antwi C, Kwiatkowska-Malina J, Szara E, Thornton S, Fenton O, Malina G (2020) Efficacy of woodchip biochar and Brown coal waste as stable sorbents for abatement of bioavailable cadmium, lead and zinc in soil. Water Air Soil Pollut 231:1–17

Annabi M, Le Bissonnais Y, Le Villio-Poitrenaud M, Houot S (2011) Improvement of soil aggregate stability by repeated applications of organic amendments to a cultivated silty loam soil. Agric Ecosyst Environ 144:382–389

Arif M, Jan T, Riaz M, Fahad S, Adnan M, Ali K, Mian IA, Khan B, Rasul F (2020) Biochar; a remedy for climate change. Environment, climate, plant and vegetation growth, pp 151–171

Asai H, Samson BK, Stephan HM, Songyikhangsuthor K, Homma K, Kiyono Y, Inoue Y, Shiraiwa T, Horie T (2009) Biochar amendment techniques for upland rice production in northern Laos. Field Crops Res 111:81–84

Banwart S (2011) Save our soils. Nature 474:151–152

Barrow C (2012) Biochar: potential for countering land degradation and for improving agriculture. Appl Geogr 34:21–28

Barskov S, Zappi M, Buchireddy P, Dufreche S, Guillory J, Gang D, Hernandez R, Bajpai R, Baudier J, Cooper R, Sharp R (2019) Torrefaction of biomass: a review of production methods for biocoal from cultured and waste lignocellulosic feedstocks. Renew Energ 142:624–642

Beesley L, Moreno-Jiménez E, Gomez-Eyles JL (2010) Effects of biochar and greenwaste compost amendments on mobility, bioavailability, and toxicity of inorganic and organic contaminants in a multi-element polluted soil. Environ Pollut 158:2282–2287

Benedetti V, Patuzzi F, Baratieri M (2018) Characterization of char from biomass gasification and its similarities with activated carbon in adsorption applications. Appl Energy 227:92–99

Biederman LA, Harpole WS (2013) Biochar and its effects on plant productivity and nutrient cycling: a metaanalysis. GCB Bioenergy 5(2):202–214

Blanco-Canqui H (2017) Biochar and soil physical properties. Soil Sci Soc Am J 81:687–711

Bot A, Benites J (2005) The importance of soil organic matter: key to drought-resistant soil and sustained food production. FAO UN, Rome, p 78

Briggs C, Breiner JM, Graham RC (2012) Physical and chemical properties of Pinus ponderosa charcoal. Soil Sci 177:263–268

Brodowski S, Amelung W, Haumaier L, Abetz C, Zech W (2005) Morphological and chemical properties of black carbon in physical soil fractions as revealed by scanning electron microscopy and energy-dispersive X-ray spectroscopy. Geoderma 128:116–129

Burrell LD, Zehetner F, Rampazzo N, Wimmer B, Soja G (2016) Long-term effects of biochar on soil physical properties. Geoderma 282:96–102

Chan KY, Van Zwieten L, Meszaros I, Downie A, Joseph S (2008) Using poultry litter biochars as soil amendments. Soil Res 46:437

Chathurika JS, Kumaragamage D, Zvomuya F, Akinremi OO, Flaten DN, Indraratne SP, Dandeniya WS (2016) Woodchip biochar with or without synthetic fertilizers affects soil properties and available phosphorus in two alkaline, chernozemic soils. Can J Soil Sci 96: 472–484

Choi JH, Kim SS, Ly HV, Kim J, Woo HC (2017) Effects of water-washing Saccharina japonica on fast pyrolysis in a bubbling fluidized-bed reactor. Biom Bioener 98:112–123

Colombo C, Palumbo G, He JZ, Pinton R, Cesco S (2013) Review on iron availability in soil: interaction of FE minerals, plants, and microbes. J Soils Sediments 14:538–548

Compant S, Clément C, Sessitsch A (2010) Plant growth-promoting bacteria in the rhizo- and endosphere of plants: their role, colonization, mechanisms involved and prospects for utilization. Soil Biol Biochem 42:669–678

De Meyer A, Poesen J, Isabirye M, Deckers J, Raes D (2011) Soil erosion rates in tropical villages: a case study from Lake Victoria basin, Uganda. Catena 84:89–98

DeLuca TH, Gundale MJ, MacKenzie MD, Jones DL (2015) Biochar effects on soil nutrient transformations. In: Biochar for environmental management, pp 453–486

Demirbas A (2004) Combustion characteristics of different biomass fuels. Prog Energy Combust Sci 30:219–230

Diatta AA, Fike JH, Battaglia ML, Galbraith JM, Baig MB (2020) Effects of biochar on soil fertility and crop productivity in arid regions: a review. Arab J Geosci 13:1–7

El-Naggar A, Lee SS, Awad YM, Yang X, Ryu C, Rizwan M, Rinklebe J, Tsang DC, Ok YS (2018) Influence of soil properties and feedstocks on biochar potential for carbon mineralization and improvement of infertile soils. Geoderma 332:100–108

El-Naggar A, El-Naggar AH, Shaheen SM, Sarkar B, Chang SX, Tsang DC, Rinklebe J, Ok YS (2019) Biochar composition-dependent impacts on soil nutrient release, carbon mineralization, and potential environmental risk: a review. J Environ Manag 241:458–467

Enders A, Hanley K, Whitman T et al (2012) Characterization of biochars to evaluate recalcitrance and agronomic performance. Bioresour Technol 114:644–653

Fang Y, Singh B, Singh BP, Krull E (2013) Biochar carbon stability in four contrasting soils. Eur J Soil Sci 65:60–71

FAO (2009) How to feed the world in 2050. High-Level Expert Forum, Rome
FAO (2012) Current world fertilizer trends and outlook to 2016. Food and Agriculture Organization of the United Nations Rome
Foley JA, DeFries R, Asner GP, Barford C, Bonan G, Carpenter SR, Chapin FS, Coe MT, Daily GC, Gibbs HK (2005a) Global consequences of land use. Science 309:570–574
Foley JA, DeFries R, Asner GP, Barford C, Bonan G, Carpenter SR, Chapin FS, Coe MT, Daily GC, Gibbs HK, Helkowski JH, Holloway T, Howard EA, Kucharik CJ, Monfreda C, Patz JA, Prentice IC, Ramankutty N, Snyder PK (2005b) Global consequences of land use. Science 309: 570–574
Gabhane JW, Bhange VP, Patil PD, Bankar ST, Kumar S (2020) Recent trends in biochar production methods and its application as a soil health conditioner: a review, vol 2. SN Appl Sci
Gao L, Wang R, Shen G, Zhang J, Meng G, Zhang J (2017) Effects of biochar on nutrients and the microbial community structure of tobacco-planting soils. J Soil Sci Plant Nutr 17:884–896
Gholizadeh M, Li C, Zhang S, Wang Y, Niu S, Li Y, Hu X (2020) Progress of the development of reactors for pyrolysis of municipal waste. Sustain Energy Fuels 4:5885–5915
Glaser B, Birk JJ (2012) State of the scientific knowledge on properties and genesis of anthropogenic dark earths in Central Amazonia (Terra preta de Índio). Geochimica et Cosmochimica Acta 82:39–51
Glaser B, Lehmann J, Zech W (2002) Ameliorating physical and chemical properties of highly weathered soils in the tropics with charcoal - a review. Biol Fertil Soils 35:219–230
Gomez JD, Denef K, Stewart CE, Zheng J, Cotrufo MF (2013) Biochar addition rate influences soil microbial abundance and activity in temperate soils. Eur J Soil Sci 65:28–39
Grossman JM, O'Neill BE, Tsai SM, Liang B, Neves E, Lehmann J, Thies JE (2010) Amazonian Anthrosols support similar microbial communities that differ distinctly from those extant in adjacentunmodified soils of the same mineralogy. Microbiol Ecol 60:192–205
Gruhn P, Goletti F, Yudelman M (2000) Integrated nutrient management, soil fertility, and sustainable agriculture: current issues and future challenges. Intl Food Policy Res Inst
Gul S, Whalen JK, Thomas BW, Sachdeva V, Deng H (2015) Physico-chemical properties and microbial responses in biochar-amended soils: mechanisms and future directions. Agric Ecosyst Environ 206:46–59
Hale S, Hanley K, Lehmann J, Zimmerman A, Cornelissen G (2011) Effects of chemical, biological, and physical aging as well as soil addition on the sorption of pyrene to activated carbon and biochar. Environ Sci Technol 45:10445–10453
Hass A, Gonzalez JM (2014) Biochar, fertilizers: components, uses in agriculture and environmental impacts
Hassan M, Liu Y, Naidu R, Parikh SJ, Du J, Qi F, Willett IR (2020) Influences of feedstock sources and pyrolysis temperature on the properties of biochar and functionality as adsorbents: a meta-analysis. Sci Total Environ 744:140714
Henao J, Baanante CA (1999) Estimating rates of nutrient depletion in soils of agricultural lands of Africa. International Fertilizer Development
Ippolito JA, Cui L, Kammann C, Wrage-Mönnig N, Estavillo JM, Fuertes-Mendizabal T, Cayuela ML, Sigua G, Novak J, Spokas K, Borchard N (2020) Feedstock choice, pyrolysis temperature and type influence biochar characteristics: a comprehensive meta-data analysis review. Biochar 2:421–438
Jafri N, Wong W, Doshi V, Yoon L, Cheah K (2018) A review on production and characterization of biochars for application in direct carbon fuel cells. PSEP 118:152–166
Jatav HS, Singh SK, Rajput VD, Minkin T (2021) Sustainable soil fertility management. Nova Science Publishers
Jeffery S, Verheijen F, Van der Velde M, Bastos A (2011) A quantitative review of the effects of biochar application to soils on crop productivity using meta-analysis. Agric Ecosyst Environ 144:175–187
Jeyasubramanian K, Thangagiri B, Sakthivel A, Dhaveethu Raja J, Seenivasan S, Vallinayagam P, Madhavan D, Malathi Devi S, Rathika B (2021) A complete review on biochar: production,

property, multifaceted applications, interaction mechanism and computational approach. Fuel 292:120243
Jianping Z (2006) Soil erosion in Guizhou province of China: a case study in Bijie prefecture. Soil Use Manag 15:68–70
Joseph S, Graber E, Chia C, Munroe P, Donne S, Thomas T, Nielsen S, Marjo C, Rutlidge H, Pan G, Li L, Taylor P, Rawal A, Hook J (2013) Shifting paradigms: development of high-efficiency biochar fertilizers based on nanostructures and soluble components. Carbon Manag 4:323–343
Jouhara H, Ahmad D, Van den Boogaert I, Katsou E, Simons S, Spencer N (2018) Pyrolysis of domestic based feedstock at temperatures up to 300 °C. Therm Sci Eng Prog 5:117–143
Jung K, Kim K, Jeong T, Ahn K (2016) Influence of pyrolysis temperature on characteristics and phosphate adsorption capability of biochar derived from waste-marine macroalgae (*Undaria pinnatifida* roots). Bioresour Technol 200:1024–1028
Kammann C I, Schmidt H, Messerschmidt N, Linsel S, Steffens D, Müller C, Koyro H, Conte P, Joseph S (2015) Plant growth improvement mediated by nitrate capture in co-composted biochar. Sci Rep 5
Karhu K, Mattila T, Bergström I, Regina K (2011) Biochar addition to agricultural soil increased CH_4 uptake and water holding capacity – results from a short-term pilot field study. Agric Ecosyst Environ 140:309–313
Karimi A, Moezzi A, Chorom M, Enayatizamir N (2020) Application of biochar changed the status of nutrients and biological activity in a calcareous soil. J Soil Sci Plant Nutr 20:450–459
Kätterer T, Roobroeck D, Andrén O, Kimutai G, Karltun E, Kirchmann H, Nyberg G, Vanlauwe B, Röing de Nowina K (2019) Biochar addition persistently increased soil fertility and yields in maize-soybean rotations over 10 years in sub-humid regions of Kenya. Field Crops Res 235:18–26
Kavitha B, Reddy PV, Kim B, Lee SS, Pandey SK, Kim K (2018) Benefits and limitations of biochar amendment in agricultural soils: a review. J Environ Manag 227:146–154
Kim J, Sparovek G, Longo RM, De Melo WJ, Crowley D (2007) Bacterial diversity of Terra preta and pristine forest soil from the western Amazon. Soil Biol Biochem 39:684–690
Kloss S, Zehetner F, Wimmer B, Buecker J, Rempt F, Soja G (2013) Biochar application to temperate soils: effects on soil fertility and crop growth under greenhouse conditions. J Plant Nutr Soil Sci 177:3–15
Kookana RS, Sarmah AK, Van Zwieten L, Krull E, Singh B (2011) Biochar application to soil: agronomic and environmental benefits and unintended consequences. In: Donald LS (ed) Advances in agronomy. AP, pp 103–143
Kumar P, Lai L, Battaglia ML, Kumar S, Owens V, Fike J, Galbraith J, Hong CO, Farris R, Crawford R, Crawford J, Hansen J, Mayton H, Viands D (2019) Impacts of nitrogen fertilization rate and landscape position on select soil properties in switchgrass field at four sites in the USA. Catena 180:183–193
Kuzyakov Y, Subbotina I, Chen H, Bogomolova I, Xu X (2009) Black carbon decomposition and incorporation into soil microbial biomass estimated by 14C labeling. Soil Biol Biochem 41: 210–219
Laird DA, Fleming P, Davis DD, Horton R, Wang B, Karlen DL (2010a) Impact of biochar amendments on the quality of a typical midwestern agricultural soil. Geoderma 158:443–449
Laird D, Fleming P, Wang B, Horton R, Karlen D (2010b) Biochar impact on nutrient leaching from a midwestern agricultural soil. Geoderma 158:436–442
Lal R (2009) Soils and sustainable agriculture: a review. Sustain Agric:15–23
Lal R (2015) Sequestering carbon and increasing productivity by conservation agriculture. J Soil Water Conserv 70:55A–62A
Lehmann J, Joseph S (2009) Biochar for environmental management: science and technology
Lehmann J, Rondon M (2006) Bio-char soil management on highly weathered soils in the humid tropics. Bio Approaches to Sus Soil Syst 517–529

Lehmann J, Silva Júnior JPD, Steiner C, Nehls T, Zech W, Glaser B (2003) Nutrient availability and leaching in an archaeological Anthrosol and a Ferralsol of the Central Amazon basin: fertilizer, manure, and charcoal amendments. Plant Soil 249:343–357

Lehmann J, Gaunt J, Rondon M (2006) Bio-char sequestration in terrestrial ecosystems – a review. Mitig Adapt Strateg Glob Chang 11:403–427

Lehmann J, Rillig MC, Thies J, Masiello CA, Hockaday WC, Crowley D (2011) Biochar effects on soil biota – a review. Soil Biol Biochem 43:1812–1836

Liang B, Lehmann J, Sohi SP, Thies JE, O'Neill B, Trujillo L, Gaunt J, Solomon D, Grossman J, Neves EG, Luizão FJ (2010) Black carbon affects the cycling of non-Black carbon in soil. Org Geochem 41:206–213

Lima HN, Schaefer CE, Mello JW, Gilkes RJ, Ker JC (2002) Pedogenesis and pre-colombian land use of "Terra Preta Anthrosols" ("Indian Black earth") of western Amazonia. Geoderma 110:1–17

Liu E, Yan C, Mei X, He W, Bing SH, Ding L, Liu Q, Liu S, Fan T (2010) Long-term effect of chemical fertilizer, straw, and manure on soil chemical and biological properties in Northwest China. Geoderma 158:173–180

Liu Z, Hoekman SK, Balasubramanian R, Zhang F (2015) Improvement of fuel qualities of solid fuel biochars by washing treatment. Fuel Process Technol 134:130–135

Liu Z, Dugan B, Masiello CA, Barnes RT, Gallagher ME, Gonnermann H (2016) Impacts of biochar concentration and particle size on hydraulic conductivity and DOC leaching of biochar–sand mixtures. J Hydrol 533:461–472

Liu Z, Dugan B, Masiello CA, Gonnermann HM (2017) Biochar particle size, shape, and porosity act together to influence soil water properties. PLoS One 12:0179079

Major J, Rondon M, Molina D, Riha SJ, Lehmann J (2010) Maize yield and nutrition during 4 years after biochar application to a Colombian savanna oxisol. Plant Soil 333:117–128

Major J, Rondon M, Molina D, Riha SJ, Lehmann J (2012) Nutrient leaching in a colombian Savanna Oxisol amended with biochar. J Environ Qual 41:1076–1086

Martinsen V, Alling V, Nurida N, Mulder J, Hale S, Ritz C, Rutherford D, Heikens A, Breedveld G, Cornelissen G (2015) PH effects of the addition of three biochars to acidic Indonesian mineral soils. Soil Sci Plant Nutr 61:821–834

Mašek O, Brownsort P, Cross A, Sohi S (2013) Influence of production conditions on the yield and environmental stability of biochar. Fuel 103:151–155

Mclennon E, Solomon JK, Neupane D, Davison J (2020) Biochar and nitrogen application rates effect on phosphorus removal from a mixed grass Sward irrigated with reclaimed wastewater. Sci Total Environ 715:137012

Murtaza G, Ahmed Z, Usman M, Tariq W, Ullah Z, Shareef M, Iqbal H, Waqas M, Tariq A, Wu Y, Zhang Z, Ditta A (2021) Biochar induced modifications in soil properties and its impacts on crop growth and production. J Plant Nutr 44(11):1677–1691

Nath H, Sarkar B, Mitra S, Bhaladhare S (2022) Biochar from biomass: a review on biochar preparation its modification and impact on soil including soil microbiology. Geomicrobiology J 39:373–388

Ndor E, Jayeoba O, Asadu C (2015) Effect of biochar soil amendment on soil properties and yield of sesame varieties in Lafia, Nigeria. Am J Exp Agric 9:1–8

Nidheesh P, Gopinath A, Ranjith N, Praveen Akre A, Sreedharan V, Suresh Kumar M (2021) Potential role of biochar in advanced oxidation processes: a sustainable approach. Chem Eng J 405:126582

Novak J, Busscher W, Watts D, Laird D, Ahmedna M, Niandou M (2010) Short-term CO_2 mineralization after additions of biochar and switchgrass to a typic Kandiudult. Geoderma 154:281–288

Novotny EH, Maia CM, Carvalho MT, Madari BE (2015) Biochar: Pyrogenic carbon for agricultural use - a critical review. Rev Bras Cienc 39:321–344

O'Neill B, Grossman J, Tsai MT, Gomes JE, Lehmann J, Peterson J, Neves E, Thies JE (2009) Bacterial community composition in Brazilian Anthrosols and adjacent soils characterized using culturing and molecular identification. Microb Ecol 58:23–35

Oguntunde PG, Fosu M, Ajayi AE, Van de Giesen N (2004) Effects of charcoal production on maize yield, chemical properties, and texture of soil. Biol Fertil Soils 39:295–299

Omondi MO, Xia X, Nahayo A, Liu X, Korai PK, Pan G (2016) Quantification of biochar effects on soil hydrological properties using meta-analysis of literature data. Geoderma 274:28–34

Omotade I, Momoh S, Oluwafemi B, Agboola E (2020) Comparative analysis of nutrients composition in biochar produced from different feedstocks at varying pyrolysis temperature. Environ Sci Technol 3:64–70

Ouyang L, Tang Q, Yu L, Zhang R (2014) Effects of amendment of different biochars on soil enzyme activities related to carbon mineralisation. Soil Res 52:706

Palm CA, Gachengo CN, Delve RJ, Cadisch G, Giller KE (2001) Organic inputs for soil fertility management in tropical agroecosystems: application of an organic resource database. Agric Ecosyst Environ 83:27–42

Pandit NR, Mulder J, Hale SE, Martinsen V, Schmidt HP, Cornelissen G (2018) Biochar improves maize growth by alleviation of nutrient stress in a moderately acidic low-input Nepalese soil. Sci Total Environ 625:1380–1389

Pender J (2009) The world food crisis, land degradation, and sustainable land management

Pietikäinen J, Kiikkilä O, Fritze H (2000) Charcoal as a habitat for microbes and its effect on the microbial community of the underlying humus. Oikos 89:231–242

Purakayastha T, Bera T, Bhaduri D, Sarkar B, Mandal S, Wade P, Kumari S, Biswas S, Menon M, Pathak H, Tsang DC (2019) A review on biochar modulated soil condition improvements and nutrient dynamics concerning crop yields: pathways to climate change mitigation and global food security. Chemosphere 227:345–365

Rashid M, Hussain Q, Khan KS, Al-Wabel MI, Afeng Z, Akmal M, Ijaz SS, Aziz R, Shah GA, Mehdi SM, Alvi S, Qayyum MF (2020) Prospects of biochar in alkaline soils to mitigate climate change. In: Environment, climate, plant and vegetation growth. Springer, Cham, pp 133–149

Razzaghi F, Obour PB, Arthur E (2020) Does biochar improve soil water retention? A systematic review and meta-analysis. Geoderma 361:114055

Rillig MC, Mummey DL (2006) Mycorrhizas and soil structure. New Phytol 171:41–53

Rondon MA, Lehmann J, Ramírez J, Hurtado M (2006) Biological nitrogen fixation by common beans (*Phaseolus vulgaris* L.) increases with bio-char additions. Biol Fertil Soils 43:699–708

Rosset P, Collins J, Lappe FM, (2000) Lessons from the green revolution. Third World Resurgence, pp 11–14

Schmidt H, Pandit B, Martinsen V, Cornelissen G, Conte P, Kammann C (2015) Fourfold increase in pumpkin yield in response to low-dosage root zone application of urine-enhanced biochar to a fertile tropical soil. Agriculture 5:723–741

Schwartz MW, Hoeksema JD, Gehring CA, Johnson NC, Klironomos JN, Abbott LK, Pringle A (2006) The promise and the potential consequences of the global transport of mycorrhizal fungal inoculum. Ecol Lett 9:501–515

Seleiman MF, Alotaibi MA, Alhammad BA, Alharbi BM, Refay Y, Badawy SA (2020) Effects of ZnO nanoparticles and biochar of rice straw and cow manure on characteristics of contaminated soil and sunflower productivity, oil quality, and heavy metals uptake. Agronomy 10:790

Shackley S, Carter S, Knowles T, Middelink E, Haefele S, Haszeldine S (2012) Sustainable gasification–biochar systems? A case-study of rice-husk gasification in Cambodia, part II: field trial results, carbon abatement, economic assessment, and conclusions. Energy Policy 41:618–623

Shafer AB, Wolf JB, Alves PC, Bergström L, Bruford MW, Brännström I, Zieliński P (2015) Genomics and the challenging translation into conservation practice. Trends Ecol Evol 30:78–87

Singh B, Singh BP, Cowie AL (2010) Characterisation and evaluation of biochars for their application as a soil amendment. Soil Res 48:516

Sohi SP (2012) Carbon storage with benefits. Science 338:1034–1035
Sohi S, Lopez-Capel E, Krull E, Bol R (2009) Biochar, climate change and soil: a review to guide future research. CSIRO
Sohi S, Krull E, Lopez-Capel E, Bol R (2010) A review of biochar and its use and function in soil. Adv Agron:47–82
Song D, Tang J, Xi X, Zhang S, Liang G, Zhou W, Wang X (2018) Responses of soil nutrients and microbial activities to additions of maize straw biochar and chemical fertilization in a calcareous soil. Eur J Soil Biol 84:1–10
Spokas K, Koskinen W, Baker J, Reicosky D (2009) Impacts of woodchip biochar additions on greenhouse gas production and sorption/degradation of two herbicides in a Minnesota soil. Chemosphere 77:574–581
Srivastava AK (2009) Integrated nutrient management: concept and application in citrus: citrus II. Tree For Sci Biotechnol 3:32–58
Steiner C, Das KC, Garcia M, Förster B, Zech W (2008a) Charcoal and smoke extract stimulate the soil microbial community in a highly weathered xanthic Ferralsol. Pedobiologia 51:359–366
Steiner C, Glaser B, Geraldes Teixeira W, Lehmann J, Blum WE, Zech W (2008b) Nitrogen retention and plant uptake on a highly weathered central Amazonian Ferralsol amended with compost and charcoal. J Plant Nutr Soil Sci 171:893–899
Sun F, Lu S (2013) Biochars improve aggregate stability, water retention, and pore-space properties of clayey soil. J Plant Nutr Soil Sci 177:26–33
Tan ZX, Lal R, Wiebe KD (2005) Global soil nutrient depletion and yield reduction. J Sustain Agric 26:123–146
ten Berge HFM, van Loon MP, Vonk WJ (2022) Nutrient cycling indicators and their relation with nutrient use efficiency in agro-food systems. In: Book of abstracts XXI international N workshop 2022, pp 299–299
Thines K, Abdullah E, Mubarak N, Ruthiraan M (2017) Synthesis of magnetic biochar from agricultural waste biomass to enhancing route for wastewater and polymer application: a review. Renew Sustain Energy Rev 67:257–276
Thomas SC, Gale N (2015) Biochar and forest restoration: a review and meta-analysis of tree growth responses. New Forests 46:931–946
Tian S, Tan Z, Kasiulienė A, Ai P (2017) Transformation mechanism of nutrient elements in the process of biochar preparation for returning biochar to soil. Chin J Chem Eng 25:477–486
Tilman D, Balzer C, Hill J, Befort BL (2011) Global food demand and the sustainable intensification of agriculture. Proc Natl Acad Sci 108:20260–20264
Uchimiya M, Wartelle LH, Klasson KT, Fortier CA, Lima IM (2011) Influence of pyrolysis temperature on biochar property and function as a heavy metal sorbent in soil. J Agric Food Chem 59:2501–2510
UNEP (2012) Avoiding future famines: strengthening the ecological Foundation of Food Security through sustainable food systems; United Nations Environment Program (UNEP): Nairobi
Van der Voort TS, Hagedorn F, McIntyre C, Zell C, Walthert L, Schleppi P, Feng X, Eglinton TI (2016) Variability in C contents of soil organic matter at the plot and regional scale across climatic and geologic gradients. Biogeosciences 13:3427–3439
Vanlauwe B, Bationo A, Chianu J, Giller K, Merckx R, Mokwunye U, Ohiokpehai O, Pypers P, Tabo R, Shepherd K, Smaling E, Woomer P, Sanginga N (2010) Integrated soil fertility management. Outlook Agric 39:17–24
Vijay V, Shreedhar S, Adlak K, Payyanad S, Sreedharan V, Gopi G, Sophia van der Voort T, Malarvizhi P, Yi S, Gebert J, Aravind P (2021) Review of large-scale biochar field-trials for soil amendment and the observed influences on crop yield variations. Front Energy Res 9:710766
Wang D, Zhang W, Hao X, Zhou D (2013) Transport of biochar particles in saturated granular media: effects of pyrolysis temperature and particle size. Environ Sci Technol 47:821–828
Wang J, Xiong Z, Kuzyakov Y (2015) Biochar stability in soil: meta-analysis of decomposition and priming effects. GCB Bioenergy 8:512–523

Wang D, Fonte SJ, Parikh SJ, Six J, Scow KM (2017) Biochar additions can enhance soil structure and the physical stabilization of C in aggregates. Geoderma 303:110–117

Wang D, Jiang P, Zhang H, Yuan W (2020) Biochar production and applications in agro and forestry systems: a review. Sci Total Environ 723:137775

Wardle DA, Nilsson M, Zackrisson O (2008) Fire-derived charcoal causes loss of forest humus. Science 320:629–629

Warnock DD, Lehmann J, Kuyper TW, Rillig MC (2007) Mycorrhizal responses to biochar in soil – concepts and mechanisms. Plant Soil 300:9–20

Winsley P (2007) Biochar and bioenergy production for climate change mitigation. N Z Sci Rev 64: 5–10

Woolf D, Amonette JE, Street-Perrott FA, Lehmann J, Joseph S (2010) Sustainable biochar to mitigate global climate change. Nat Commun 1:56

Xie Y, Wang L, Li H, Westholm LJ, Carvalho L, Thorin E, Yu Z, Yu X, Skreiberg Ø (2022) A critical review on production, modification and utilization of biochar. J Anal Appl Pyrolysis 161:105405

Yadav V, Khare P (2020) Impact of pyrolysis techniques on biochar characteristics: application to soil. J Soil Sci Environ Manage 33–52

Yao Y, Gao B, Zhang M, Inyang M, Zimmerman AR (2012) Effect of biochar amendment on sorption and leaching of nitrate, ammonium, and phosphate in a sandy soil. Chemosphere 89: 1467–1471

You S, Ok YS, Chen SS, Tsang DC, Kwon EE, Lee J, Wang C (2017) A critical review on sustainable biochar system through gasification: energy and environmental applications. Bioresour Technol 246:242–253

Zhai L, CaiJi Z, Liu J, Wang H, Ren T, Gai X, Xi B, Liu H (2015) Short-term effects of maize residue biochar on phosphorus availability in two soils with different phosphorus sorption capacities. Biol Fertil Soils 51:113–122

Zhang H, Voroney R, Price G (2015) Effects of temperature and processing conditions on biochar chemical properties and their influence on soil C and N transformations. Soil Biol Biochem 83: 19–28

Zhang C, Lin Y, Tian X, Xu Q, Chen Z, Lin W (2017) Tobacco bacterial wilt suppression with biochar soil addition associates to improved soil physiochemical properties and increased rhizosphere bacteria abundance. Appl Soil Ecol 112:90–96

Zhang M, Song G, Gelardi DL, Huang L, Khan E, Mašek O, Parikh SJ, Ok YS (2020) Evaluating biochar and its modifications for the removal of ammonium, nitrate, and phosphate in water. Water Res 186:116303

Zheng H, Wang Z, Deng X, Herbert S, Xing B (2013) Impacts of adding biochar on nitrogen retention and bioavailability in agricultural soil. Geoderma 206:32–39

Zimmerman AR (2010) Abiotic and microbial oxidation of laboratory-produced black carbon (biochar). Environ Sci Technol 44:1295–1301

Chapter 14
Phytohormones as Stress Mitigators in Plants

Hunny Waswani and Rajiv Ranjan

Abstract The sessile nature of plants makes them susceptible to a wide variety of abiotic and biotic stresses. Abiotic stress conditions incorporate factors like drought, heat, cold, salinity, whereas biotic stresses emerge mainly from microorganisms, insects and from nematodes. Well-developed mechanisms have been evolved by plants to perceive abiotic and biotic stresses that affect plants. Through flaws in the genetic control of cellular pathways, abiotic stresses have a detrimental effect on the morphology and physiology of plants. Tolerating stresses that arise from changes in metabolism is a sophisticated process that plants employ to mitigate the impact of stresses. They play an important role in both stimulating plant defense mechanisms against stresses and regulating plant metabolism, making phytohormones some of the most important growth regulators. Supplementing with exogenous phytohormones has been used to enhance growth and metabolism in stressful environments. Recent studies have demonstrated that phytohormones made by microorganisms associated with roots may prove to be significant metabolic engineering targets for promoting host tolerance to abiotic stresses. Numerous genetic and biochemical techniques have been used to identify the biosynthesis pathways for phytohormones, here, we present the state of our understanding on how phytohormones help plants exposed to diverse stressors increase their ability to withstand abiotic stress and their defence mechanisms. The aim of this paper is to examine recent developments in understanding the mechanisms by which phytohormones increase plant stress tolerance, especially in crop plants. This chapter does this so by highlighting significant morpho-physiological characteristics of plants that can be used to determine the beneficial impacts of phytohormones on stress tolerance.

Keywords Reactive oxygen species · Reactive nitrogen species · Salicylic acid · Jasmonic acid · Abscisic acid · Gibberellic acid · Ethylene · Cytokinin · Onithine decarboxylase · Arginine decarboxylase · Heavy metals/metalloids

H. Waswani · R. Ranjan (✉)
Plant Molecular Biology Laboratory, Department of Botany, Dayalbagh Educational Institute, Dayalbagh, Agra, India

M. Hasanuzzaman et al. (eds.), *Mineral Biofortification in Crop Plants for Ensuring Food Security*, https://doi.org/10.1007/978-981-99-4090-5_14

14.1 Introduction

Environmental alterations either due to hasty seasonal changes in weather or due to evolutionary activities give rise to serious survival challenges to plants. Adverse environmental conditions are faced by plants since they are sessile in nature (Zhu 2016). Amongst variable changes in environment factors like, drought, high temperature or low temperature, water deficit, high salinity, water logging, high salinity, solar radiations, heavy metals, are become detrimental for growth and optimum productivity of plant (Johnson et al. 2022). This leads to biomass decrement and grain yields at a worldwide level. Drastic environment conditions cause profuse types of stresses. There have been precise definitions of "stress" in mechanics; however, it has been given a wide range of meanings in biology. Since many of these definitions extend the physical meaning to encompass a wide range of environmental factors "unfavourable" for living organisms, the term "stress" has become synonymous with those factors. Under natural conditions plants are frequently exposed to different types of stress. Any abiotic or biotic constraint that affects the process of photosynthesis and reduces the plants' ability to convert carbon dioxide (CO_2) into biomass may be defined as stress, or more broadly, any external factor that negatively impacts the plant's growth. The basic assimilation mechanisms (CO_2 and mineral uptake) and plant survival, crop production, growth (biomass accumulation), and stress are all related to general growth and development. Abiotic and biotic stress are the two main types into which stress may be categorised. Any physical or chemical element that the environment may impose on the plant is considered an abiotic stress. Water stress, reduced soil availability of critical nutrients (or, conversely, the build-up of harmful ions during salt stress), temperature extremes, or abundant light, especially when photosynthesis is inhibited, are examples of abiotic stress (Hasanuzzaman et al. 2013; Akter et al. 2014; Vardhini and Anjum 2015; Ghassemi et al. 2018; Sytar et al. 2019). The effects of extreme temperatures on salt stress can be exacerbated by droughts, which are frequently linked with salt stress. Biotic stress may be a biological factor (pest, pathogen, or herbivore attack). Plants adapt specific defence systems to deal with biotic and abiotic challenges in their natural soil environments. To each given stimulus, numerous cellular signalling pathways are triggered, the growth and development of plants can be affected by even very low concentrations of these hormones. Physiological changes occur in plants by the occurrence of abiotic and biotic stress is below mentioned in Fig. 14.1. The synthesis of phytohormones is accelerated by these signals. There are many biotic and abiotic elements that influence the number of hormones, which can vary substantially. These hormones may have a major impact on plant growth and development even at extremely low concentrations. These biochemical compounds, known as phytohormones, are created by plants and it is well known that signalling molecules called phytohormones have a wide range of biological and physiological effects on plants. The detrimental effects of numerous environmental conditions, both biotic and abiotic, on plant growth and development are also significantly reduced by these hormones. The significance of

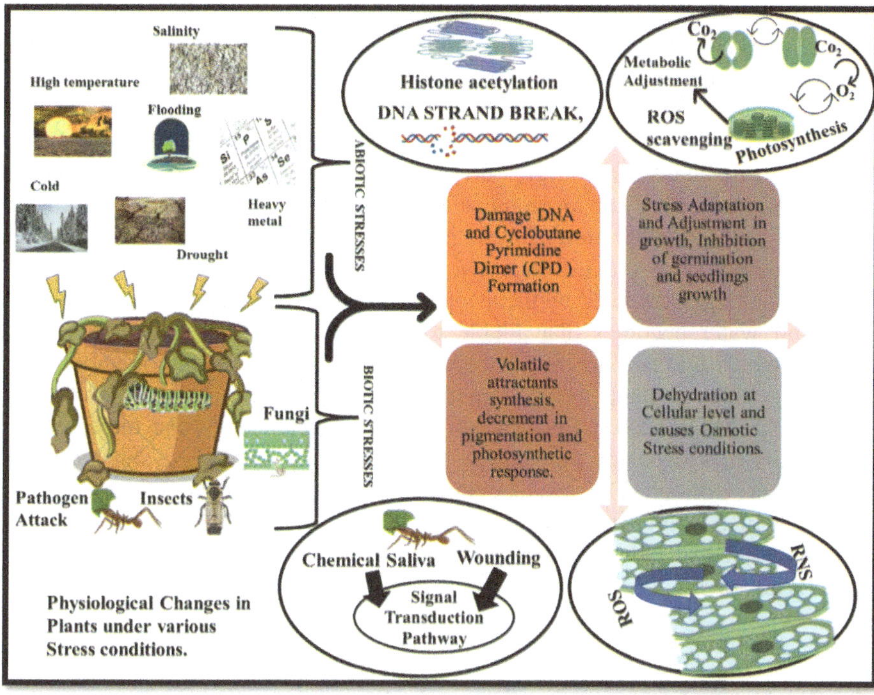

Fig. 14.1 Physiological changes in plants under biotic and abiotic stress conditions

phytohormones in plant stress tolerance is highlighted in this chapter. The following sections provide a review and discussion of phytohormone sources, metabolism, and physiological effects on plant growth, particularly in stressful environments.

14.2 Abiotic Stresses

14.2.1 Salinity Stress

Arid and semiarid regions often struggle with soil salinity as one of the environmental stresses, because of salinity, soil quality is degraded, cultivation area is reduced, and crop yield is minimized (Sadiq et al. 2002). There has been a 50% reduction in the yields of major crops in arid and semiarid regions due to salinity. A primary impact of salinity on plants is growth inhibition, which can be attributed to its negative effects on photosynthesis and cellular disruption as well as oxidative disintegration. Almost all physiological and morphological plants processes are affected by salinity. There are several ions that negatively impact plant growth in salty conditions but sodium interferes more pronouncedly with potassium uptake, leading to potassium deficiency in plants (Nawaz and Ashraf 2010). Increased Na/K ratios in plants cause certain nutritional and cellular imbalances, such as lower essential nutrients and soluble sugars, while also lowering the vital cell organelles membrane stability. As a result, plant metabolic activities are also disrupted by higher chloride concentrations, which affect certain enzyme activities as well as Na^+, Cl^-, Example- due to the elevation of chloride ion in the cytoplasm plants lose its ability to maintain osmotic pressure (Misra and Saxena 2009). As a result of decreased production of certain amino acids and nitrogenous bases needed for RNA and DNA synthesis, inhibition of these processes has been observed in salty environments (Chen et al. 2003; Song et al. 2006). It is widely believed that plants' production is largely dictated by the availability of water. In the presence of salinity stress, soil water becomes unavailable to plants due to increased soil osmotic potential, which makes extracting water from soil difficult, causing metabolic and cellular processes of plants to be disrupted, resulting in improper plant growth and development (Munns 2005). Because, soil osmotic potential increases under salinity stress, soil water becomes unavailable to plants and removing water from soil becomes more difficult. This results in the disruption of plant cellular and metabolic activities, which lead to improper plant growth. As a result of altered physiological processes related to growth and reproduction under water deficit conditions, plant growth is adversely affected (Manivannan et al. 2008).

14.2.2 Drought Condition/Drought Stress

Promptly plants respond to water stress, in fact, even a short-term drought at any growth stage can negatively impact a plant's whole life cycle. Drought affects almost every aspect of a physiological state, plant's morphology and biochemistry, posing for adverse limitations of crop productions (Aroca 2012). As some literature reveals that drought stress majorly affects key procedures regarding division of cell, water linking, uptake of nutrients, energy transfer, carbon fixation, nutrient assimilation, and photosynthesis (Yamane et al. 2003; Gomes et al. 2010; Asrar and Elhindi 2011). Reduction in cytoplasm water contents restricts following processes, elongation, cell division and differentiation due to foremost decrement in, metabolic activity, turgor pressure and inhibition of energy transfer. The cell multiplication inhibition negatively affects the reproductive and vegetative growth due to degrading biomass accumulation leading to shoot growth and stunted root, fruit development and poor flowering (Asrar and Elhindi 2011). Beneath conditions like water-deficit, major plants exhibit restricted uptake of nutrients and enhance nutrient deficiency. The roots and shoots of plants show significant reductions in macro- and micronutrients under water shortage, especially nitrogen, potassium, and phosphorus (Subramanian et al. 2006; Asrar and Elhindi 2011).

14.2.3 Temperature Stress (High/Low Temperature)

Every chemical, physical, and biological process in living cells is affected by temperature. Irreversible crop damages can cause by slight increment and decrement in temperature. Highly elevated temperature is the most extreme form of temperature stress, which challenges plant's survival in extreme climatic conditions. Responses are shown at cellular and molecular level by plants under high temperature stress. Adverse effect scan be seen on the growth, biochemistry, phenology, anatomy, and physiology of plant (Wahid et al. 2012). As with other stress from germination to seed production, high temperatures can influence plants' growth at any stage. It has been reported that changes in protein expression profile are a contributing factor to the inhibition of germination caused by high temperatures (Ren et al. 2009). Plants are more sensitive at germination stage, to change in temperature which adversely inhibits the development and emergence of seedling (Egli et al. 2005). Enzymes are degraded and denatured by the adverse effect of high temperature which ultimately inhibits the germination and growth of embryo's (Wahid et al. 2012). The photosynthesis process is also adversely affected by heat stress since high temperatures destroy the internal ground tissue known as mesophyll cell by deform and devour the chloroplasts, lamella and stroma severely inhibiting photosystem II activity (Carpentier 1999). Low temperature-Injuries caused by low temperatures, which are slightly above freezing point, are known as chilling, while injuries caused by freezing are caused by freezing temperatures which cause solutions in plants to freeze, crystallize, and cease working completely, resulting in the rupture of membrane structures and complete cessation of biochemical machinery. Due to disturbances in biochemical mechanisms, (ROS) are also produced that cause oxidative stress, since low temperatures prevent oxidoreductive enzymes from functioning. Specifically, catalase inhibition leads to the accumulation of free radicals and hydrogen peroxide (Los and Murata 2004; Sun et al. 2010).

14.2.4 Toxic/Heavy Metal

In the present era, environmental pollution is becoming increasingly severe due to rapid industrialization and urbanization. There are a lot of toxic pollutants in the environment, especially heavy metals, and studies have shown these pollutants are present in every ecosystem (Wei and Yang 2010; Azizullah et al. 2011). It is becoming increasingly difficult to grow crops in such toxic environment, as heavy metals/toxic metal(loid)s incorporating, Cr, Cd, As, Pb and Hg, are un-essential for plant growth (Bandara et al. 2020). Their contamination is released into soil and water not only in the vicinity of industrial locations as well as in remote are as heavy metal stress becoming a serious issue. The misuse of chemicals for crop production has severely contaminated the soil with different hazardous metals (Hjortenkrans et al. 2006; Nada et al. 2007). These hazardous metals pose mutagenic, cytogenic and genotoxic effects on plants. Some of the heavy metals are diligently up taken by plants roots and via various processes it transferred into the food chain which results in serious issues of health in humans as well as in animals (Flora et al. 2008; Balali-Mood et al. 2021; Dai et al. 2022).

14.3 Biotic Stress

Biotic or biological stress in plants caused by the living organism like bacteria, nematodes, fungi, viruses' arachnids, nematodes, and weeds. These agents which cause biotic stress deprives nutrients of plants and leads to the death of plant (Madani et al. 2019). Due to pre-harvest and post-harvest loses, biological stress become dominant. Because plants have evolved sophisticated strategies to deal with biotic stresses, they can survive without an adaptive immune system. It is the genetic code contained in a plant's genome that regulates its defence mechanisms against these stresses. Many genes in the genome of a plant are responsible for resistance to these stresses. This discussion shows that plants face numerous stresses in soil environment. Various plant physiological processes are negatively affected by all these stresses, resulting in decreased growth and development of plants Interactions of these stresses may increase their intensity, and their impact may vary depending on the species and growth stage of the plant. To properly grow and develop plants, it may be beneficial to control the negative effects of one stress at a certain growth stage (Gull et al. 2019; Khare et al. 2020).

14.4 Mitigation of Plant Stresses Through Phytohormones

Promoting plant growth and development with plant growth regulators is a highly effective approach (Bano and Yasmeen 2010). Phytohormones have proven to be effective in increasing crop production both under normal and stressful conditions due to their ability to promote growth (Afzal et al. 2005).

To conquer the losses in yield due to abiotic and biotic stresses, mechanisms of mitigation, avoidance and tolerance are needed by plants to overcome various stresses. Plant responses to biotic and abiotic stress are significantly influenced by growth hormones, such as GA and abscisic acid (Lee and Park 2010). Seed priming with phytohormones, which boosts antioxidant enzyme activity, lowers oxidative damage, and shields plants from diverse stimuli, can improve plant growth (Ayyaz et al. 2022).

There is evidence that crops can be made more heat tolerant by the application of exogenous ABA. It is believed that ABA improves thermotolerance by increasing levels of other molecules, such as nitric oxide (NO) (Ding et al. 2010). Abiotic stress responses and plant responses to pathogen infections are also closely related to (SA). In addition to promoting stress tolerance, SA has also been shown to enhance plant

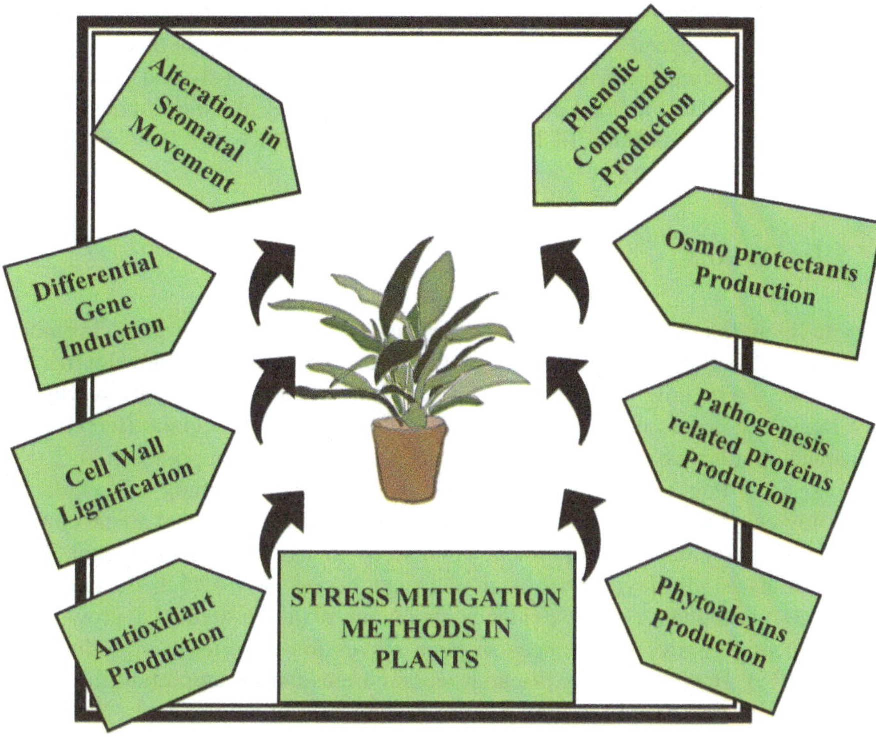

Fig. 14.2 Stress mitigation methods in plants

growth under salt and osmotic stress (Krantev et al. 2006). It is well known that SA is involved in a wide range of plant reactions to stress, including UV-B, ozone, and pathogen attack. When plants are stressed, SA also activates the antioxidant system. Accumulation of polyamines and particularly putrescine accumulate because of potassium deficit, osmotic stress, low pH, nutritional deficiency, or light exposure. According to Minocha et al. (2014), enhanced ADC or (ODC) activity is associated with putrescine buildup during environmental stress. According to Chen et al. (2006), In a variety of crops jasmonic acid (JA) is also known to increase thermotolerance. Plants adapt various mitigation methods to mitigate stress which is below mentioned in Fig. 14.2.

14.5 Phytohormones and Plant Stress

The ability of plants to maintain their growth under stress is greatly enhanced with the use of endogenous and exogenous phytohormones. Phytohormones can also be applied exogenously to improve plant tolerance to adverse environmental conditions, in addition to endogenous hormones (Fan et al. 2014; Afzal et al. 2005). There have been several reports demonstrating the positive effects of plant growth and development when phytohormones are applied endogenously and exogenously (Fässler et al. 2010; Shaddad et al. 2013). The actions of phytohormones in plant stress are presented in Table 14.1.

14.5.1 Auxins

The auxin indole-3 acetic acid (IAA) first identified and a key plant hormone, research revealed that it may have a part in stress adaptation and development in plants and shows how it plays a crucial function in controlling plant growth and developmental processes as root expansion, cell elongation, vascular differentiation, and apical dominance (Sachs 2005; Asgher et al. 2015; Lymperopoulos et al. 2018). Auxins, according to studies, collaborate with CKs in a variety of cellular or physiological processes activities such cell growth, cell division, and apical control, leaf growth, and embryo development when seeds are maturing (Tromas et al. 2009; Jurado et al. 2010). Tryptophan is used to synthesize IAA, which is chemically related to tryptophan. Strong evidence in favour of auxin-mediated growth and developmental control via changes in gene expression patterns was produced by Ljung (2013). However, many research reports are available supporting the role of auxins in improving and mediating plant tolerance to abiotic stresses. Numerous reports are available depicting various modulations in the transport, metabolism, synthesis, and auxins mechanism of action after plant exposure to stresses (Ljung 2013; Kazan 2013). IAA significantly decreased in rice plants after being subjected to salinity stress.

According to Iqbal and Ashraf (2013), this difference in IAA might also cause growth regulation by increasing other phytohormones like ABA. An association between auxin signalling and salt stress was discovered by Jung and Park (2011). This association resulted from auxin participation in modifying the membrane-bound transcription factor NTM2 (Park et al. 2011); over-expression investigations on the *NTM2 IAA30* gene further confirmed these involvements, however the precise mechanism by which IAA-induced salinity mitigation works is yet unknown. Since (Hu et al. 2013) found that heavy metals negatively affect auxin production, auxins play a significant role in improving heavy metal tolerance, whether directly or indirectly.

The addition of a small amount of IAA (1010 M) reduced the harmful effects of Pb on sunflower plant growth and promoted increases in diameter, root volume and surface area (Fässler et al. 2010). Auxins have the potential to improve phytoextraction of metals since IAA enhanced shoot biomass and Pb and Zn accumulation in plant tissue. As a result of gene expression, aluminium inhibited the transport and synthesis of IAA from shoot to root in *Medicago sativa*. However, exogenous administration of IAA could partially alleviate aluminium stress by increasing the expression of AUX1 and PIN2 genes (Wang et al. 2016).

Table 14.1 Representing various phytohormones and their mechanisms of action in plant stresses

Hormone	Nature	Active Form	Effects	Stress	Mechanisms	References
Ethylene	Alkenes (a class of hydrocarbon)	Methionine	Fruit/flowering ripening; responses in various stresses; germination of seeds	Biotic stress (Pathogens, microorganisms)	Integrated stress response; interactions with jasmonic acid; regulates SA/JA antagonism; influence plant defense genes such as THI2.1 (thionin), Plant Defensin1.2 (PDF1.2)	Großkinsky et al. (2016), Iqbal et al. (2017), Iqbal et al. (2022)
Jasmonic acid	Sesquiterpene lactones	JA-Ile	Inhibition of root growth; plant defense followed for herbivores insects; responses for necrotrophic pathogens	Biotic stress	An integrated stress response, to mediate stress it interacts with many other hormones	Yang et al. (2019) Liu et al. (2019), Yang et al. (2019), Wang et al. (2020), Wang et al. (2021)
				Abiotic stress	Activates antioxidant system, regulates closing and opening of stomata; brings accumulation of soluble sugars and amino acids	
Salicylic acid	Organic phenol compound	SA	Growth and development of plant including germination of seed, flowering, fruit yield, senescence. Vegetative growth, closure of stomata,	Biotic stress	Mediates system acquired resistance SAR; For pathogen specific response JA/SA antagonism is perfect; In resistance to aphids JA/SA together involved in MAPK signalling	Rajjou et al. (2006), Li et al. (2006), Gao et al. (2015), Großkinsky et al. (2016), Lefevere et al. (2020), Kaur et al. (2022), Shaukat et al. (2022)
				Abiotic stress	Increment in antioxidant activity	

(continued)

Table 14.1 (continued)

Hormone	Nature	Active Form	Effects	Stress	Mechanisms	References
Gibberellins	Diterpene	GA_1, GA_2	In stresses, fruit growth, floral development, elongation of stem, growth of root, germination of seed	Abiotic stress	Degradation of DELLA proteins	Robert-Seilaniantz et al. (2007), Alonso-Ramírez et al. (2009), Alonso-Ramírez et al. (2009), Großkinsky et al. (2016), Abbas et al. (2022)
				Biotic stress	Initiation of defence pathway dependent upon salicylic acid; Act on SA/JA signalling relative strength	
Auxins	Derivative of tryptophan	IAA	Elongation of cell; Apical dominance	Biotic stress	Interaction with different phytohormones.	Ghanashyam and Jain (2009), Shi et al. (2014) Gao et al. (2015)
					Contribution with system acquired resistance	
				Abiotic stress (drought stress)	Root architecture modulation, reactive oxygen species metabolism, expression of abscisic acid (ABA) responsive genes	
Abscisic acid	Isoprenoid compounds	ABA	Seed germination, maturation; storage, dessication tolerance; Stomatal closure; leaf senescence; root, shoot growth	Biotic stress	Influences plant defence central backbone of SA/JA/ET	Großkinsky et al. (2016)
				Abiotic stress	Closure of stomata; depletion in ROS levels	
CK's	Derivative of isopentenyladenine	*trans-zeatin*	Division of cell; delays senescence, growth of root	Abiotic stress	Stimulating cell division, regulation of shoot meristem, leaf and shoot growth	Jiang et al. (2013), O'Brien and Benková (2013), Zwack and Rashotte (2015), Großkinsky et al. (2016)

Using auxins as priming sources had a beneficial outcome. After seed priming with IAA, which led to ionic homeostasis and the stimulation of SA biosynthesis, Iqbal and Ashraf (2007) observed a considerable alleviation of salt stress-induced adverse effects in wheat. These results suggest that auxin and SA may interact to influence how plants respond to tolerance in plants. Salt inhibits the synthesis of IAA, although the exogenous administration of SA showed successful in reducing harmful effects by significantly alleviating the inhibition caused by salinity (Fahad and Bano 2012).

14.5.2 Cytokinin

The CK in plants play a crucial role in maintaining cellular proliferation and differentiation, in regulating various aspects of plant growth and development, such as, root formation, apical dominance, chloroplast development stomatal behaviour, and cell division, as well as preventing senescence (Schaller et al. 2014). So, they prevent premature leaf senescence by inhibiting cellular proliferation and differentiation. The xylem sap of stay-green plants was found to be more CK-rich, which led to an increased mitigation to drought under stress conditions, particularly water stress at grain filling. In comparison to wild-type cassava plants, transgenic cassava plants that expressed CK showed greater tolerance to drought (Borrell et al. 2022; Zhang et al. 2010). Overexpressed genes involved in CK biosynthesis have been linked to stress mitigation. For example, in field analysis the isopentenyl transferase (ipt) gene has been validated (Peleg and Blumwald 2011). ABA-induced stomatal closure reduces carbon assimilation and uptake when CK is reduced, and CK oxidase upregulation reduces carbon metabolism under stressful conditions; CK improves grain filling, so research on this topic can be beneficial to improving plant growth and yield. Presently, optimum CK concentrations are being achieved with exogenous CK application. Additionally, the accumulation of Pb and Zn in plant tissues inhibits chickpea seedling growth by inhibiting GA3 and Z concentrations, respectively. It was previously reported that kinetin stimulated chickpea growth under salt stress, while another study shown it improved eggplant's antioxidant capacity under Cd stress.

14.5.3 Abscisic Acid

By enhancing stress reactions and adaptation, ABA is recognised to play a significant function in plants, just as other phytohormones. It is a sesquiterpenoid, a class of important phytohormones involved in the control of growth, which occurs naturally. Numerous studies have supported the use of ABA in integrating signals after stress exposure and controlling subsequent downstream reactions (Wilkinson et al. 2012). ABA-induced and -mediated signalling controls the expression of stress-responsive genes under abiotic stress, which improves the induction of tolerance responses (Sah et al. 2016).

Additionally, under drought stress conditions, ABA has been shown to influence water content and root development (Cutler et al. 2010). However, a rapid increase in ABA concentrations during stressful situations might delay growth and alter the body's reaction to stress (Asgher et al. 2015). Exogenous ABA may, however, play a beneficial role in counteracting the negative impacts of many stresses, including, chilling (Nayyar et al. 2005), salinity (Parwez et al. 2022), cold stress (Li et al. 2014) and drought (Bano et al. 2012). According to Bano et al. (2012), exogenous administration of ABA improved the antioxidant system and relative water content of wheat, protecting it from drought-induced oxidative damage.

Exogenous ABA treatment has been suggested as a useful method for reducing stress and increasing stress tolerance. Mora-Herrera and Lopez-Delgado (2007) found that the antioxidant enzyme peroxidase significantly increased during ABA application, which reduced the formation of free radicals and improved stress tolerance in *Solanum tuberosum*. Exogenous ABA treatment under drought stress circumstances caused a considerable alteration in the tea proteome, including modifications to proteins involved in transport, carbon metabolism, and stress tolerance, according to Zhou et al. (2014). According to some theories, ABA keeps levels of other hormones like ethylene stable, which keeps *Zea mays* shoot and root growth active (Spollen et al. 2000). Stresses cause an increase in ABA accumulation and production in plant tissue. The ability of ABA to operate as an anti-transpirant following the induction of stomatal closure and restriction of canopy growth is its most crucial function, in addition to its role in signalling (Wilkinson and Davies 2002). Application of exogenous ABA to rice seedlings exposed to drought improved photosynthetic capacity and stomatal regulation under both normal and stressful conditions, suggesting that these genes are involved in the induction of photosystem II after exogenous ABA application. This was accomplished by up regulating the expression of the, *OsNCED2, OsNCED3, OsNCED4, OsNCED5 and OsPsbD1, OsPsbD2* genes.

ABA facilitates efficient water and nutrient uptake when plants are under stress by developing deeper roots and changing other essential root changes (Spollen et al. 2000; Vysotskaya et al. 2009). Further, ABA activates the antioxidant system and increases the concentration of compatible osmolytes in the body, which maintains the water balance in the body. This results in improved drought tolerance and the maintenance of tissue turgor potential. ABA also maintains the hydraulic conductivities of root and shoot to better exploit soil water content. Zhou et al. (2005) showed that in *Stylosanthes guianensis*, ABA-induced antioxidant defence was mediated by enhanced nitric oxide generation. According to Guajardo et al. (2016), ABA treatment boosted antioxidant enzyme activity, which decreased oxidative stress-induced damage and increased desiccation tolerance. According to Cabot et al. (2009), ABA administered exogenously decreased the buildup of salt and chloride in citrus plants. In a different investigation, common bean plants under salt stress received ABA therapy, which boosted plant growth, nutrient uptake, and nitrogen fixing (Khadri et al. 2006).

14.5.4 Gibberellic Acid

Gibberellin, a plant's primary growth regulator, is involved in seed dormancy, lateral branch development, and the appearance of floral organs (Olszewski et al. 2002). GAhas a clear ameliorative effect against salinity, according to the literature that is currently accessible.

GA promoted growth and development in plants under varied abiotic stress conditions (Ahmad et al. 2010). GA-treated tomato plants that were grown in salty circumstances displayed reduced stomatal resistance and improved water uptake (Maggio et al. 2010). Inducing effective absorption and ion partitioning within the plant system, gibberellic acid promotes development and maintains the metabolism of plants under both normal and stressful circumstances (Iqbal and Ashraf 2013). GA has been linked to better germination and growth under salt stress conditions, according to numerous research (Tuna et al. 2008; Ahmad et al. 2010; Manjili et al. 2012). Gibberellins can also interact with other phytohormones, evoking significant reactions and mediating tolerance processes to improve stress tolerance. Other hormones, such auxin, can also be used to stimulate the production of gibberellins (Wolbang et al. 2004). Enhanced ABA catabolism results from increased gibberellic acid production.

A rise in the osmotic component was also noted in plants under salt stress, and GA treatment boosted their level even more. Plants' osmotic stress was altered by endogenous GA administration, and tissue water content was maintained (Ahmad et al. 2010). Tuna et al. (2008) for maize and Manjili et al. (2012) for wheat both noted these effects. Additionally, gibberellic acid improved development under stress by increasing antioxidant enzyme activity and decreasing (ROS) levels (Manjili et al. 2012). Additionally, exogenous GA administration reduces the effects of salinity on germination and growth in *Arabidopsis thaliana* by facilitating higher SA synthesis, which results in increased action of isochorismate synthase 1 (Alonso-Ramírez et al. 2009). The same study also showed that Arabidopsis's ability to tolerate salt was improved by over-expressing the gibberellin-responsive gene from *Fagus sylvatica*.

14.5.5 Salicylic Acid

Another significant phytohormone with a phenolic character is SA, which plays a significant role in modulating the activity of antioxidant enzymes to help plants tolerate stress (Ahmad et al. 2011; da Silva et al. 2017). Feng et al. (2022) for salt stress, Senaratna et al. (2000) for water stress, and Ahmad et al. (2011) for heavy metal stress reported the relief of diverse abiotic stresses by application of SA. Through stress-activated signal pathways and response mechanisms, SA modifies several physiological processes crucial to plant stress tolerance (Ahmad et al. 2011; Janda et al. 2012; Khan et al. 2014). There are various findings on the easing effect of SA on plants, including wheat, fava beans, and maize (Azooz et al. 2011). A study by Azooz et al. (2011) found that SA reduced the negative effects on growth, biomass accumulation, and the antioxidant system of *Vicia faba* plants that had been treated with seawater, and that organic osmolytes such as proline and free amino acids could be effectively accumulated. (Khan et al. 2014) showed that SA treatment resulted in a decrease in endogenous levels of ethylene in salt stressed *Vigna radiata*. The findings demonstrated that SA treatment facilitated effective Na sequestration and partitioning of harmful ions. Proline and ABA have been synthesized and accumulated with increased frequency. The findings demonstrated that SA treatment facilitated effective Na sequestration and partitioning of harmful ions. Wheat seedlings under salinity stress have been shown to produce more proline and ABA, leading to enhance production and growth (Shakirova et al. 2003).

Khan et al. (2014) found that treating seeds with SA significantly lessened the effects of salt stress on the *Vigna radiata* plant. Plants treated with SA exhibited improved growth in terms of cell division, biomass accumulation, antioxidant enzyme activity, photosynthetic rate (da Silva et al. 2017). Salinity stress altered the rate of photosynthesis and membrane stability in barley plants; however, SA treatment counteracted these negative effects of salinity stress (Janda et al. 2012). In the same study, da Silva et al. (2017) reported accelerated sesame plant growth when drought stress was present by using SA at 105 M. SA treatments in maize reduced the accumulation of Na in plant tissues and counteracted the harmful effects of salt on plants (Gunes et al. 2007). Additionally, SA maintains the transpiration rate, suppresses lipid peroxidation, enhances membrane stability, and reduces electrolyte leakage (Azooz et al. 2011; Stevens et al. 2006). According to Tang et al. (2017), SA treatment reduces water stress by keeping ROS levels low. According to several studies, SA therapy boosted the production and activity of the enzyme known as mitochondrial alternative oxidase (AOX), which is crucial for abiotic stress tolerance (Zhang et al. 2010).

14.6 Conclusion

The script mentioned above demonstrates the significance of plant growth regulators in encouraging plant development. Plants experience a variety of biotic and abiotic stresses in the soil environment, which have an impact on several physiological processes. The synthesis of hormones is one of the techniques that plants adopt to deal with these challenges. By speeding plant processes, these phytohormones contribute significantly to the growth and development of plants. Several phytohormones interact with each other, and that interaction can be either positive or negative according to the chapter mentioned above. Plant growth is impaired by unfavourable phytohormone reactions caused by high concentrations that directly damage or interfere with certain processes or interfere with the production of other hormones. This increased amount of growth hormone may be the result of environmental factors or the infected strains' excessive synthesis of a particular hormone.

Researchers' work here suggests that a few factors should be considered for improving plant growth and development under normal and stressful conditions, including the concentration of a hormone, its use for a specific purpose, as well as the application of phytohormone-producing strains. We need to conduct research on the gross root level to clear up our understanding of these aspects. There should be more research on the rate and timing of phytohormone application, their stability, and their bioavailability in soil environments, in addition to using biotechnological and molecular approaches, genetically engineered plants could also be created that can synthesize substances that enable them to cope with adverse soil conditions and maintain their growth. Plants with transgenic DNA would be able to grow under a wide range of conditions with minimal yield losses.

Acknowledgments We are grateful to Director, Dayalbagh Educational Institute, Dayalbagh, Agra for encouragement and kind support.

References

Abbas HM, Askri SM, Ali S, Fatima A, Qamar MT, Xue SD, Muhammad Z, Akram W, Zhong YJ (2022) Mechanism associated with brassinosteroids crosstalk with gibberellic acid in plants. In: Khan MTA, Yusuf M, Qazi F, Ahmad A (eds) Brassinosteroids signalling: intervention with phytohormones and their relationship in plant adaptation to abiotic stresses. Springer, Singapore, pp 101–115

Afzal I, Basra SA, Iqbal A (2005) The effects of seed soaking with plant growth regulators on seedling vigor of wheat under salinity stress. J Stress Physiol Biochem 1:6–15

Ahmad P, Jaleel CA, Salem MA, Nabi G, Sharma S (2010) Roles of enzymatic and nonenzymatic antioxidants in plants during abiotic stress. Crit Rev Biotechnol 30:161–175

Ahmad P, Nabi G, Ashraf M (2011) Cadmium-induced oxidative damage in mustard [*Brassica juncea* (L.) Czern. & Coss.] plants can be alleviated by salicylic acid. S Afr J Bot 77:36–44

Akter N, Rafiqul Islam M, Abdul Karim M, Hossain T (2014) Alleviation of drought stress in maize by exogenous application of gibberellic acid and CK. J Crop Sci Biotechnol 17:41–48

Alonso-Ramírez A, Rodríguez D, Reyes D, Jiménez JA, Nicolás G, López-Climent M, Gómez-Cadenas A, Nicolás C (2009) Evidence for a role of gibberellins in salicylic acid-modulated early plant responses to abiotic stress in Arabidopsis seeds. Plant Physiol 150:1335–1344

Aroca R (2012) Plant responses to drought stress. From morphological to molecular features:1–5

Asgher M, Khan MI, Anjum NA, Khan NA (2015) Minimising toxicity of cadmium in plants—role of plant growth regulators. Protoplasma 252:399–413

Asrar AW, Elhindi KM (2011) Alleviation of drought stress of marigold (*Tagetes erecta*) plants by using arbuscular mycorrhizal fungi. Saudi J Biol Sci 18:93–98

Ayyaz A, Shahzadi AK, Fatima S, Yasin G, Zafar ZU, Athar HU, Farooq MA (2022) Uncovering the role of melatonin in plant stress tolerance. Theor Exp Plant Physiol 34:335–346

Azizullah A, Khattak MN, Richter P, Häder DP (2011) Water pollution in Pakistan and its impact on public health—a review. Environ Int 37:479–497

Azooz MM, Youssef AM, Ahmad P (2011) Evaluation of salicylic acid (SA) application on growth, osmotic solutes and antioxidant enzyme activities on broad bean seedlings grown under diluted seawater. Int J Plant Physiol Biochem 3:253–264

Balali-Mood M, Naseri K, Tahergorabi Z, Khazdair MR, Sadeghi M (2021) Toxic mechanisms of five heavy metals: mercury, lead, chromium, cadmium, and arsenic. Front Pharmacol 12:643972

Bandara T, Franks A, Xu J, Bolan N, Wang H, Tang C (2020) Chemical and biological immobilization mechanisms of potentially toxic elements in biochar-amended soils. Crit Rev Environ Sci Technol 50:903–978

Bano A, Yasmeen S (2010) Role of phytohormones under induced drought stress in wheat. Pak J Bot 42:2579–2587

Bano A, Ullah F, Nosheen A (2012) Role of abscisic acid and drought stress on the activities of antioxidant enzymes in wheat. Plant Soil Environ 58:181–185

Borrell AK, Wong AC, George-Jaeggli B, van Oosterom EJ, Mace ES, Godwin ID, Liu G, Mullet JE, Klein PE, Hammer GL, McLean G (2022) Genetic modification of *PIN* genes induces causal mechanisms of stay-green drought adaptation phenotype. J Exp Bot 73:6711–6712

Cabot C, Sibole JV, Barceló J, Poschenrieder C (2009) Abscisic acid decreases leaf Na^+ exclusion in salt-treated Phaseolus vulgaris L. J Plant Growth Regul 28:187–192

Carpentier RO (1999) Effect of high-temperature stress on the photosynthetic apparatus. Handbook of Plant Crop Stress 2:337–348

Chen GP, Ma WS, Huang ZJ, Xu T, Xue YB, Shen YZ (2003) Isolation and characterization of TaGSK1 involved in wheat salt tolerance. Plant Sci 16:1369–1375

Chen P, Yu S, Zhan Y, Kang XL (2006) Effects of jasmonate acid on thermotolerance of grape seedlings. J Shihezi Univ Nat Sci 24:87–91

Cutler SR, Rodriguez PL, Finkelstein RR, Abrams SR (2010) Abscisic acid: emergence of a core signaling network. Annu Rev Plant Biol 2:651–679

Dai ZH, Guan DX, Bundschuh J, Ma LQ (2022) Roles of phytohormones in mitigating abiotic stress in plants induced by metal (loid) s As, Cd, Cr, Hg, and Pb. Crit Rev Environ Sci Technol 1–21

Ding W, Song L, Wang X, Bi Y (2010) Effect of abscisic acid on heat stress tolerance in the calli from two ecotypes of *Phragmites communis*. Biol Plant 54:607–613

Egli DB, TeKrony DM, Heitholt JJ, Rupe J (2005) Air temperature during seed filling and soybean seed germination and vigor. Crop Sci 45:1329–1335

Fahad S, Bano A (2012) Effect of salicylic acid on physiological and biochemical characterization of maize grown in saline area. Pak J Bot 44:1433–1438

Fan Y, Zhu M, Shabala S, Li CD, Johnson P, Zhou MX (2014) Antioxidant activity in salt-stressed barley leaves: evaluating time-and age-dependence and suitability for the use as a biochemical marker in breeding programs. J Agron Crop Sci 200:261–272

Fässler E, Evangelou MW, Robinson BH, Schulin R (2010) Effects of indole-3-acetic acid (IAA) on sunflower growth and heavy metal uptake in combination with ethylene diamine disuccinic acid (EDDS). Chemosphere 80:901–907

Feng Y, Wang X, Du T, Shu Y, Tan F, Wang J (2022) Effects of salicylic acid concentration and post-treatment time on the direct and systemic chemical defense responses in maize (*Zea mays* L.) following exogenous foliar application. Molecules 27:6917

Flora SJ, Mittal M, Mehta A (2008) Heavy metal induced oxidative stress and its possible reversal by chelation therapy. Indian J Med Res 128:501–523

Gao QM, Zhu S, Kachroo P, Kachroo A (2015) Signal regulators of systemic acquired resistance. Front Plant Sci 6:228

Ghanashyam C, Jain M (2009) Role of auxin-responsive genes in biotic stress responses. Plant Signal Behav 4:846–848

Ghassemi S, Farhangi-Abriz S, Faegi-Analou R, Ghorbanpour M, Lajayer BA (2018) Monitoring cell energy, physiological functions and grain yield in field-grown mung bean exposed to exogenously applied polyamines under drought stress. J Soil Sci Plant Nutr 18:1108–1125

Gomes FP, Oliva MA, Mielke MS, Almeida AA, Aquino LA (2010) Osmotic adjustment, proline accumulation and cell membrane stability in leaves of *Cocos nucifera* submitted to drought stress. Sci Hortic 126:379–384

Großkinsky DK, van der Graaff E, Roitsch T (2016) Regulation of abiotic and biotic stress responses by plant hormones. In: Collinge DB (ed) Plant pathogen resistance biotechnology. Wiley, New York, pp 131–154

Guajardo E, Correa JA, Contreras-Porcia L (2016) Role of abscisic acid (ABA) in activating antioxidant tolerance responses to desiccation stress in intertidal seaweed species. Planta 243: 767–781

Gull A, Lone AA, Wani NU (2019) Biotic and abiotic stresses in plants. In: Oliveira AD (ed) Abiotic and biotic stress in plants. IntechOpen, Rijeka, Croatia, pp 1–19

Gunes A, Inal A, Alpaslan M, Eraslan F, Bagci EG, Cicek N (2007) Salicylic acid induced changes on some physiological parameters symptomatic for oxidative stress and mineral nutrition in maize (*Zea mays* L.) grown under salinity. J Plant Physiol 164:728–736

Hasanuzzaman M, Gill SS, Fujita M (2013) Physiological role of nitric oxide in plants grown under adverse environmental conditions. In: Tuteja N, Singh Gill S (eds) Plant acclimation to environmental stress. Springer, New York, pp 269–322

Hjortenkrans D, Bergbäck B, Häggerud A (2006) New metal emission patterns in road traffic environments. Environ Monit Assess 117:85–98

Hu YF, Zhou G, Na XF, Yang L, Nan WB, Liu X, Zhang YQ, Li JL, Bi YR (2013) Cadmium interferes with maintenance of auxin homeostasis in Arabidopsis seedlings. J Plant Physiol 170: 965–975

Iqbal M, Ashraf M (2007) Seed treatment with auxins modulates growth and ion partitioning in salt-stressed wheat plants. J Integr Plant Biol 49:1003–1015

Iqbal M, Ashraf M (2013) Gibberellic acid mediated induction of salt tolerance in wheat plants: growth, ionic partitioning, photosynthesis, yield and hormonal homeostasis. Environ Exp Bot 86:76–85

Iqbal N, Khan NA, Ferrante A, Trivellini A, Francini A, Khan MIR (2017) Ethylene role in plant growth, development and senescence: interaction with other phytohormones. Front Plant Sci 8: 475

Iqbal N, Czékus Z, Angeli C, Bartók T, Poór P, Ördög A (2022) Fumonisin b1-induced oxidative burst perturbed photosynthetic activity and affected antioxidant enzymatic response in tomato plants in ethylene-dependent manner. J Plant Growth Regul 4:1–14

Janda K, Hideg É, Szalai G, Kovács L, Janda T (2012) Salicylic acid may indirectly influence the photosynthetic electron transport. J Plant Physiol 169:971–978

Jiang CJ, Shimono M, Sugano S, Kojima M, Liu X, Inoue H, Sakakibara H, Takatsuji H (2013) CKs act synergistically with salicylic acid to activate defense gene expression in rice. Mol Plant-Microbe Interact 26:287–296

Johnson R, Vishwakarma K, Hossen MS, Kumar V, Shackira AM, Puthur JT, Abdi G, Sarraf M, Hasanuzzaman M (2022) Potassium in plants: growth regulation, signaling, and environmental stress tolerance. Plant Physiol Biochem 172:56–69

Jung JH, Park CM (2011) Auxin modulation of salt stress signaling in Arabidopsis seed germination. Plant Signal Behav 6:1198–1200

Jurado S, Abraham Z, Manzano C, Lopez-Torrejon G, Pacios LF, Del Pozo JC (2010) The Arabidopsis cell cycle F-box protein SKP2A binds to auxin. Plant Cell 22:3891–3904

Kaur G, Tak Y, Asthir B (2022) Salicylic acid: a key signal molecule ameliorating plant stresses. Cereal Res Commun 50:1–10

Kazan K (2013) Auxin and the integration of environmental signals into plant root development. Ann Bot 112:1655–1665

Khadri M, Tejera NA, Lluch C (2006) Alleviation of salt stress in common bean (*Phaseolus vulgaris*) by exogenous abscisic acid supply. J Plant Growth Regul 25:110–119

Khan MI, Asgher M, Khan NA (2014) Alleviation of salt-induced photosynthesis and growth inhibition by salicylic acid involves glycinebetaine and ethylene in mungbean (*Vigna radiata* L.). Plant Physiol Biochem 80:67–74

Khare S, Singh NB, Singh A, Hussain I, Niharika KM, Yadav V, Bano C, Yadav RK, Amist N (2020) Plant secondary metabolites synthesis and their regulations under biotic and abiotic constraints. J Plant Biol 63:203–216

Krantev A, Yordanova R, Popova L (2006) Salicylic acid decreases cd toxicity in maize plants. Gen Appl Plant Physiol 1:45–52

Lee S, Park CM (2010) Modulation of reactive oxygen species by salicylic acid in Arabidopsis seed germination under high salinity. Plant Signal Behav 5:1534–1536

Lefevere H, Bauters L, Gheysen G (2020) Salicylic acid biosynthesis in plants. Front Plant Sci 11: 338

Li J, Brader G, Kariola T, Tapio Palva E (2006) WRKY70 modulates the selection of signaling pathways in plant defense. Plant J 46:477–491

Li X, Cai J, Liu F, Dai T, Cao W, Jiang D (2014) Exogenous abscisic acid application during grain filling in winter wheat improves cold tolerance of offspring's seedlings. J Agron Crop Sci 200: 467–478

Liu Y, Du M, Deng L, Shen J, Fang M, Chen Q, Lu Y, Wang Q, Li C, Zhai Q (2019) MYC2 regulates the termination of jasmonate signaling via an autoregulatory negative feedback loop. Plant Cell 31:106–127

Ljung K (2013) Auxin metabolism and homeostasis during plant development. Development 140: 943–950

Los DA, Murata N (2004) Membrane fluidity and its roles in the perception of environmental signals. Biochem Biophysic Acta – Biomembr 1666:142–157

Lymperopoulos P, Msanne J, Rabara R (2018) Phytochrome and phytohormones: working in tandem for plant growth and development. Front Plant Sci 9:1037

Madani B, Mirshekari A, Imahori Y (2019) Physiological responses to stress. In: Yahia EM (ed) Postharvest physiology and biochemistry of fruits and vegetables. Woodhead Publishing, pp 405–423

Maggio A, Barbieri G, Raimondi G, De Pascale S (2010) Contrasting effects of GA3 treatments on tomato plants exposed to increasing salinity. J Plant Growth Regul 29:63–72

Manivannan P, Jaleel CA, Somasundaram R, Panneerselvam R (2008) Osmoregulation and anti-oxidant metabolism in drought-stressed *Helianthus annuus* under triadimefon drenching. C R Biol 331:418–425

Manjili FA, Sedghi M, Pessarakli M (2012) Effects of phytohormones on proline content and antioxidant enzymes of various wheat cultivars under salinity stress. J Plant Nutr 35:1098–1111

Minocha R, Majumdar R, Minocha SC (2014) Polyamines and abiotic stress in plants: a complex relationship. Front Plant Sci 5:175

Misra N, Saxena P (2009) Effect of salicylic acid on proline metabolism in lentil grown under salinity stress. Plant Sci 177:181–189

Mora-Herrera ME, Lopez-Delgado HA (2007) Freezing tolerance and antioxidant activity in potato microplants induced by abscisic acid treatment. Am J Potato Res 84:467–475

Munns R (2005) Genes and salt tolerance: bringing them together. New Phytol 167:645–663

Nada E, Ferjani BA, Ali R, Bechir BR, Imed M, Makki B (2007) Cadmium-induced growth inhibition and alteration of biochemical parameters in almond seedlings grown in solution culture. Acta Physiol Plant 29:57–62

Nawaz K, Ashraf M (2010) Exogenous application of glycinebetaine modulates activities of antioxidants in maize plants subjected to salt stress. J Agron Crop Sci 196:28–37

Nayyar H, Bains TS, Kumar S (2005) Chilling stressed chickpea seedlings: effect of cold acclimation, calcium and abscisic acid on cryoprotective solutes and oxidative damage. Environ Exp Bot 54:275–285

O'Brien JA, Benková E (2013) CK cross-talking during biotic and abiotic stress responses. Front Plant Sci 4:451

Olszewski N, Sun TP, Gubler F (2002) Gibberellin signaling: biosynthesis, catabolism, and response pathways. Plant Cell 14:S61–S80

Park J, Kim YS, Kim SG, Jung JH, Woo JC, Park CM (2011) Integration of auxin and salt signals by the NAC transcription factor NTM2 during seed germination in *Arabidopsis*. Plant Physiol 156:537–549

Parwez R, Aftab T, Gill SS, Naeem M (2022) Abscisic acid signaling and crosstalk with phytohormones in regulation of environmental stress responses. Environ Exp Bot 104885

Peleg Z, Blumwald E (2011) Hormone balance and abiotic stress tolerance in crop plants. Curr Opin Plant Biol 14:290–295

Rajjou L, Belghazi M, Huguet R, Robin C, Moreau A, Job C, Job D (2006) Proteomic investigation of the effect of salicylic acid on Arabidopsis seed germination and establishment of early defense mechanisms. Plant Physiol 141:910–923

Ren CH, Bilyeu KD, Beuselinck PR (2009) Composition, vigor, and proteome of mature soybean seeds developed under high temperature. Crop Sci 49:1010–1022

Robert-Seilaniantz A, Navarro L, Bari R, Jones JD (2007) Pathological hormone imbalances. Curr Opin Plant Biol 10:372–379

Sachs T (2005) Auxin's role as an example of the mechanisms of shoot/root relations. Plant Soil 268:13–19

Sadiq M, Jamil M, Mehdi SM, Sarfraz M, Hassan G (2002) Comparative performance of brassica varieties/lines under saline sodic condition. Asian J Plant Sci 2:77–78

Sah SK, Reddy KR, Li J (2016) Abscisic acid and abiotic stress tolerance in crop plants front. Plant Sci 7:571

Schaller GE, Street IH, Kieber JJ (2014) CK and the cell cycle. Curr Opin Plant Biol 21:7–15

Senaratna T, Touchell D, Bunn E, Dixon K (2000) Acetyl salicylic acid (aspirin) and salicylic acid induce multiple stress tolerance in bean and tomato plants. Plant Growth Regul 30:157–161

Shaddad MA, Abd El-Samad HM, Mostafa D (2013) Role of gibberellic acid (GA3) in improving salt stress tolerance of two wheat cultivars. Int J Plant Physiol 5:50–57

Shakirova FM, Sakhabutdinova AR, Bezrukova MV, Fatkhutdinova RA, Fatkhutdinova DR (2003) Changes in the hormonal status of wheat seedlings induced by salicylic acid and salinity. Plant Sci 164:317–322

Shaukat K, Zahra N, Hafeez MB, Naseer R, Batool A, Batool H, Raza A, Wahid A (2022) Role of salicylic acid–induced abiotic stress tolerance and underlying mechanisms in plants. In: Aftab T, Naeem M (eds) Emerging plant growth regulators in agriculture. Academic, pp 73–98

Shi H, Chen L, Ye T, Liu X, Ding K, Chan Z (2014) Modulation of auxin content in *Arabidopsis* confers improved drought stress resistance. Plant Physiol Biochem 82:209–217

Silva AC, Suassuna JF, Melo AS, Costa RR, Andrade WL, Silva DC (2017) Salicylic acid as attenuator of drought stress on germination and initial development of sesame. Rev Bras de Eng Agricola e Ambient 21:156–162

Song JQ, Mei XR, Fujiyama H (2006) Adequate internal water status of NaCl-salinized rice shoots enhanced selective calcium and potassium absorption. Soil Sci Plant Nutr 52:300–304

Spollen WG, LeNoble ME, Samuels TD, Bernstein N, Sharp RE (2000) Abscisic acid accumulation maintains maize primary root elongation at low water potentials by restricting ethylene production. Plant Physiol 122:967–976

Stevens J, Senaratna T, Sivasithamparam K (2006) Salicylic acid induces salinity tolerance in tomato (*Lycopersicon esculentum* cv. Roma): associated changes in gas exchange, water relations and membrane stabilisation. Plant Growth Regul 49:77–83

Subramanian KS, Santhanakrishnan P, Balasubramanian P (2006) Responses of field grown tomato plants to arbuscular mycorrhizal fungal colonization under varying intensities of drought stress. Sci Hortic 107:245–253

Sun WH, Duan M, Li F, Shu DF, Yang S, Meng QW (2010) Overexpression of tomato *tAPX* gene in tobacco improves tolerance to high or low temperature stress. Biol Plant 54:614–620

Sytar O, Kumari P, Yadav S, Brestic M, Rastogi A (2019) Phytohormone priming: regulator for heavy metal stress in plants. J Plant Growth Regul 38:739–752

Tang Y, Sun X, Wen T, Liu M, Yang M, Chen X (2017) Implications of terminal oxidase function in regulation of salicylic acid on soybean seedling photosynthetic performance under water stress. Plant Physiol Biochem 112:19–28

Tromas A, Braun N, Muller P, Khodus T, Paponov IA, Palme K, Ljung K, Lee JY, Benfey P, Murray JA, Scheres B (2009) The AUXIN BINDING PROTEIN 1 is required for differential auxin responses mediating root growth. PLoS One 4:e6648

Tuna AL, Kaya C, Dikilitas M, Higgs D (2008) The combined effects of gibberellic acid and salinity on some antioxidant enzyme activities, plant growth parameters and nutritional status in maize plants. Environ Exp Bot 62:1–9
Vardhini BV, Anjum NA (2015) Brassinosteroids make plant life easier under abiotic stresses mainly by modulating major components of antioxidant defense system. Front Environ Sci 2:67
Vysotskaya LB, Korobova AV, Veselov SY, Dodd IC, Kudoyarova GR (2009) ABA mediation of shoot CK oxidase activity: assessing its impacts on CK status and biomass allocation of nutrient-deprived durum wheat. Funct Plant Biol 36:66–72
Wahid A, Farooq M, Hussain I, Rasheed R, Galani S (2012) Responses and management of heat stress in plants. In: Environmental adaptations and stress tolerance of plants in the era of climate change, pp 135–157
Wang Q, Dodd IC, Belimov AA, Jiang F (2016) Rhizosphere bacteria containing 1-aminocyclopropane-1-carboxylate deaminase increase growth and photosynthesis of pea plants under salt stress by limiting Na^+ accumulation. Funct Plant Biol 43:161–172
Wang J, Song L, Gong X, Xu J, Li M (2020) Functions of jasmonic acid in plant regulation and response to abiotic stress. Int J Mol Sci 21:1446
Wang Y, Mostafa S, Zeng W, Jin B (2021) Function and mechanism of jasmonic acid in plant responses to abiotic and biotic stresses. Int J Mol Sci 22:8568
Wei B, Yang L (2010) A review of heavy metal contaminations in urban soils, urban road dusts and agricultural soils from China. Microchem J 94:99–107
Wilkinson S, Davies WJ (2002) ABA-based chemical signalling: the co-ordination of responses to stress in plants. Plant Cell Environ 25:195–210
Wilkinson S, Kudoyarova GR, Veselov DS, Arkhipova TN, Davies WJ (2012) Plant hormone interactions: innovative targets for crop breeding and management. J Exp Bot 63:3499–3509
Wolbang CM, Chandler PM, Smith JJ, Ross JJ (2004) Auxin from the developing inflorescence is required for the biosynthesis of active gibberellins in barley stems. Plant Physiol 134:769–776
Yamane K, Hayakawa K, Kawasaki M, Taniguchi M, Miyake H (2003) Bundle sheath chloroplasts of rice are more sensitive to drought stress than mesophyll chloroplasts. J Plant Physiol 160: 1319–1327
Yang J, Duan G, Li C, Liu L, Han G, Zhang Y, Wang C (2019) The crosstalks between jasmonic acid and other plant hormone signaling highlight the involvement of jasmonic acid as a core component in plant response to biotic and abiotic stresses. Front Plant Sci 10:1349
Zhang P, Wang WQ, Zhang GL, Kaminek M, Dobrev P, Xu J, Gruissem W (2010) Senescence-inducible expression of isopentenyl transferase extends leaf life, increases drought stress resistance and alters CK metabolism in cassava. J Integr Plant Biol 52:653–669
Zhou B, Guo Z, Xing J, Huang B (2005) Nitric oxide is involved in abscisic acid-induced antioxidant activities in *Stylosanthes guianensis*. J Exp Bot 56:3223–3228
Zhou L, Xu H, Mischke S, Meinhardt LW, Zhang D, Zhu X, Li X, Fang W (2014) Exogenous abscisic acid significantly affects proteome in tea plant (*Camellia sinensis*) exposed to drought stress. Hortic Res 1:14029. https://doi.org/10.1038/hortres.2014.29
Zhu JK (2016) Abiotic stress signaling and responses in plants. Cell 167:313–324
Zwack PJ, Rashotte AM (2015) Interactions between CK signalling and abiotic stress responses. J Exp Bot 66:4863–4871

The manufacturer's authorised representative in the EU is Springer Nature Customer Service Centre GmbH, Europaplatz 3, 69115 Heidelberg, Germany. If you have any concerns regarding our products, please contact ProductSafety@springernature.com

Printed and bound by CPI Group (UK) Ltd, Croydon, CR0 4YY
30/09/2026
02225319-0011